Liquids Under Negative Pressure

NATO Science Series

A Series presenting the results of scientific meetings supported under the NATO Science Programme.

The Series is published by IOS Press, Amsterdam, and Kluwer Academic Publishers in conjunction with the NATO Scientific Affairs Division

Sub-Series

I.	**Life and Behavioural Sciences**	IOS Press
II.	**Mathematics, Physics and Chemistry**	Kluwer Academic Publishers
III.	**Computer and Systems Science**	IOS Press
IV.	**Earth and Environmental Sciences**	Kluwer Academic Publishers
V.	**Science and Technology Policy**	IOS Press

The NATO Science Series continues the series of books published formerly as the NATO ASI Series.

The NATO Science Programme offers support for collaboration in civil science between scientists of countries of the Euro-Atlantic Partnership Council. The types of scientific meeting generally supported are "Advanced Study Institutes" and "Advanced Research Workshops", although other types of meeting are supported from time to time. The NATO Science Series collects together the results of these meetings. The meetings are co-organized bij scientists from NATO countries and scientists from NATO's Partner countries – countries of the CIS and Central and Eastern Europe.

Advanced Study Institutes are high-level tutorial courses offering in-depth study of latest advances in a field.
Advanced Research Workshops are expert meetings aimed at critical assessment of a field, and identification of directions for future action.

As a consequence of the restructuring of the NATO Science Programme in 1999, the NATO Science Series has been re-organised and there are currently Five Sub-series as noted above. Please consult the following web sites for information on previous volumes published in the Series, as well as details of earlier Sub-series.

http://www.nato.int/science
http://www.wkap.nl
http://www.iospress.nl
http://www.wtv-books.de/nato-pco.htm

Series II: Mathematics, Physics and Chemistry – Vol. 84

Liquids Under Negative Pressure

edited by

A.R. Imre

KFKI Atomic Energy Research Institute,
Materials Department,
Budapest, Hungary

H.J. Maris

Department of Physics,
Brown University,
Providence, RI, U.S.A.

and

P.R. Williams

Centre for Complex Fluids Processing,
Department of Chemical and Biological Process Engineering,
University of Wales,
Swansea, U.K.

Kluwer Academic Publishers

Dordrecht / Boston / London

Published in cooperation with NATO Scientific Affairs Division

Proceedings of the NATO Advanced Research Workshop on
Liquids Under Negative Pressure
Budapest, Hungary
23–25 February 2002

A C.I.P. Catalogue record for this book is available from the Library of Congress.

ISBN 1-4020-0895-3 (HB)
ISBN 1-4020-0896-1 (PB)

Published by Kluwer Academic Publishers,
P.O. Box 17, 3300 AA Dordrecht, The Netherlands.

Sold and distributed in North, Central and South America
by Kluwer Academic Publishers,
101 Philip Drive, Norwell, MA 02061, U.S.A.

In all other countries, sold and distributed
by Kluwer Academic Publishers,
P.O. Box 322, 3300 AH Dordrecht, The Netherlands.

Printed on acid-free paper

TABLE OF CONTENTS

Part V: Cavitation during static and dynamic tensions

Part VI: Heterogeneous nucleation

INTRODUCTION

Some physical quantities (such as acceleration) may have either positive or negative values, while others (such as length) cannot be negative. For many scientists, the concept of pressure appears to belong in the latter category. Whereas it is commonly appreciated that, in decreasing the pressure within a system such as a liquid from an initial positive value one can approach P=0 by increasing the volume, it is not widely known that negative values of pressure within a liquid may be physically realisable. Indeed, thermodynamics does not exclude the possibility of the existence of P<0 states in condensed matter and the contributions to this Volume are the results of research into various aspects of the development of negative pressure in liquids. Many of the Contributors to this Volume will have had occasion to explain to colleagues outside the field that it is possible to 'stretch' a liquid; and that when suitably prepared, liquids are in fact capable of sustaining substantial levels of tension (often for significant periods of time).

This general lack of appreciation of the phenomenon of negative pressure development in liquids may stem from the fact that although Toricelli's famous experiment is widely taught, relatively few schoolchildren are told of similar work conducted by Huygens less than two decades later. We recall that in performing Toricelli's experiment, one fills a long (>76 cm) glass tube (closed at one end) with mercury: the tube is then inverted with the open end immersed in mercury. When the experiment is performed at sea level, the mercury column drops down to 76 cm – while above the mercury we have Toricelli's vacuum (practically P=0): within the liquid column the pressure changes linearly with height, from zero to atmospheric. This was the result experienced by Toricelli and that which is *usually* experienced by those who repeat the experiment. However, the mercury column is occasionally observed to 'stick' i.e. it does not fall down to 76 cm. This unexpected result was recorded several times by Huygens in experiments involving a 1.5 m long tube (see G. S. Kell: Early observations of negative pressures in liquids, *Am. J. Phys.*, 51 (1983) 1038), this work being followed by that of Boyle and Papen before further experiments were abandoned. The phenomenon was difficult to reproduce and was inexplicable given the prevailing knowledge of the liquid state. More than 150 years were to pass before Donny, in the mid-XIXth century, rediscovered and explained the phenomenon.

Donny realised that the tendency of the mercury to stick was due to adhesive forces at the tube wall which resulted in the upper part of the mercury column (i.e. that above 76 cm) being stretched. Since the hydrostatic pressure is proportional to the height of the liquid column, the pressure in the upper part of the liquid has to be below P=0. In Huygens' experiment, this pressure was negative (approximately down to −1 bar). Despite this evidence and the related contributions of several great scientists of the XIXth and XXth centuries (like Berthelot and Reynolds), knowledge of the existence and importance of negative pressures in liquids has remained relatively obscure. It was against this background that the Organisers of this NATO Advanced Research

Workshop decided to prepare the first meeting entirely dedicated to the topic of *Liquids under Negative Pressure* (February 23-25, 2002, Budapest, Hungary).

Negative pressure states for liquids are metastable states, very closely related to superheated states; in both cases the liquid is metastable with respect to the liquid-vapour phase transition, only in the second case the metastable region has been reached by heating, while in the first case it has been reached by stretching. Metastability means that for some perturbation the liquid can show a sudden liquid-vapour phase transition (boiling or cavitation) and the pressure will jump back to the vapour pressure, which is a positive quantity. But liquids can withstand negative pressure (even in the -10 to -100 MPa region) for days, therefore even a substantial experimental investigation of these liquids is possible. There is a limiting negative pressure for every particular liquid, a value where the liquid must cavitate because the stretched state would not be metastable but unstable. For regular liquids, the range of these maximal values is from -0.2 MPa (for He^3) to well above (or rather below) -100 MPa (for water).

It is no exaggeration to state that negative pressures occur in liquids all about (and sometimes within) us. Liquids (as sap) in plants experience moderate negative pressure. Negative pressures up to -100 MPa have been reported to occur within liquid inclusions in crystals (including mercury inclusions in artificial diamonds). In the animal world, negative pressure can be generated by an octopus under its liquid-filled suckers; the value of this pressure can be in the 0 to -3 bar range, causing a substantial pressure difference between the inner and outer side of the skin of its victim. Also, by suddenly pulling on a 'biological bearing' such as the finger joint, the lubricant within it may be suddenly stretched and subsequently relax by cavitation: the consequent emission of an audible 'crack' has been related, in some instances, to the occurrence of cavitation damage. Positive pressure waves (such as those generated by earthquakes) may be reflected from a liquid/solid or liquid/gas interface and can result in the development of rarefaction waves involving substantial negative pressure.

Information concerning the generation of negative pressure under laboratory conditions can be found in several books or review articles (V.P. Skripov: *Metastable Liquids*, Wiley, New York, 1974; D.H. Trevena: *Cavitation and Tension in Liquids*, Adam Hilger, Bristol, 1987; P. G. Debenedetti: *Metastable Liquids – Concepts and Principles*, Princeton University Press, Princeton, NJ, 1996, A. Imre et al, Thermodynamics of Negative Pressures in Liquids, *J. Noneq. Thermodyn.*, 23 (1998) 351; H. Maris and S. Balibar: Negative Pressures and Cavitation in Liquid Helium, *Physics Today*, 53 (2000) 29, also in www.aip.org/pt/feb00/maris.htm). There is also an Internet-site dedicated to the subject of liquids under negative pressure (www.kfki.hu/~pressure).

The purpose of this Volume is to provide an overview of recent and current research into various aspects of the behaviour of liquids under negative pressure. Part I includes papers concerning the thermodynamics of stretched liquids. Part II includes papers dealing with the physical and chemical behaviour of liquids under negative pressure. In Part III one can find papers dealing with the effect of negative pressure on the solidification of a liquid. Part IV includes papers concerning stretched helium. Part V includes papers concerning cavitation in various stretched liquids, while the last part (Part VI) is dealing with the effect of foreign substances on the cavitation. We believe that this book represents a necessary and timely overview of recent and past studies of this subject.

We thank all the contributing authors for preparing their manuscripts in a timely manner. We acknowledge the generous support of the NATO Science Programme and additional supports given by the Hungarian Academy of Science and the Chemical Research Center, Hungary. Thanks for Dr. G. Veres for his help dealing with TeX-files and for Prof. V.A. Akulichev for the photo. Further thanks to the KFKI Atomic Energy Research Institute for the technical help. We also express our gratitude to Kluwer for the publication of these contributions.

A. R. Imre
Budapest
Hungary

H. J. Maris
Providence, RI
USA

P. R. Williams
Swansea, Wales
UK

NATO ARW Liquids Under Negative Pressure
Budapest, Hungary, February 23-25, 2002

Sitting: L.P.N. Rebelo, J. Szydlowski, K. Martinás, A.R. Imre, V. Vinogradov, V.A. Akulichev and V. Bulanov

Standing: D.M. Heyes, Y. Ohde, Z.P. Visak, H.J. Maris, J. Navarro, P. Jedlovszky, E. Piorkowska, S. Sastry, A. Galeski, H.E. Stanley, S. Balibar, V.K. Kedrinskii, A. Geiger, R.J. Speedy, A. Besov, M.Z. Faizullin, R.L. Williams, F. Caupin and S.J. Rzoska

Absent: A.A. Angell, D.M. Ceperley, M. Barranco, J. Boronat, G. Házi, W.D. Machin, L. Smeller and P.R. Williams

PARTICIPANTS

Co-Directors

P. Rhodri Williams
Centre for Complex Fluids Processing,
Dept. of Chem. and Biological Process
Engineering, University of Wales,
Swansea, Singleton Park, Swansea, United
Kingdom SA2 8PP

Attila R. Imre
 KFKI Atomic Energy Research Institute,
Materials Department, 1525 Budapest,
POB. 49, Hungary

Organizing Committee

Valery K. Kedrinskii
Lavrentyev Institute of Hydrodynamics
630090, Novosibirsk, pr. Ac. Lavrentyeva, 15, Russia

Humprey J. Maris
Department of Physics, Box 1843, Brown University
Providence, RI 02912, USA

Katalin Martinás
Eötvös University, Department of Atomic Physics
1117 Budapest Pázmány Péter sétány 1/A, Hungary

Keyspeakers

V.A. Akulichev
Pacific Oceanological Institute, Far Eastern Branch, Russian Academy of Sciences
43 Baltiyskaya Street, Vladivostok 690041, Russia

C. Austen Angell
Arizona State University, Department of Chemistry
P.O. Box 871604, Tempe, Arizona 85287-1604, USA

Sebastian Balibar
Ecole Normale Supérieure, Paris France, Laboratoire de Physique Statistique
24, rue Lhomond 75231 Paris Cedex 05 - France

xiv

Andrzej Galeski
Centre of Molecular and Macromolecular Studies, Polish Academy of Sciences
90-363 Lodz, ul. Sienkiewicza 112, Poland

Yoshihito Ohde
Department of Systems Engineering, Nagoya Institute of Technology,
Gokiso, Shouwa-ku, Nagoya, 466-8555, Japan

Robin J. Speedy
Apt. 504, 120 Courtenay Place, Wellington 6001, New Zealand

H. Eugene Stanley
Boston University, Center for Polymer Studies, Department of Physics
590 Commonwealth Avenue, Boston, Massachusetts 02215 USA

Further Participants

Manuel Barranco
Departament ECM , Facultat de Fisica , Universitat de Barcelona
08028 Barcelona, Spain

A. Besov
Lavrentyev Institute of Hydrodynamics
630090, Novosibirsk, pr. ac. Lavrentyeva, 15, Russia

V. Bulanov
Institute of Marine Technology Problems, Far Eastern Branch, Russian Academy of
Sciences, 5a Sukhanov Street, Vladivostok 690960, Russia

Jordi Boronat
Quantum Monte Carlo group, Departament de Fisica i Enginyeria Nuclear, Campus
Nord B4--B5 , Universitat Politecnica de Catalunya, E--08034 Barcelona, Spain

Frederic Caupin
Laboratoire de Physique Statistique de l'Ecole Normale Supérieure
24 rue Lhomond, 75231 Paris Cedex 05, France

David M. Ceperley
University of Illinois, Department of Physics
405 N. Matthews Av., Urbana, Il, 61801

M. Faizullin
Institute of Thermophysics Urals Branch of the Russian Academy of Sciences,
Pervomaiskaya Str.91,GSP-828, Ekaterinburg,620219, Russia

Alfons Geiger
Universität Dortmund, Physikalische Chemie PC2a
Raum C1-05-104, Otto-Hahn-Str. 6, D-44221 Dortmund, Germany

Gábor Házi
KFKI Atomic Energy Research Institute, Simulator Development Laboratory
1525 Budapest, POB. 49, Hungary

David M. Heyes
University of Surrey, Department of Chemistry
Guildford, Surrey , GU2 5XH, UK

Pál Jedlovszky
Eötvös University, Department of Colloid Chemistry
1117 Budapest Pázmány Péter sétány 1/A, Hungary

William D. Machin
Memorial University of New Foundland , Department of Chemistry
St. John's, NF, Canada A1B 3X7

Jesus Navarro
IFIC (CSIC-Universitat Valencia) , Edificio Institutos Paterna
Apdo. 2085 , 46071 Valencia, Spain

Eva Piorkowska
Centre of Molecular and Macromolecular Studies, Polish Academy of Sciences
90-363 Lodz, ul. Sienkiewicza 112, Poland

Luis Paulo N. Rebelo
Instituto de Tecnologia Química e Biológica, ITQB 2, Universidade Nova de Lisboa
Av. da República, Apartado 127, 2780-901 Oeiras, Portugal

Sylwester J. Rzoska
Dept. of Biophysics and Molecular Physics, Institute of Physics, Silesian University
ul. Uniwersytecka 4, 40-007 Katowice, Poland

Srikanth Sastry
Theoretical Sciences Unit, Jawaharlal Nehru Centre For Advanced Scientific Research,
Jakkur Campus, Bangalore 560064, India

László Smeller
Institute of Biophysics & Radiation Biology, Semmelweis Medical University
H-1444 Budapest, Puskin u. 9. POB 263, Hungary

xvi

Jerzy Szydlowski
Chemistry Department, Warsaw University
Zwirki i Wigury 101, 02-089 Warsaw, Poland

V. Vinogradov
Institute of Thermophysics Urals Branch of the Russian Academy of Sciences,
Pervomaiskaya Str.91,GSP-828, Ekaterinburg,620219, Russia

Zoran P. Visak
Instituto de Tecnologia Química e Biológica, ITQB 2, Universidade Nova de Lisboa
Av. da República, Apartado 127, 2780-901 Oeiras, Portugal

R.L. Williams
Mail: Centre for Complex Fluids Processing, Dept. of Chem. and Biological Process
Engineering, University of Wales, Swansea, Singleton Park, Swansea, United Kingdom
SA2 8PP

LIMITS OF STABILITY FOR LIQUIDS UNDER TENSION

ROBIN J. SPEEDY
504-120 Courtenay Place, Wellington, New Zealand
robin.speedy@clear.net.nz

Abstract. Two distinct thermodynamic limits of stability for a liquid are discussed. One is the familiar spinodal line where $(\partial P / \partial V)_T \to 0$ when the liquid is stretched on an isotherm or warmed on a sub-critical isobar. The other is an ideal glass transition line where the configurational entropy $S_c \to 0$ when the liquid is cooled on an isobar. Thermodynamics requires that these two lines have the same slope where they meet and, if the spinodal has a monotonic slope, the intersection occurs at the maximum tension the liquid can sustain. A generalized van der Waals model is used to illustrate the results and some implications for the interpretation of simulation data for stretched and supercooled water are noted.

1. Introduction

Liquids can be studied in superheated, supercooled and stretched states where they are metastable [1] with respect to boiling, freezing or cavitation. In practice, the range of observation is limited by crystal or vapor nucleation or by a glass transition. These kinetic limits depend on variable factors like the sample size and observation times.

Thermodynamic bounds 1 on the metastability of a liquid are more speculative, partly because their existence and location can be inferred only by extrapolation beyond the kinetic limits that pre-empt them. The two main contenders are the spinodal lines [1,2] of mean field theories, where the compressibility and heat capacity diverge and an ideal glass transition line [3,4] where the entropy of the liquid tends to that of an individual glass. While these lines are hypothetical constructs, beyond the reach of real experiments, they may help to unify the interpretation of diverse phenomena.

The relation between the two lines, their place in the overall phase diagram and their connection to density maxima, is the subject of this paper. Section 2 summarizes thermodynamics near a spinodal. Stillinger [5] has argued against the notion of an ideal glass transition, so section 3 provides a brief discussion to justify retaining the concept. Section 4 gives explicit results for a simple liquid, modeled by a generalized van der Waals equation [6] and shows that the spinodal and ideal glass transition lines meet with the same slope at the maximum negative pressure the liquid can sustain. Section 5 shows, by a thermodynamic argument, that the spinodal and ideal glass transition lines always have the same slope where they meet.

Water is our most studied liquid but there is not yet a consensus on the interpretation of its unusual properties or how it behaves beyond the range of current measurements. Section

1

A.R. Imre et al. (eds.), Liquids Under Negative Pressure, 1–12.
© 2002 *Kluwer Academic Publishers. Printed in the Netherlands.*

6 notes thermodynamic constraints on the interpretation of recent simulation results for stretched and supercooled water

2. Spinodals

An analytic equation for the pressure, $P(V,T)$, as a function of volume, V, and temperature, T, of coexisting phases must approximate the pressure in the coexistence region by "loops" where the isotherms have maxima and minima. Spinodals lines1 are the loci, $P_S(T)$, of the extrema where $(\partial P / \partial V)_T = 0$. They bound a region where the stability condition $(\partial P / \partial V)_T < 0$ is violated and where a single homogeneous phase is thermodynamically unstable. Skripov [7] and coworkers pioneered the exploration of behaviour near the spinodal in superheated liquids and showed that real liquids behave, in the kinetically accessible range, as though the spinodal is real.

From the above definition of a spinodal, the isothermal compressibility

$$\kappa_T \equiv -(\partial \ln V / \partial P)_T \to \infty \text{ as } P \to P_S(T) \tag{1}$$

diverges there.

In the pressure-temperature plane, a spinodal line is an envelope [1,7] of lines of constant volume, lines of constant entropy, S, lines of constant enthalpy, H, and, wherever the slope of the spinodal, dP_S / dT, is finite1

$$(\partial P / \partial T)_V \to (\partial P / \partial T)_S \to (\partial P / \partial T)_H \to dP_S / dT \text{ as } P \to P_S(T). \tag{2}$$

For instance, where an isochor meets a spinodal line the two lines have the same slope, and, because $(\partial P / \partial T)_V = \alpha / \kappa_T$, the isobaric expansivity

$$\alpha \equiv (\partial \ln V / \partial T)_P \to (dP_S / dT)\kappa_T \text{ as } P \to P_S(T). \tag{3}$$

The sign of α is therefore the same as the sign of dP_S / dT near the spinodal. [1,2,7]

3. The ideal glass transition

Kauzmann [8] showed that the excess entropy of a liquid, over that of the crystal,

$$S_{ex}(P,T) \equiv S_{liquid}(P,T) - S_{crystal}(P,T), \tag{4}$$

decreases rapidly on cooling and extrapolates to zero at a temperature T_0 not far below the kinetic glass transition temperature, T_g. His discussion [8] of this apparent paradox suggests "a lower metastable limit" which might be associated either with the inevitability of freezing or with the intervention of a thermodynamic glass transition. Gibbs and DiMarzio [3] developed a lattice model with an ideal glass transition, where the entropy of a liquid tends to that of an ideal glass. DiMarzio [9] noted that some liquids do not have a crystalline phase

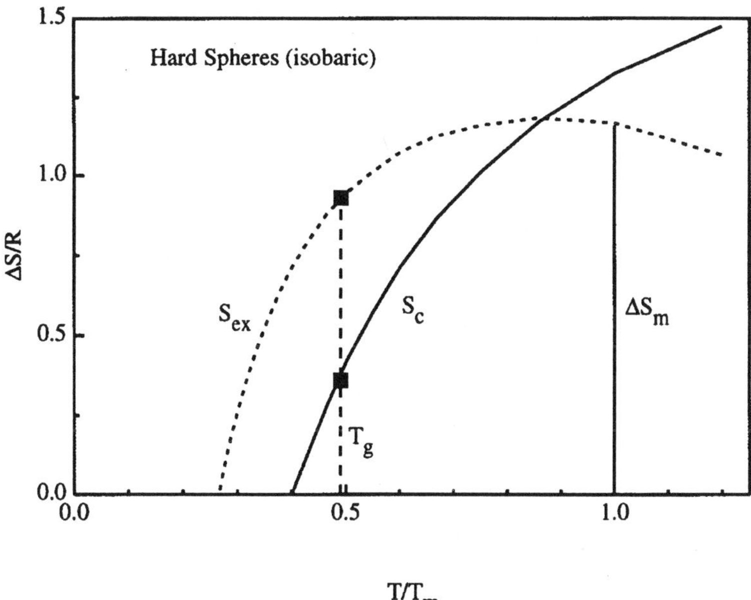

Figure 1.   The configurational entropy, S_c, and the excess entropy, S_{ex}, for hard spheres, plotted against the temperature relative to the melting [10] temperature, T/T_m. The temperature on an isobar is $(yPV/RT)^{-1}$ where y is the packing fraction. The entropy of the crystal is from eq. (2) of ref. [11]. The entropy of the liquid, from eqs. (4) and (5) of ref. [12], is extrapolated when $T/T_m < 0.56$. The same extrapolation is used to calculate S_c and S_{ex} so it contributes no error to their difference. The entropies of glasses that have the same pressure and density as the liquid at the same temperature, from eqs. (1) and (6) of ref. [13], are measured near T_g but depend on assumptions and extrapolations above and below T_g. The assumptions [13] on which the the extrapolation of the hard sphere glass properties depend have been tested for hard discs. [14] Squares show values of S_c and S_{ex} at T_g, where they are established with minimal extrapolation.

so freezing cannot always be invoked to stop the rapid decrease in the liquid entropy as it is cooled below T_g.

Adam and Gibbs [4] argued that the viscosity of a liquid varies as

$$\eta = \eta_0 \exp\{C/TS_c\} \tag{5}$$

where η_0 and C are approximately constant and S_c is the configurational entropy, which counts the number of distinct configurations, or glasses, that the liquid samples. To estimate S_c they supposed that the entropy of an individual glass is about the same as that of a crystal, so that

$$S_c(P,T) \approx S_{ex}(P,T). \tag{6}$$

4

Angell [15,16] shows that, for a wide range of liquids, the temperature where S_{ex} extrapolates to zero is close to the temperature where relaxation times extrapolate to infinity, consistent with eqs. (5) and (6). However, eq. (6) is a poor approximation. [17] For instance17, $S_{ex} \approx 2.5 S_c$ near T_g in ethylbenzene [18] and figure 1 shows that $S_{ex} \approx 2.6 S_c$ near T_g for hard spheres.

The properties of glasses depend on their history and estimates of S_c for real materials vary widely because they require assumptions and extrapolations that are difficult to test [17]. However, the entropy of glasses formed with different histories can be measured in computer simulation experiments [13,19,20,21,22] and the studies that have tested [13,14,23,24,25] Adam and Gibbs' theory using directly measured values of S_c support eq. (5).

Stillinger [5] presents a persuasive argument against the idea of an ideal glass transition. He shows that when S_c is defined in terms of the number of inherent structures that a liquid samples it cannot go to zero at a finite temperature. The following discussion aims to overcome Stillinger's objection so that the concept of an ideal glass transition can be retained. The main assumption is that one can define perfect and imperfect glasses, according to whether they contain point defects. If S_c is defined to count only perfect glasses, rather than all inherent structures, then it can go to zero at a finite temperature. Only the most numerous [26] point defects, simple independent vacancies, are considered because this is sufficient for the purposes of the discussion.

A vacancy can be created in a crystal of N particles by selecting a molecule in the interior and moving it to the surface. N is taken to be large so that surface effects are negligible. The enthalpy change, h_v, required to create a vacancy is about equal to the enthalpy of sublimation [26]. The number of ways of distributing n_v independent vacancies over $N + n_v$ lattice sites,

$$\Omega_v = (N + n_v)! / N! n_v!, \tag{7}$$

provides the main [26] contribution to the entropy of vacancy formation and the equilibrium concentration of vacancies in a crystal, obtained by minimizing the free energy $n_v h_v - k_B T \ln\{\Omega_v(n_v)\}$ with respect to n_v, is [26]

$$n_v / N = \exp\{-h_v / k_B T\}, \tag{8}$$

where k_B is the Boltzmann constant. For simple crystals near their melting temperatures this concentration is very small [26], typically in the range 10^{-10} to 10^{-18}.

To apply similar ideas to glasses we need to define vacancies in a glass. Following Stillinger [5], we refer to any structure that is mechanically stable at absolute zero as an inherent structure, and we confine attention to amorphous inherent structures. A perfect glass is defined as an inherent structure that, like a perfect crystal, contains no vacancies. A vacancy is said to exist at a specified site within an inherent structure if the enthalpy can be reduced by transferring a molecule from the surface to the site. The inherent structure is a perfect glass when no such site is available. Each imperfect glass belongs to the perfect glass that results from filling its vacancies.

Let the number of structurally distinct perfect glasses with enthalpy z at absolute zero be $N_g(z)$. The Gibbs free energy of the liquid can then be expressed as

$$G_l(P,T) = G_g(P,T,z,n_v) - k_B T \ln\{N_g(z)\Omega_v(n_v)\} \qquad (9$$

where $G_g(P,T,z,n_v)$ is the free energy of an individual glass [17] and $k_B \ln\{N_g(z)\Omega_v(n_v)\}$ is the entropy due to the multiplicity [5] of inherent structures. The equilibrium values of the internal parameters, z and n_v, in eq. (9) are those that minimize [27], $G_l(P,T)$ so

$$(\partial G_g / \partial z)_{P,T,n_v} - k_B T(\partial \ln\{N_g(z)\} / \partial z)_{P,T,n_v} = 0 \qquad (10$$

and

$$(\partial G_g / \partial n_v)_{P,T,z} - k_B T(\partial \ln\{\Omega(n_v)\} / \partial n_v)_{P,T,z} = 0 . \qquad (11$$

The last derivative

$$(\partial \ln\{\Omega(n_v) / \partial n_v\}_{P,T,z} \approx -\ln\{n_v / N\} \approx h_v / k_B T , \qquad (12$$

diverges as T and n_v go to zero but this divergence has no effect on the terms in equation (10). In particular, the divergence has no implications for configurational entropy, defined in terms of the number of perfect glasses, by

$$S_c(z) = k_B \ln\{N_g(z)\} . \qquad (13$$

In Stillinger's argument [5] the configurational entropy, $k_B \ln\{N_g(z)\Omega_v(n_v)\}$, counts all the inherent structures, so it can go to zero only as T and n_v go to zero. Re-defining S_c in terms of perfect glasses allows us to retain the useful concept of ideal glass transition, where $S_c(z(T)) \to 0$ at $T > 0$. In practice the difference is negligible. For instance, the enthalpy of evaporation of glassy water [28] near $150 K$ is 46.9 kJ/mol which gives, from eq. (8), $n_v / N < 10^{-16}$, corresponding to an entropy difference of $k_B \ln\{\Omega_v(n_v)\} < 10^{-15} R$. $R = Nk_B$ is the gas constant.

4. Generalized van der Waals model

The familiar van der Waals' equation, $P = RT/(V - Nb) - a(N/V)^2$, was generalised by Longuet-Higgins and Widom [6], who replaced the repulsive term, $Nk_B T/(V - Nb)$ by the pressure of hard spheres, P_{HS}. They [6] showed that the model gives a good account of melting and of the properties of Argon at its triple point. The constants b and a measure the particle size and the strength of the mean field potential.

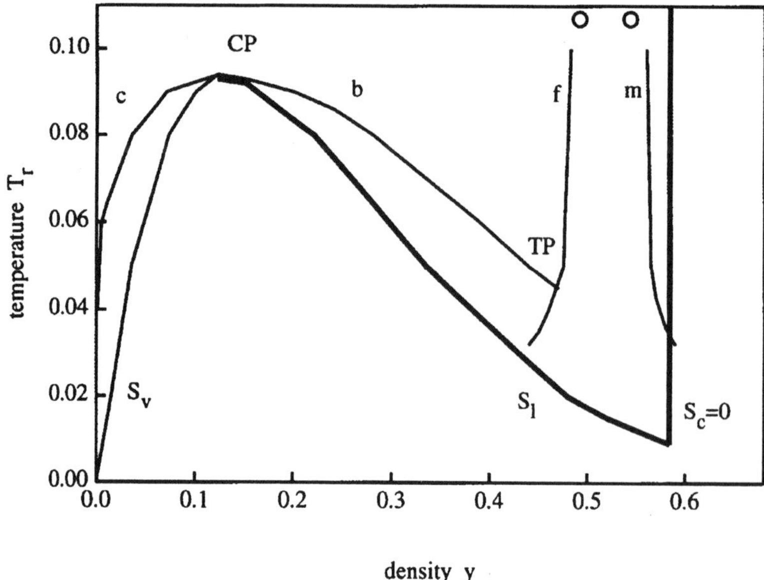

Figure 2. Temperature-density diagram for the Hard sphere model with van der Waals attractions. *CP* is the gas-liquid critical point. *TP* is the triple point. The vapor condensation line is *c*. The liquid boiling line is *b* and freezing line *f*. The fcc crystal melting line is m. S_v is the vapor spinodal. S_l is the liquid spinodal. $S_c = 0$ is the ideal glass transition line. Circles show the freezing and melting densities in the high temperature limit, that is, for hard spheres. The stability limits for the liquid, the bold lines, are shown in the *P,T* plane in figure 3.

The pressure of the hard sphere model with van der Waals mean field attractions (HSvdW) has the form

$$P_r = f(y)T_r - y^2 \qquad (14$$

where $P_r \equiv PV_0^2 / N^2 a$ and $T_r \equiv RTV_0 / N^2 a$ are a dimensionless reduced pressure and temperature, $f(y) = yP_{HS}V / RT$, where the density $y = V_0 / V$, $V_0 = N\pi\sigma^3 / 6$ is the volume of the spheres, σ is the diameter of a sphere.

Figure 2 shows the temperature-density diagram for the HSvdW model. The free energy of the gas, liquid and crystalline phases was calculated from empirical equations for the pressure and entropy of the hard sphere fluid [12] and face-centered-cubic crystal [11]. Hoover and Ree [10] first measured the entropy of a hard sphere crystal to prove that it melts and to locate the melting transition thermodynamically. The entropy of the hard sphere fluid or crystal, $S_{HS}(y,T)$, is obtained by thermodynamic integration [10,11,12]. The entropy of the HSvdW model is the same as that of hard spheres at the same density and temperature, because the mean field potential has no effect on the motion of the spheres. The energy of the model is $U = (3/2)RT - aV_0 y$ and the Gibbs free energy, $G = U + PV - TS$, is given by

$$G / RT = 3/2 + f(y)/y - 2y/T_r - S_{HS}(y,T)/R. \tag{15}$$

The coexistence lines in figure 2 were calculated numerically by iterating P_r and T_r to find where two phases have the same $G(P_r, T_r)$. The spinodal lines in figure 2 were located from the condition

$$(\partial P_r / \partial y)_{T_r} = 0 = T_r df(y)/dy - 2y. \tag{16}$$

Hard sphere crystals also tend towards a spinodal instability [11], near the density $y = 0.47$.

The hard sphere fluid has a kinetic glass transition near the density [13,29] $y_{gt} = 0.57$. The measured pressure of the kinetic glasses fits the empirical equation [13,20] $PV/RT = 1 + 2.8y/(y_0 - y)$ and the measured entropy, relative to an ideal gas, is $\Delta_{ig}^{glass} S/R = 2.8\ln(y_0 - y) - 0.25$, with the (reproducible [13,30,31,32]) limiting density $y_0 \approx 0.648 \pm 0.001$. The ideal glass is described by the same equations with the limiting density shifted to $y_0 = 0.657$. This locates the ideal glass transition [13] at density $y_{gt}^0 = 0.586$, where the ideal glass has the same pressure and entropy as the fluid. The mean field potential does not affect the entropy or the dynamics so the HSvdW model has an ideal glass transition at the same density.

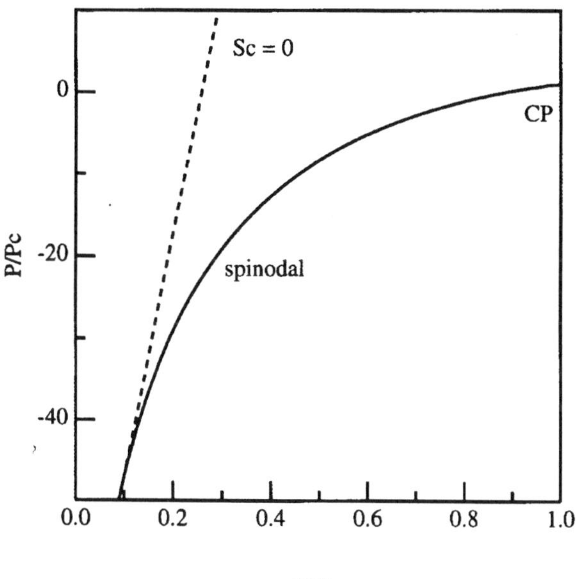

Figure 3. The spinodal and ideal glass transition lines for the Hard sphere model with van der Waals attractions, shown in the pressure-temperature plane. The estimated coordinates of the gas-liquid critical point are $T_c = RTV_0/N^2 a = 0.094$, $P_c = PV_0^2/N^2 a = 0.0043$ and $y_c = 0.12$.

Figure 3 shows the relation between the spinodal and ideal glass lines in the pressure–temperature plane. The ideal glass line is straight. A short extrapolation of the hard sphere

8

fluid equation of state [12] gives $f(y_{gt}^0) = 14.13$ and eq (14) then yields $P_r = 14.13 T_r - 0.34$ along the line where $S_c(P,T) = 0$. In the HSvdW model, the line $S_c(P,T) = 0$ is an isochor so eq. (2) requires that it meet the spinodal with same slope, as shown in figure 3.

Sastry [23] computed the locus of the spinodal and ideal glass transition lines for a Lennard-Jones model. In a temperature-density diagram [23] the line $S_c = 0$ has a positive slope, much lower than the infinite slope shown in figure 2. This difference is expected because the Lennard-Jones model has a softer repulsive potential and the effective density decreases with temperature [33]. When Sastry's results are plotted in the pressure-temperature plane [34] they are similar to figure 3. First, the pressure and temperature where Sastry's lines meet [34] $T = 0.11 T_c, P = -34 P_c$, are close to the values $T = 0.09 T_c, P = -49 P_c$ shown in figure 3, which suggests that the HSvdW model provides semi-quantitative predictions for simple liquids. Second, although the line $S_c(P,T) = 0$ is not an isochor in the Lennard Jones model, it meets the spinodal with same slope. La Nave et al, [35] studied orthoterphenyl, represented by three fused Lennard Jones atoms, and their results [36] also suggest that the two lines meet with the same slope. The coincidence of the slopes is explained in section 5.

Stillinger, Debenedetti and Truskett [37] present a different interpretation of the low temperature limit for a supercooled liquid. They do not consider an ideal glass transition and instead regard the "Kauzmann" line, where the excess entropy $S_{ex}(P,T) = 0$, as the low temperature limit. Figure 1 shows evidence that the ideal glass transition pre-empts the Kauzmann line in hard spheres, and the same conclusion was reached for ethylbenzene and toluene [17,18].

5. Coincidence of the slopes

Figure 3 shows that the ideal glass transition line, where $S_c(P,T) = 0$, and the spinodal line, $P_s(T)$, meet with the same slope. For the HSvdW model this is because the line $S_c(P,T) = 0$ is an isochor but that does not explain why the two lines seem to meet with the same slope [34,36] in other models. This section explains the apparent coincidence.

From the differential of the entropy $S(P,T)$

$$dS = (\partial S / \partial T)_P \, dT + (\partial S / \partial P)_T \, dP \tag{17}$$

the slope of a line of constant entropy is

$$(\partial P / \partial T)_S = -\frac{(\partial S / \partial T)_P}{(\partial S / \partial P)_T} = \frac{C_p}{TV\alpha} \tag{18}$$

where $C_p = T(\partial S / \partial T)_P$ is the isobaric heat capacity and $\alpha = -(\partial S / \partial P)_T / V$ is the isobaric expansivity.

The configurational entropy can be defined [17] as

$$S_c(z(P,T)) \equiv S_l(P,T) - S_g(P,T,z) \tag{19}$$

where $S_l(P,T)$ is the entropy of the liquid and $S_g(P,T,z)$ is the entropy of a glass of the type [27] $z(P,T)$ that has the same volume and enthalpy as the liquid [27] at P,T. Davies and Jones [27] show that two internal parameters, $z = \{z_1, z_2\}$, are needed to characterize glasses, but that complication is not important in the present context. The slope of a line of constant S_c is

$$(\partial P / \partial T)_{S_c} = -\frac{(\partial S_c / \partial T)_P}{(\partial S_c / \partial P)_T} = \frac{\Delta C_p}{TV\Delta\alpha} \tag{20}$$

where Δ indicates the difference between liquid and glass properties, $\Delta C_p = C_{p,l} - C_{p,g}$ and $\Delta\alpha = \alpha_l - \alpha_g$. The differences arise [27] because $z(P,T)$ varies in the liquid but is fixed in a glass. Near the liquid spinodal $C_{p,l}$ and α_l diverge so $\Delta C_p \approx C_{p,l}$, $\Delta\alpha \approx \alpha_l$ and eqs. (2), (18) and (20) imply that

$$(\partial P / \partial T)_{S_c} \to (\partial P / \partial T)_S \to dP_s / dT \text{ as } P \to P_s(T). \tag{21}$$

The ideal glass transition line, where $S_c(P,T) = 0$ is constant, therefore meets the spinodal with the same slope, as shown for the special case of the HSvdW model in figure 3. If the slope of the spinodal is monotonic, $dP_s / dT > 0$, then the intersection occurs at the maximum possible tension.

6. Water

Stability limits for stretched and supercooled water, from recent simulation studies, are shown in figure 4. An intriguing feature of figure 4 is that the locus of the density maxima, where $\alpha = 0$, extrapolates to meet the ideal glass transition line near the point where its slope changes sign. Another feature is that a simple linear extrapolation of the spinodal and ideal glass lines would have them meet at right angles, which is inconsistent with eq. (21).

Equation (21) requires that the spinodal and ideal glass transition lines meet with the same slope and figure 5 sketches two ways in which this requirement might be met. In figure 5(a) the spinodal goes through a pressure minimum [1,2] and turns up to meet the line $S_c(P,T) = 0$. Equation (3) requires [2] that the line $\alpha = 0$ ends at the spinodal where its slope, dP_s / dT, changes sign. However, the slope, $(\partial P / \partial T)_{\alpha=0}$, of the line $\alpha = 0$ must be negative [38] where it meets the spinodal in figure 5(a) while simulations [40,39] suggest that it is positive, as shown in figure 4, and there is no sign from the simulations that the spinodal goes through a minimum, as shown in figure 5(a).

Figure 5(b) shows a more likely scenario where the line $S_c(P,T) = 0$ goes through a temperature maximum and turns back to meet the spinodal with the same slope. Figure 4 suggests that the line $\alpha = 0$ may end on the line $S_c(P,T) = 0$ where the slope $(\partial P / \partial T)_{S_c}$ diverges, and it is drawn that way in figure 5(b). This speculation requires that

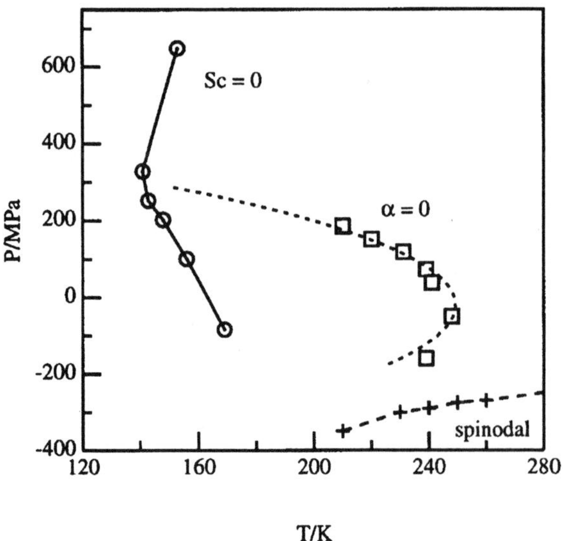

Figure 4. Properties of simulated simple point charge model for water [40]. Squares locate density maxima, where the isobaric expansivity $\alpha = 0$, and circles locate the (extrapolated) line where $S_c(P,T) = 0$, from ref. [40]. Crosses show the spinodal line from ref. [41].

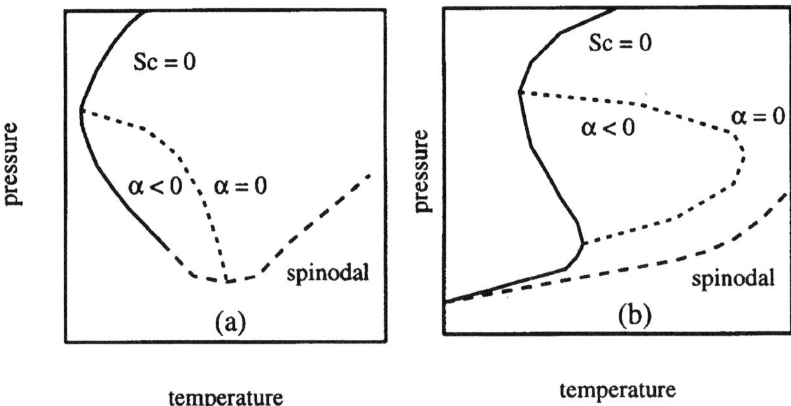

Figure 5. Possible loci of the stability limits and density maxima in water that are consistent with figure 4 and with equation (21).

$\alpha_g = 0$ at those points, because eq. (19) implies that $\Delta\alpha = \alpha_l - \alpha_g = 0$ where the slope $(\partial P / \partial T)_{S_c}$ diverges.

The region where $\alpha < 0$ must be enclosed by the line $\alpha = 0$ and the stability limits, otherwise it would be possible for α to change sign without going through zero. A third possibility, not shown, is that the line $\alpha = 0$ is a closed loop surrounding a region where $\alpha < 0$.

7. Conclusions

The main theme of the paper is that the appropriate thermodynamic limits of stability for a liquid are the spinodal line and the ideal glass transition line. The idea of a spinodal limit dates back to van der Waals and Kauzmann [8] alluded to the idea of a thermodynamic glass transition as a low temperature limit in 1948, so these are old ideas.

One reason for the lack of progress with the ideal glass transition concept since the early work [3,4] is the difficulty of measuring the entropy of real glasses with substantially different histories and entropies. Little is known about how the entropy of real glasses, $S_g(P,T,z)$, or the configurational entropy $S_c(z)$ (eq. (19)), varies with z, so extrapolations to find where $S_c \to 0$ and to establish the properties of ideal glasses are unreliable [17]. However, the entropy of glasses can be measured in computer simulation experiments [13,19,20,21,22,23,24,25,35,40] and these studies offer increasing support for the ideal glass transition concept. As noted in section 4, the present interpretation agrees with Sastry's interpretation [23,34] of the stability limits in a Lennard Jones liquid.

A new result, Equation (21), requires that the spinodal and ideal glass transition lines have the same slope where they meet and imposes a condition on the extrapolation of the lines to extreme tensions.

Acknowledgement

I thank P. G. Debenedetti, S. Sastry and F. Sciortino for helpful discussions and for sharing unpublished results.

References

1. Debenedetti, P. G. (1996) *Metastable Liquids* Princeton University Press: Princeton, NJ, USA.
2. Speedy, R. J. (1982) The Stability Limit Conjecture: An Interpretation of the Properties of Water, *J. Phys. Chem* 86, 982.
3. Gibbs, J. H. and DiMarzio, E. A. (1958) Nature of the Glass transition and the Glassy Sate, *J. Chem. Phys.* 28, 373 and 807.
4. Adam, G. and Gibbs, J. H. (1965) On the Temperature Dependence of Cooperative Relaxation Properties in Glass-Forming Liquids, *J. Chem. Phys.* 43, 139.
5. Stillinger, F. H. (1988) Supercooled Liquids, Glass Transitions, and the Kauzmann Paradox, *J. Chem. Phys.* 88, 7818.
6. Longuet-Higgins, H. C. and Widom, B. (1964) A Rigid Sphere Model for the Melting of Argon. *Mol. Phys.* 8, 549.
7. Skripov, V. P. (1980) Metastable States of Water-Superheating, Supercooling and the boundary of Thermodynamic Stability. *Water and Steam. Their Properties and Current Industrial Applications. Proceedings of the 9th International Conference on the Properties of Steam.* J. Straub and K. Sheffler, eds., Pergamon Press: Oxford.
8. Kauzmann, W. (1948) The Nature of the Glassy State and the Behavior of Liquids at Low Temperatures, *Chem. Rev.* 43, 219.
9. DiMarzio, E. A. (1981) Equilibrium Theory of Glasses, *Ann. N.Y. Acad. Sci.* 371, 1.
10. Hoover, W. G. and Ree, F. H. (1968) Melting Transition and Communal Entropy for Hard Spheres, *J. Chem. Phys.* 49, 3609.
11. Speedy, R. J. (1998) Pressure and entropy of hard-sphere crystals, *J. Phys.:Condens. Matter* 10, 4387.

12. Speedy, R. J. (1997) Pressure of the metastable hard-sphere fluid, *J. Phys.:Condens. Matter* 9, 8591.
13. Speedy, R. J. (1998) The hard sphere glass transition, *Mol. Phys.* 95, 169.
14. Speedy, R. J. (2001) Configurational Entropy and diffusion in a hard disc fluid, *J. Chem. Phys.* 114, 9069.
15. Angell, C. A. (1985) *Relaxations in Complex Systems*, K. Ngai and G. B. Wright (eds.). National Technical Information Series, U.S. Department of Commerce, Springfield, VA, USA. Page 1.
16. Angell, C. A. (1997) Entropy and Fragility in Supercooled Liquids, *J. Res. Natl. Inst. Stand. Technol* 102, 171.
17. Speedy, R. J. (2001) Estimates of the Configurational Entropy of a Liquid, *J. Phys. Chem. B* 105, 11741.
18. Yamamuro, O., Tsukushi, I., Lindqvist, A.; Takahara, S., Ishikawa, M. and Matsuo, T. J. (1998) Calorimetric Study of Glassy and Liquid Toluene and Ethylbenzene, *J. Phys. Chem. B* 102, 1605.
19. Speedy, R. J. (1993) The entropy of a glass, *Mol. Phys.* 80, 1105.
20. Speedy, R. J. and Debenedetti, P. G. (1996) The distribution of tetravalent network glasses, *Mol. Phys.* 88, 1293.
21. Büchner, S. and Heuer, A. (1999) The potential energy landscape of a model glass former: thermodynamics, anharmonicities, and finite size effects, *Phys. Rev. E* 60, 6507.
22. Sciortino,F., Kob, W. and Tartaglia, P. (1999) Inherent Structure Entropy of Supercooled Liquids, *Phys. Rev. Lett.* 83, 3214.
23. Sastry, S. (2000) Liquid Limits: The Glass Transition and Liquid-Gas Spinodal Boundaries of Metastable Liquids, *Phys. Rev. Lett.* 85, 5590.
24. Sastry, S. (2001) Fragility, Configurational Entropy and the Potential Energy Landscape of Glass Forming Liquids, *Nature* 409, 164.
25. Scala, A., Starr, F., La Nave, F., Sciortino. F. and Stanley, H. E. (2000) Configurational Entropy and Diffusivity of Spercooled Water, *Nature* 406, 166.
26. Berry, R. S., Rice, S. A. and Ross, J. (1980) *Physical Chemistry*, John Wiley & Sons, Inc. New York. Section 22.4.
27. Davies, R. O. and Jones, G. O. (1953) On the Thermodynamic and Kinetic Properties of Glasses, *Adv. Phys.* 2, 370.
28. Speedy, R. J., Debenedetti, P. G., Smith, R. S., Huang, C. and Kay, B. D. (1996) The evaporation rate, free energy and entropy of amorphous water at 150 K, *J. Chem. Phys.* 105, 240.
29. Woodcock, L. V. (1981) Glass transition in the Hard-Sphere Model and Kauzmann's Paradox, *Ann. N.Y. Acad. Sci.* 371, 274.
30. Jodrey, W. S. and Tory, E. M. (1985) Computer simulation of close random packing of equal spheres, *Phys. Rev. A* 32, 2347.
31. Speedy, R. J. (1994) Quench rate independence of the hard sphere glass transition, Mol. Phys. 83, 591.
32. Speedy, R. J. (1994) On the reproducibility of glasses, *J. Chem. Phys.* 100, 6684.
33. Speedy, R. J., Preilmeier, F. X., Vardag, T., Lang, E. W. and Lüdemann, H.-D. (1989) Diffusion in simple fluids, *Mol. Phys.* 66, 577.
34. Sastry, S. (2002) personal communication.
35. La Nave, E, Mossa, S. and Sciortino, F. (2002) Potential Energy Landscape Equation of State, *Cond-mat*/0202039
36. Sciortino, F. (2002) personal communication.
37. Stillinger, F. H., Debenedetti, P. G. and Truskett, T. (2001) The Kauzmann Paradox Revisited, *J. Phys. Chem. B* 105, 11809.
38. D'Antonio, M. C. and Debenedetti, P. G. (1987) Loss of tensile strength in liquids without property discontinuities: A thermodynamic analysis, *J. Chem. Phys.* 86, 2229.
39. Poole, P. H., Sciortino, F., Essmann, U. and Stanley, H. E. (1993) Spinodal of liquid water, *Phys. Rev. E* 48, 3799.
40. Scala, A., Starr, F. W., La Nave, E., Stanley, H. E. and Sciortino, F. (2000) Free energy surface of supercooled water, *Phys. Rev. E* 62, 8016.
41. Netz, P. A., Starr, F. W., Stanley, H. E. and Barbosa, M. C. (2001) Static and dynamic properties of stretched water, *J. Chem. Phys.* 115, 344.

LIMITING SUPERHEAT OF AQUEOUS SOLUTIONS AT NEGATIVE PRESSURES

V.E. VINOGRADOV AND P.A. PAVLOV

Institute of Thermal Physics, Ural Branch of the Russian Academy of Sciences

Pervomaiskaya Str.91, GSP-828, Ekaterinburg, 620219,Russia

Abstract. The temperature of the limiting superheat for water-ethyl alcohol, water-isobutyl alcohol and water-acetone solutions has been measured in the region of negative pressures to -10 MPa at water concentration from 0 to 100%. Dependencies of the limiting-superheat temperature for pure liquids and solutions on the value of the negative pressure $T^*(P)_J$ have been obtained in the region of temperatures exceeding the boiling temperature of pure components at atmospheric pressure. For organic liquids one can observe a continuous transition of the dependence $T^*(P)_J$ from the region of positive into the region of negative pressures. Experimental results within the limits of experimental error agree with calculations by the formulae of homogeneous nucleation theory. In the case of water, the slope of the curve $T^*(P)_J$ increases in passing from positive to negative pressures. The discrepancy between experimental data and theory increases with the liquid stretch. Additions of water into organic liquids increase the slope of curves $T^*(P)_J$ in the region of negative pressures when a certain concentration of water is achieved. Analysis of the results has shown that it happens when the concentration of water in a solution exceeds that of the azeotropic compound.

1. Introduction

Tensile strengths of water measured in different laboratories of the world often differ by 1-2 orders [1-8]. Part of these results is shown with dots in Fig.1. Here solid lines show the line of liquid–vapor equilibrium (binodal) and the boundary of the liquid thermodynamic stability (spinodal). Calculation of the spinodal in the region of negative pressures is not an easy task and requires a special discussion. In our work we used the simplest Furth formula [9], which works fairly well in the region of positive pressures and is in good agreement with more correct calculations:

$$P_{Sp} = P_S - C \cdot \sigma^{3/2} \cdot (k \cdot T)^{-3/2}, \tag{1}$$

where PS is the vapor pressure on the saturation line at the experiment temperature; C=1.32; σ is the liquid surface tension at the liquid- vapor interface; T is the liquid temperature in Kelvin degrees; k is the Boltzmann constant. The dashes show the lines of limiting superheats of water for different nucleation rates calculated with the help of formulas of the classical homogeneous nucleation theory [9].

$$J = N_1 B \exp(-W_k / kT). \tag{2}$$

A.R. Imre et al. (eds.), Liquids Under Negative Pressure, 13–22.
© 2002 *Kluwer Academic Publishers. Printed in the Netherlands.*

14

Here N_l is the number of molecules in a unit of the liquid volume; B is the kinetic coefficient; W_K is the work of formation of a critical bubble:

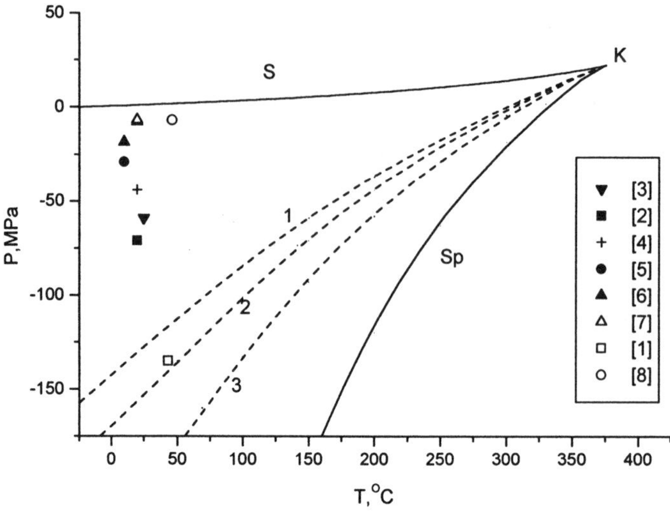

Figure 1. Limiting stretches of water from data of different authors [Refs. 1-8]. K: critical point; S: liquid-vapor equilibrium line; Sp: spinodal calculated with the help of Furth formula (1); dashed lines show the lines of limiting superheats of water for different nucleation rates calculated with the help of formulas of the classical homogeneous nucleation theory (2-4): 1: J=10^0 m^{-3}s^{-1}, 2: J=10^{12} m^{-3}s^{-1}, 3: J=10^{24} m^{-3}s^{-1}.

$$W_k = \frac{16\pi\sigma^3}{3(P_S - P')^2(1 - v'/v'')^2},$$

(3)

where P' is the pressure in the liquid; v', v'' are the specific volumes of the liquid and the vapor respectively. The kinetic coefficient B has been calculated by the formula for a viscous liquid [12]:

$$B = \frac{N_1 r_k}{2\eta}\sqrt{kT\sigma},$$

(4)

where $r_k = 2\sigma/(P_S-P')$ is radius of a critical bubble; η is the liquid viscosity.

It is known that at positive pressures also water can be superheated much less than all other liquids. To a great extent it is connected with the fact that water contains a great number of ready boiling centres and weak spots. By some data the density of nucleation centres in water reaches 10^{12} m^{-3}. Therefore, the values of limiting superheats and stretches of water considerably depend on the methods of measurements. In particular, water can be superheated to theoretical values only by the method of pulse heating [11].

In this method at the expense of fast conversion of a liquid into the metastable state boiling on easily activated evaporation centres has no time to develop to the beginning of thermofluctuation boiling.

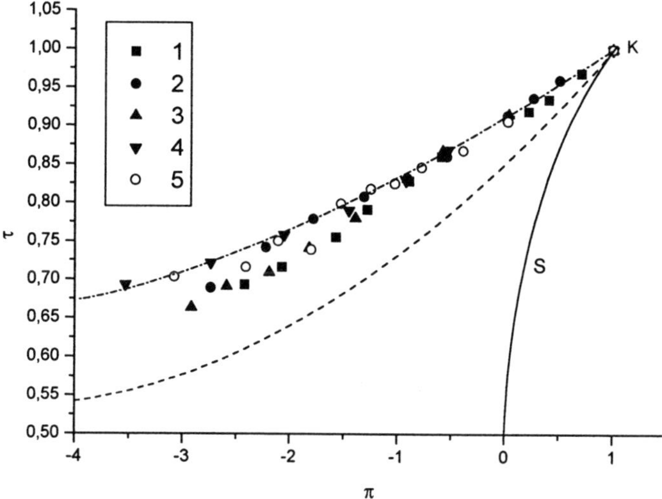

Figure 2. Limiting-superheat temperature of organic liquids as a function of pressure in reduced coordinates: $\pi = P/P_K$, $\tau = T/T_K$. 1: benzene, 2: toluene, 3: n-heptane, 4: n-hexane, 5: diethyl ether. K: critical point; S: liquid-vapor equilibrium line; dashed line: spinodal calculated from the Van der Waals equation; dash-dot line: boundary of limiting superheats calculated with the help of formulas of the classical homogeneous nucleation theory (2-4) for $J=10^{24}$ m^{-3}s^{-1}.

The paper presents dependencies of the temperature of the limiting superheat of water and its solutions with ethyl alcohol, isobutyl alcohol and acetone obtained in the transition region from positive to negative pressures by using a combination of the methods of pulse heating on thin wire heaters and the method of pulse stretch during the reflection of a pressure wave from a free surface.

2. Experimental setup and methods of conducting experiments

The scheme of the experimental setup is given in Fig.3. A liquid was superheated on a thin platinum wire (diameter: 20 μm, length: 10mm) immersed by 3-5 mm into the liquid under investigation. The height of the liquid column in the chamber was 40-45 mm. The chamber internal diameter was equal to 50 mm. Previously the liquid in the chamber was degassed by boiling and pumping-out with a vacuum pump. The wire (2) included in the bridge circuit was heated by rectangular current pulses of duration 15-25 μs. The liquid boiling-up on the wire was accompanied by the appearance of a high-frequency component on the heating pulse. The boiling-up pulse is caused by the change of the heat transfer from the wire as a result of origination on it of a vapor film.

16

The boiling-up signal was amplified and delivered to an oscillograph. A typical form of the boiling-up pulse is shown in Fig.4.

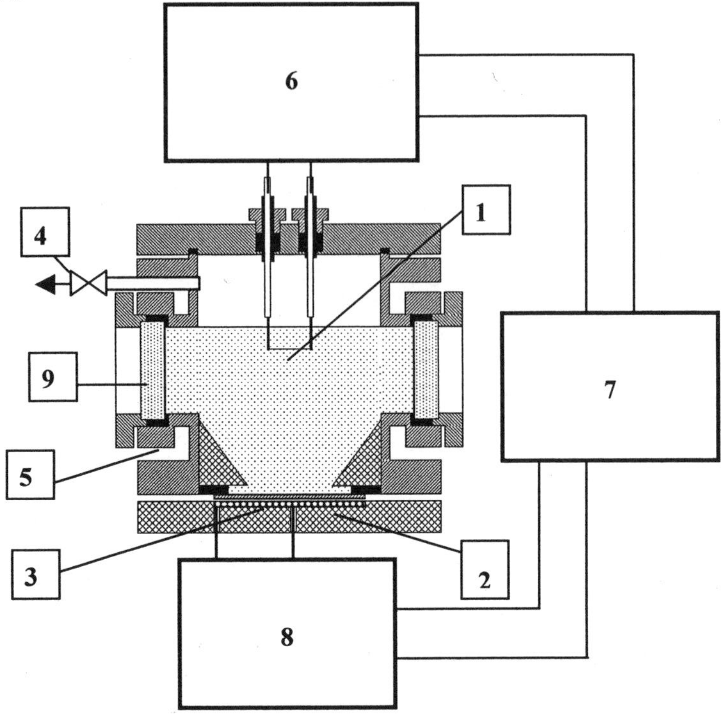

Figure 3. Block-diagram of the experimental setup. 1: wire, 2: coil, 3: membrane, 4: locking valve, 5: chamber, 6: unit of wire heating and temperature measurement, 7: control unit, 8: coil power supply unit, 9: window.

The wire temperature at the moment of the beginning of fluctuation boiling-up was determined by the position of the beginning of a boiling-up pulse with the help of the bridge circuit. The error of determining the temperature did not exceed 3K. The method of measuring the temperature of liquid boiling-up on thin wire heaters are described in detail in Ref. [11].

A negative pressure in a liquid was created during the reflection of a short pressure wave (~3 μs) from a free liquid surface. The pressure was formed by a duralumin membrane (3) when a low-inductance capacitor was discharged into a flat coil (2). The pressure pulse amplitude was measured with a piezoceramic pressure transducer. Before the beginning of experiments the setup was calibrated. The error of measuring the pressure was 5%.

The pressure and the heating pulses were started from a generator and coordinated in time in such a way that the moment of liquid boiling-up on the wire coincided with the passage through it of the maximum of a negative-pressure pulse. If necessary, the wire was photographed using pulse lighting with a flash bulb of pulse duration 1μs. In this case the setup working cycle was started from the camera shutter. This technique made it possible to realize experimentally nucleation rates between 10^{24} and 10^{26} m^{-3}s^{-1}.

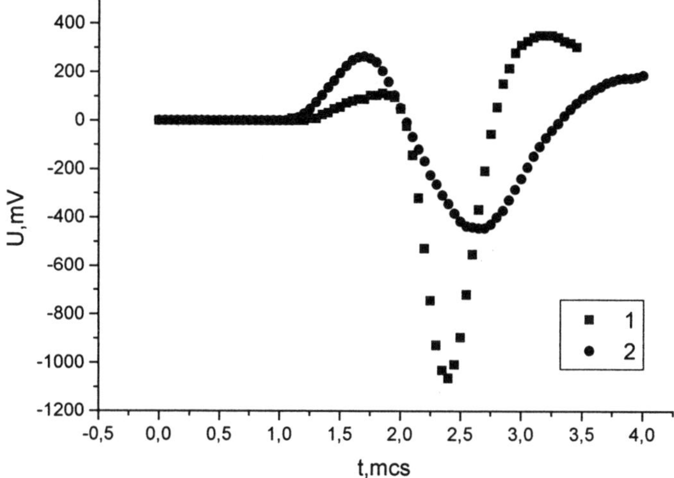

Figure 4. Shape of a boiling-up pulse in acetone. 1: at atmospheric pressure; 2: at negative pressure P= -3.0 MPa.

3. Results of experiments and their discussion

Dependencies of the temperature of the limiting superheat $T^*(P)_J$ of solutions on the value of the negative pressure in a liquid have been obtained. The results of experiments on ethanol and its solutions with water are shown in Figure 5. Here one can also find the experimental values of limiting-superheat temperatures for ethanol at positive pressures from reference [11]. For pure ethyl alcohol (azeotropic mixture of alcohol with water of volume concentration 96%) the dependence $T^*(P)_J$ has the same form as for other pure organic liquids. Within experimental error, it is in agreement with calculation by formulae of the classical homogeneous nucleation theory. Additions of water to alcohol increase the liquid limiting-superheat temperature at atmospheric pressure. At the same time one can observe an increase in the inclination of the curves $T^*(P)_{J,x}$ in going from positive to negative pressures.

Therefore there is a region of concentrations where the limiting-superheat temperature of a high-boiling solution is lower than that of a low-boiling one. It is clearly demonstrated in Fig. 6, which gives concentration dependences of the limiting-superheat temperature for an ethyl alcohol-water solution at constant pressures. Similar dependencies have been obtained for solutions of isobutyl alcohol in water (Fig.7). Experimental data on pure isobutyl alcohol are shown with an approximation curve not to shade the figure. The dashed lines that connect the critical points and the limiting-superheat temperatures at atmospheric pressure indicate arbitrarily the boundaries of the limiting superheat at positive pressures (similarly to the experimental dependence for pure ethyl alcohol in Fig. 5).

18

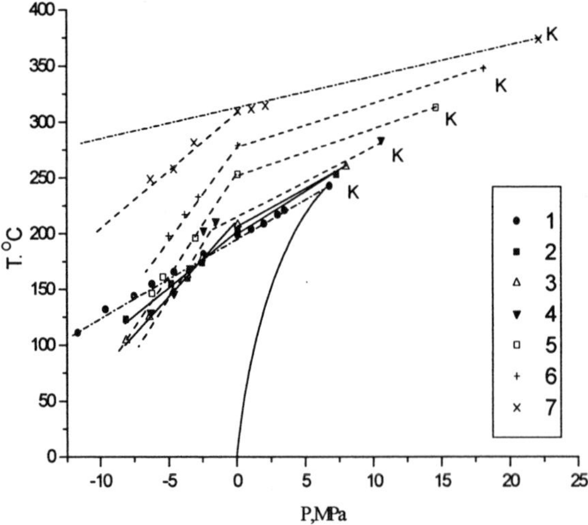

Figure 5. Temperature of the attainable superheat of ethyl alcohol and solution with water as a function of pressure. Dots in the figure show experimental temperatures of the limiting superheat: 1: "pure" ethyl alcohol (azeotropic mixture, water mole fraction x=0.11); 2: x=0.16; 3: x=0.31; 4: x=0.58; 5: 0.72; 6: 0.92; 7: water; the rest: critical points of solutions. The solid line in figure shows the liquid-vapor equilibrium line for pure ethyl alcohol; dashed lines: linear approximation of experimental data; dash-dot lines: calculated boundaries of limiting superheats.

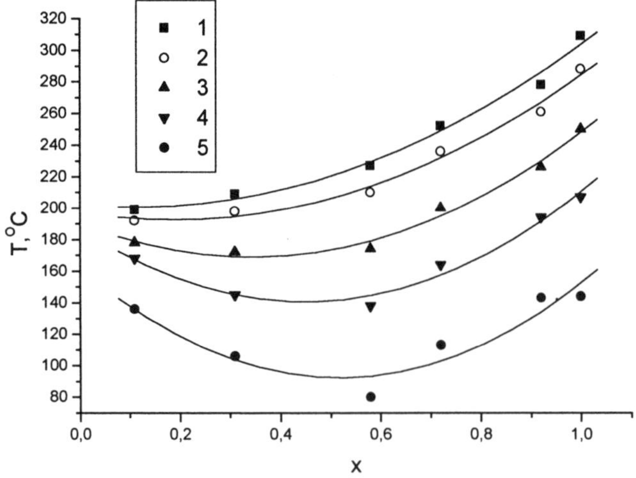

Figure 6. Temperature of the limiting superheat of solutions of ethyl alcohol with water as a function of water concentration at different pressures: 1: P=0.1 MPa, 2: P= -1.0 MPa, 3: P= -3.0 MPa, 4: P= -5.0 MPa, 5: P= -8.0 MPa.

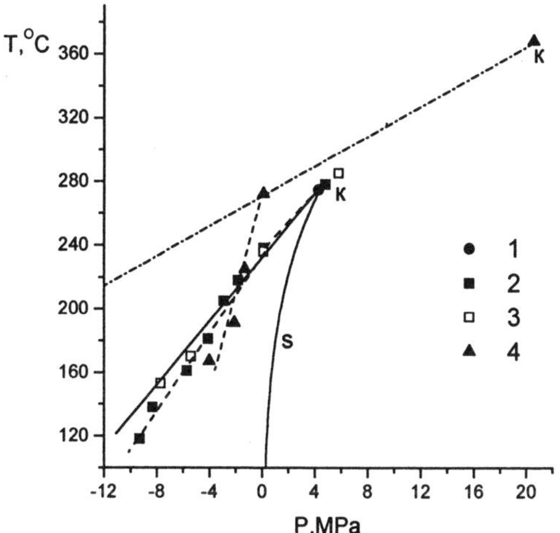

Figure 7. Boundary of limiting superheats for solutions of water in isobutyl alcohol. Dots: experiment, dashed lines: approximation of experimental data. 1: pure isobutyl alcohol (the line of limiting superheats is shown with a solid curve); 2: water weight concentration of 5%; 3: 12%; 4: 95%. K: critical points; S: liquid-vapor equilibrium line for pure isobutyl alcohol.

It should be noted that for water-isobutyl alcohol solutions there is a region of stratification at water concentrations from 8 to 92%. In the vicinity of this region dependences of the limiting-superheat temperature for a solution are the same as for pure isobutyl alcohol. This is evidently caused by a shift of the solution stratification line on the wire. As a result, isobutyl alcohol envelops the wire, and we observe the boiling-up of pure isobutyl alcohol. Figure 8 gives the results of experiments on water-acetone solutions. For pure acetone one can observe a smooth extension of the curve $T^*(P)_J$ in going from the region of positive pressures to region of stretches. However, at stretches of more than -4.0 MPa an increase in the slope of the curve $T^*(P)_J$ is observed. Additions of water to acetone smooth out the dependence $T^*(P)_J$ increasing the temperature of the acetone limiting superheat at high negative pressures. The dependence $T^*(P)_{x,J}$ becomes similar to good agreement with calculation by the homogeneous nucleation theory. Such a form of the curves is observed up to a water volume concentration of 60%. At higher water concentrations the slope of the curves $T^*(P)_{x,J}$ for solutions of water with acetone increases in the region of negative pressures, just as was the case with alcoholic solutions of water. Moreover, an increase in the slope of the curve $T^*(P)_{x,J}$ for acetone-water solutions begins even in the region of positive pressures.

Figure 9 shows dependencies of the limiting-superheat temperature of solutions on the concentration of water at constant pressures $T^*(x)_{J,P}$. One should pay attention to a very weak concentration dependence of the limiting-superheat temperature in the region of concentrations from 10 to 60%. Besides, for solutions of acetone in water, as distinct from solutions of alcohols, no decrease in the temperature of the liquid limiting

20

superheat was observed with an increase in the water concentration in the range of parameters under investigation.

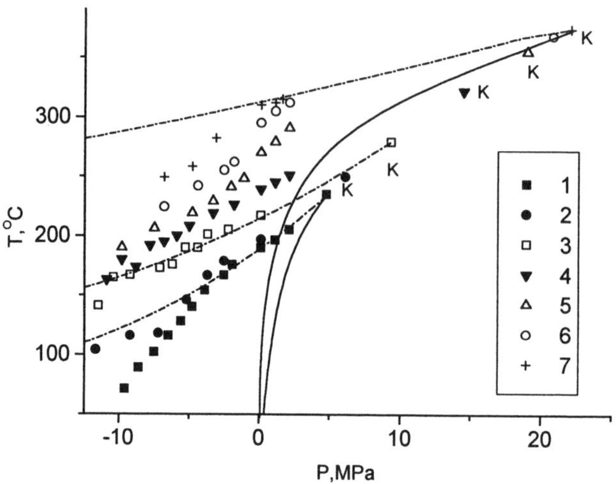

Figure 8. Boundary of limiting superheats for solutions of acetone with water. 1: acetone, 2: 10% of water in acetone, 3: 30% of water in acetone, 4: 60% of water in acetone, 5: 85% of water in acetone, 6: 95% of water in acetone, 7: water. Solid lines: liquid-vapor equilibrium lines for acetone and water; dashed lines: calculation by formulae of the homogeneous nucleation theory for $J=10^{24}$ m^{-3}s^{-1}.

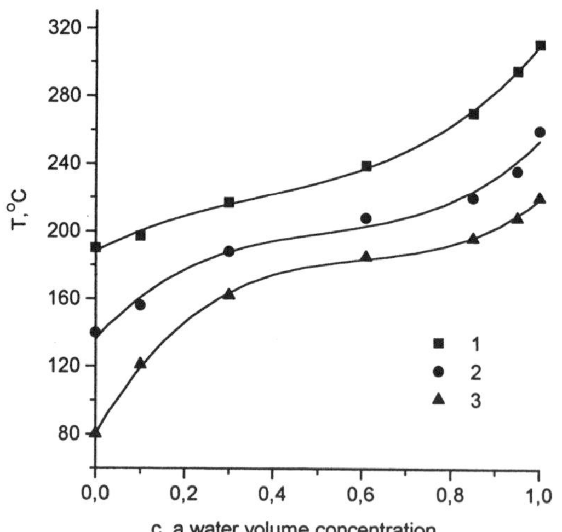

Figure 9. Temperature of the limiting superheat for solutions of acetone with water as a function of water concentration at different pressures. 1: P=0.1 MPa, 2: P= -5.0 MPa, 3: P= -9.0 MPa.

The obtained experimental results show that dependences of the limiting-superheat temperature for aqueous solutions of ethyl alcohol, isobutyl alcohol and acetone in the region of negative pressures at water concentrations lower than that of azeotropic composition have the same form as in the case of organic liquids investigated previously. Changes in the slope of the curves $T^*(P)_{x,J}$ during the entry into the region of negative pressures take place only at water concentrations exceeding the concentration of azeotropic composition. Respectively, for ethyl alcohol at water concentrations of more than 4%, for acetone at water concentrations exceeding 60%. In the case of isobutyl alcohol any additions of water change the slope of the curves $T^*(P)_{x,J}$ during the entry into the region of negative pressures. For isobutyl alcohol there are no azeotropic solutions with water. We do not understand as yet the dependence of the limiting-superheat temperature for pure acetone at negative pressures exceeding -4 MPa. We cannot explain this anomaly as well as its disappearance during the dissolution of water in acetone.

Thus, from our experiments it follows that the dependence of the limiting-superheat temperature for water on the value of the negative pressure differs qualitatively from similar dependences for organic liquids. Peculiar features of water tell on the limiting-superheat temperature for solutions of organic liquids only in the case when in a solution there is an excess of free water molecules.

Acknowledgement

The work has been done with a financial support of the Russian Foundation of Fundamental Investigations (projects № 01-02-16966, 00-15-96719)

References

1. Zheng Q, Durben D.J., Wolf G.H., Angell C.A. (1991) Liquids at large negative pressures: water at the homogeneous nucleation limit, *Science* **254.** 829-832.
2. Williams P. R... Williams P. M. Brown S. W. J., Temperley H. N. V. (1999) On the tensile strength of water under pulsed dynamic stressing, *Proc. R. Soc. Lond.* A **455**, 3311—3323.
3. Wurster C., Kohler M., Pecha R., Eisenmenger W., Suhr D., Irmer U., Brummer F., Hulser D. (1995) Negative pressure measurements of water using the glass fiber optic hydrophone , *Ultrason. World Congr., Berlin, Sept., 3-7, 1995:* Proc. Pt 2, Duisburg, pp. 635-638.
4. Bogach A.A.,Utkin A.V. (2000) Water strength under pulsed stretch, *Appl. Mech Tech. Phys.* (in Russian) **41**, 198-205.
5. Briggs L.J. (1950) Limiting negative pressure of water. *J. Appl. Phys.* **21**, 721-722.
6. Henderson S. J. and Speedy R.J. (1987) Temperature of maximum density in water at negative pressure. *J. Phys. Chem.* **91**, 3062-3068.
7. Besov A.C., Kedrinsky V.K., Morozov N.F., Petrov U.V., Utkin A.A. (2001) About analogy of an incipient state of fracture of solid bodies and fluids at an impulse loading. *Dokl. Acad. Nauk* (in Russian) **378**, 333-335.
8. Ohde Y., Ikemizu M., Okamoto H., Mosokawa W., Ando T. (1988) The two-stage increase in negative pressure with repeated cavitation for water in a metal Berthelot tube. *J. Phys.* D **21**, 1540-1542.
9. Skripov V.P.(1974) *Metastable liquids*, Wiley, New York.
10. Skripov V.P., Sinitsyn E.N.,Pavlov P.A. et al.(1988) *Thermophysical properties of liquids in metastable state*, Gordon and Breach Sci Publ., New York, London.
11. Pavlov P.A.(1988) *Boiling-up dynamics of highly superheated liquids*, Ural Scientific Centre of the USSR Academy of Sciences, Sverdlovsk.

12. Deryagin B.V., Prokhorov A.V., Tunitsky N.N. (1977) Statistical Thermodynamics of Formation of a New Phase. 2, Theory of Boiling-Up of Volatile Liquids, *Zh. Exp. Teor. Fiz.*(in Russian) **73**, 1831-1848.
13. Apfel R.E.(1971) Tensile strength of superheated n-hexane droplets. *Nature* **233**, 119-121.
14. Vinogradov V.E., Pavlov P.A.(2000) Boundary of limiting superheats for n-heptane, ethanol, benzene and toluene in the region of negative pressures. *High Temp.* **38**, 379-383.

CLASSICAL THERMODYNAMICS OF STATES WITH NEGATIVE ABSOLUTE TEMPERATURE OR WITH NEGATIVE ABSOLUTE PRESSURE

KATALIN MARTINÁS[1] AND ATTILA R. IMRE[2]
[1] *Eötvös University, Department of Atomic Physics, 1117 Budapest Pázmány Péter sétány 1/A, Hungary*
[2] *KFKI Atomic Energy Research Institute, Materials Department, H-1525 Budapest, Hungary*
E-mail: martinas@ludens.elte.hu

Abstract. Traditionally thermodynamics considers pressure defined by other disciplines, or the simple kinetic definition is accepted. Negative pressure states are often considered as impossible ones by thermodynamicians. In this paper we would like to show, that classical thermodynamics gives a good approach to investigate the properties of negative pressure states. Such kind of state is not just possible but in spite of the metastability it can be observed experimentally.

1. Introduction

We can meet with physical quantities which cannot be negative (like mass, volume, length, etc.) while some others can be negative and positive too (like charge, acceleration, etc.). But in some case the decision is not as easy as in these cases; let's see for example two thermodynamical intensives, the temperature and pressure. In elementary and middle school (primary schools) the temperature can be positive and negative too - in Celsius and Fahrenheit scale. Sometimes in the high- (or secondary) school we should learn about the absolute (Kelvin) scale - and around that time we should consider the temperature as an always positive quantity. Finally during the further studies, we should hear about the possibility of absolute negative temperatures (that funny thing which can be reached across the infinite).

But what about pressure? Although nobody had ever taught us any absolute pressure scales - that was simply natural for us, without any teaching. Additionally we have some simplified ideas about the pressure. The pressure comes from the collision of the particles with the wall. Particles push the wall, instead of pulling, therefore the pressure is positive. Further, pressure decreases with decreasing particle density; the zero pressure means vacuum, which is the BIG NOTHING. What can we found below it? It has to be nothing!

Time to time we could hear about negative pressure. For example, doctors talk about the negative pressure in the lung. That negative pressure is not really negative, just smaller that the atmospheric one. Can we say, that the pressure is always positive? When somebody is talking about negative pressure, he just means something smaller than 1 bar but bigger than zero?

The answer is definitely no. For example Landau in his famous theoretical physics

A.R. Imre et al. (eds.), Liquids Under Negative Pressure, 23–31.

24

books [1] mentions, that p<0 states can exist with limited stability - and he is talking about absolute pressure. Axiomatic thermodynamics (like Callen's [2],) cannot rule out such kind of state. In nuclear physics, QCD, astrophysics, etc., we can also meet with the pressure as a negative quantity - but just in some exotic states [3]. But what about "normal" states? Being positive pressure is "pushing", we have to accept that "pulling" (at least the big ones) are negative pressures. One cannot pull a gas – i.e. cannot generate negative pressure in it – but it is possible to pull a solid body. It is also possible to pull a liquid – although it is a little bit more difficult than to pull a solid. Pushing and pulling solid or liquid bodies we have to deal with a 3*3 pressure tensor ($\mathbf{P}$) rather than a scalar. We can define a scalar pressure as $(1/3)\mathrm{Tr}\mathbf{P} = (1/3)(p_{xx}+p_{yy}+p_{zz})$. In thermodynamics we have scalar pressure; it means that all off-diagonal elements of our pressure tensor is zero and diagonal elements are equal, i.e. $p = p_{xx} = p_{yy} = p_{zz}$. In liquids, this is the so-called hydrostatic pressure. It is possible to pull a liquid by a piston or by ultrasound or by shearing it between two rotating plate [4, 5 and references therein] but in that cases the pressure will not be scalar, therefore thermodynamics (at least classical one) cannot say anything about it. But it is also possible to generate negative hydrostatic pressure [4, 5, 6 and references therein] in liquids. Although the negative pressure states are not stable [7], they can live for considerable long times. It is easy to generate them in a laboratory, and it is also possible to find them in the nature, for example in liquid inclusions or in the xylem of trees [4,5,6 and references therein].

Therefore we should check carefully, what we need to fix an absolute pressure scale and how can we go below its zero point.

2. Pressure in Thermodynamics

In thermodynamics pressure is defined as the partial derivative of internal energy by the volume

$$-p = \partial U/\partial V \tag{1}$$

or in entropic representation.

$$p/T = \partial S/\partial V \tag{2}$$

Condition for local stability is that the second partial derivative of entropy is negative,

$$\partial^2 S/\partial V^2 < 0 \tag{3}$$

that is

$$\partial(p/T)/\partial V < 0 \tag{4}$$

if $T > 0$ then $\partial p/\partial V > 0$ states do not exist.

Thermodynamics requires that pressure must be a decreasing function of volume at constant energy. A central theorem of thermodynamics is the existence of absolute entropy scale. It means that to every state of a system belongs a unique value of entropy. **At the**

first glance that theorem implies that there is an absolute scale for pressure. Thermodynamics does not exclude the existence of $p< 0$ states, simply it does not tell anything about them. Nevertheless, that conclusion is valid, if and only if, the existence of absolute pressure scale is not a hidden (already built in) assumption in the derivation of the absolute entropy scale.

In this paper we will address the problem of the uniqueness of the thermodynamic pressure scale. Volume is a positive quantity. Natural selection for the scale is as follows: in the zero volume limit pressure tends to infinity, expressing the unavailability of the zero volume state; on the other end of the scale: pressure goes to zero when volume tends to infinity. At the first glance that definition excludes the possibility of negative absolute pressure states, as it would require volume, which is larger then infinity. Nevertheless, it is not the case. This definition does not exclude directly the existence of negative pressure. The statement that pressure goes to zero, when volume tends to infinity implies that a finite system with infinite volume has a zero pressure. On the other hand it does not mean that only systems in an infinite volume may have a zero pressure state. If there is an upper bound for volume, then the upper 'natural' zero point definition is meaningless.

Thermodynamic theories formulate postulates (laws) which ensure the existence of an entropy function, and the properties of entropy function define the pressure scale. We will investigate the question, whether the pressure is an external parameter or internal. In the first case the pressure scale is not a problem of thermodynamics. If the pressure is really an internal parameter, then its properties are constrained by thermodynamics. All traditional thermodynamic constructions are based primarily on the concept of heat. The basic postulates may be stated as follows:

> Zeroth Law of thermodynamics: If two (thermodynamic) systems are in thermal equilibrium with a third they are in thermal equilibrium with each other.

> First Law of thermodynamics: Conservation law of internal energy (U) says

$$dU = dQ + dL \tag{5}$$

> where dQ is the heat flow and dL is the work. If a system experiences an adiabatic process (dQ=0), then the work done on the system depends only on its initial and final states.

> Second Law of thermodynamics: No process is possible whose sole result is the transfer of heat from a cooler to a hotter body. (Clausius). It is impossible to convert an amount of heat completely into work by a cyclic process, without simultaneously producing other changes (Kelvin principle in Planck's formulation).

> Third Law of thermodynamics: By no finite series of processes is the absolute zero temperature attainable.

These postulates are sufficient to prove that in quasi-static reversible processes the heat dQ always can be written as:

$$dQ = T\,dS \tag{6}$$

where S is the entropy function and T the absolute temperature, that is, heat has an integrating multiplier.

There are many approaches to do that job. Best-known formulations of

26

thermodynamics are associated with the founding fathers of thermodynamics – Clausius [8], Kelvin [9], and Planck [10].

In the Clausius-Kelvin- Planck-Truesdell construction of thermodynamics work and heat are considered as primitive concepts, defined outside thermodynamics. Here first the absolute temperature scale is constructed. Entropy is obtained by a process of integration. It is shown that Second Law demands for the following inequality:

$$dQ/T < 0 \tag{7}$$

In that approach work is an external parameter. It is defined in mechanics, so the problem of negative absolute pressure is not a question of thermodynamics. It is a problem of the discipline defining the mechanical work. Usually we assume, that work is uniqelly defined . Nevertheless it was shown [11], that mechanics gives a unique definition for work only in case of a mass point. In continuum mechanics there is a freedom in the division of energy into work and heat. The problem here is, that without the unique definition of heat, that approach must not be applied.

Another treatment of the foundation of thermodynamics was given by Julius Farkas, Carathéodory [12] and reformulated later by Born [13,14] and Landsberg [15]. Here heat is considered as a Pfaffian expression, and it is shown that the Second Law ensures the existence of an integrating factor to heat. A basic theorem of thermodynamics is the proof of absolute entropy scale. Julius Farkas (1896) and Caratheodory (1909) were the firsts to show, that the Zeroth, First and Second Law of thermodynamics are sufficient to prove the existence of an entropy function, in the form

$$dS = dQ/T \tag{8}$$

where T is the absolute temperature function. There is only one freedom, namely to select the unit of temperature. The most general form of entropy function is

$$S^* = KS + S_0 \tag{9}$$

where S=S(U,V,N,..) (U is the internal energy, V is volume, N is for mole numbers, ... means the further variables). K is an arbitrary constant. The Third Law of thermodynamics allows to choose S_0 such that

$$S = 0 \tag{10}$$

at zero temperature, so S_0 is also defined.

At the first glance this is a proof of the absolute pressure scale:

$$p = T^* \partial S^* / \partial V = T \partial S / \partial V \tag{11}$$

Pressure does not depend on the choice of K. There is a uniquely defined pressure scale in thermodynamics. A negative absolute pressure state means that the increase of volume results in the decrease of entropy, it may or may not exist.

That conclusion is the result of the distinguished role of adiabatic processes. In reality, there are only two processes when the heat/work is uniquely defined, namely

    - no work process - only internal energy changes. When all the extensive parameters, but the internal energy are fixed: it is the original Joule experiment.

    - no heat process (adiabatic work) - occurs in the original Joule experiment: a body (considered as a mass point) is lowered in a gravitational field.

In all the other experiments heat and work appear simultaneously, and only the internal energy change can be measured

$$dU = dQ + dL \tag{12}$$

and the adiabatic work has the property that in an isothermal Environment one must not get effective work in a cyclic process (Moutier-theorem),

$$\int dL \geq 0 \tag{13}$$

It is easy to show that if dL_{Ad} is an adiabatic work expression, satisfying equation, then

$$dL^* = dL + p_o\, dV \tag{14}$$

is also a valid "adiabatic work" expression, satisfying equation 13.

Fowler [16] was the first to notice that problem. He showed that the statistical physical definition of entropy, logically founded on its increasing property, is not unique. "The identification of S and klogW is based on an analogy, correct enough so far as it goes, but insufficiently deep. For it is tacitly assumed that the entropy is the only function of the state of the assembly which has this increasing property."

He showed that any function defined by

$$S^* = klog\ W + KU \tag{15}$$

where W is the thermodynamic probability, K is an arbitrary constant, and U is the internal energy, has the same increasing property, and a maximum at the same position, as statistical entropy. We have no a priori reason for preferring one value of K to any other. He called S* "ekaentropy"[#], and concluded that one needs an additional postulate in statistical physics to arrive to the thermodynamic entropy. This postulate can be in the form

$$dS = d\ Q/T \tag{16}$$

or as a further postulate, we have to require in statistical derivation of the entropy that "in a quasi-static adiabatic process the entropy does not change".

Fowler's conclusion was: "the use of functions with the increasing property can apparently never lead to precise results without an appeal to dQ. If this appeal has to be made in any case, the method of approach by the increasing property loses any possible advantage over the classical method." Nevertheless, he did not notice that the problem of uniqueness is an

---

[#] In the original Fowler's paper [16] ekaentropy was marked by "P", but to avoid confusion (p is pressure), we are using S* as ekaentropy.

28

inherent problem in thermodynamics.

Problem: Specification of an adiabatic process is hard without the concept of entropy.
Solution: Postulate the properties of the adiabatic processes. As it is done in Lieb, E.H. and J. Yngvason papers [17, 18] without a receipt how to establish it.
Result: Ekaentropy family

$$S^* = A\,S + Y_{1o}U + Y_{2o}\,V + S_o \tag{17}$$

where A, S_o, Y_{1o} are constants, S is entropy, U is internal energy, V is volume and S^* is ekaentropy. Y_{io} modifies the zero point of the entropic intensive parameters. If Y_k^i is the entropic intensive, then all

$$Y^*{}_k^i = Y_k^i + Y_{ko} \tag{18}$$

can be selected as ekaentropic intensive parameters.
For temperature:

$$1/T = \partial S/\partial U \tag{19}$$

and

$$1/T^* = \partial S^*/\partial U = 1/T + Y_{1o} \tag{20}$$

that is the new temperature scale:

$$T^* = T/(1 + Y_{1o}\,T) \tag{21}$$

similarly, the new thermodynamic pressure scale is

$$p^* = p\,/((1 + Y_{1o}\,T) + Y_{2o}\,/Y_{1o} = p\,T^*/T + p_o \tag{22}$$

Ekaheat:

$$Q^* = T^*\,dS^* \tag{23}$$

In a reversible ekaadiabatic process ($Q^* = 0$) ekaentropy does not change ($dS^* = 0$). Ekawork is defined, as:

$$W^* = dU - Q^* \tag{24}$$

The relation of ekaheat and heat:

$$Q^* = Q + (1 - p_o\,/p)/(1 + TY_{1o}\,)dW \tag{25}$$

and

$$W^* = W - (1 - p_o/p)/(1 + TY_{1o})dW \qquad (26)$$

where $p_o = Y_{2o}/Y_{1o}$.

Each choice (ekaheat, ekawork, ekaentropy) gives a valid thermodynamics, only the state equations will be modified. Thermodynamics is invariant under this transformation. That invariance property is the consequence of the conservation of internal energy.

Ekaentropy will be entropy if $Y_{io} = 0$. It is sufficient to fix it for one system, most practically for ideal gas.

In gases entropy takes the absolute maximum value as a function of the relevant extensive (energy, volume) go to infinity. It gives us two new thermodynamic postulates:

$$\lim_{U \to \infty} \partial S/\partial U = 0 \qquad (27)$$

and

$$\lim_{V \to \infty} \partial S/\partial V = 0 \qquad (28)$$

These new postulates forbid negative absolute temperature and pressure states only in gases, that is in systems without upper bound on energy and volume.

3. Postulates of thermodynamics

Together with these two new postulates, we have six postulates of thermodynamics.
   1. Perpetual mobile of first type does not exist.
   2. Perpetual mobile of second type does not exist.
   3. Inaccessibility of U_{min} (inaccessibility of $1/T = \infty$).
   4. Inaccessibility of V_{min} (inaccessibility of $p/T = \infty$).
   5. Inaccessibility of $1/T = 0$.
   6. Inaccessibility of $p/T = 0$.
Postulate 3-4 follows from the fact that infinite work would be needed to reach that states. Postulate 5 and 6 reflects the fact of impossibility of infinite energy or infinite volume state, that is impossible to reach in that systems the absolute maximum entropy state.

4. Negative pressure and negative temperature

In case of upper bound on energy or volume Postulate 5 and 6 do not hold. There is no restriction for the accessibility of $1/T = 0$ or $p/T = 0$ states, then p or T can be negative. For example postulate 6 does not hold in condensed matters, i.e. in liquids and solids pressure can be negative.

Although negative pressure and negative temperature can be introduced in thermodynamics in similar way, there are some crucial differences between these two states. Probably the biggest difference that while negative pressures are <u>lower</u> than positive pressures, negative temperatures are <u>higher</u> than positive temperatures. An other important difference that while negative temperature states can be seen only in laboratories (lasers, nuclear spins), negative pressure states can be find in the Nature (xylem of trees). The differences and similarities between negative temperature and negative pressure can be seen in tabulated form in Table 1.

Table 1. Properties of $T^{<0}$ and $p^{<0}$ states

$T^{<0}$	$P^{<0}$
$T^{<0} > T^{>0}$	$P^{<0} < P^{>0}$
Stability: $\partial^2 S/\partial U^2 < 0$	Stability: $\partial^2 S/\partial V^2 < 0$
metastable state	metastable state
non-equilibrium state	non-equilibrium state
can be seen only in sub-systems (where the temperature of the whole system is positive)	can be seen in whole closed systems
few examples in laboratories (laser, nuclear spin)	common in nature (trees, liquid inclusions)

5. Conclusion

In this paper we demonstrated that negative pressure and temperature – although they are forbidden in some approach - are not impossible in classical thermodynamics. They can be introduced by the similar way; but while there are some similarities between them, there are also some crucial differences. Forbidding negative pressure in liquids by some thermodynamic approach is the result of some "error" of the approach. Negative pressure can be find in Nature, for example the sap-transport of trees is helped by negative pressure (-2 to -3 bar); i.e. trees know thermodynamics better that some of us.

Acknowledgment

This work was partially supported by the Hungarian Research Foundation (OTKA) under contract number F034333 and T029542. One of the authors (A.R.I.) was supported also by the Bolyai Research Fellowship.

References

1.Landau, L. and Lipschitz, A. (1976) *Statistical Physics*, (in Hungarian) Műszaki Kiadó, Budapest
2. Callen, H. (1960) *Thermodynamics*, Wiley and Sons, New York
3. Lukács, B., Martinás, K. (1990) Thermodynamics of Negative Absolute Pressures, *Acta Phys. Pol.*, **B21**, 177-183
4. Trevena, D.H. (1987) *Cavitation and Tension in Liquids*, Adam Hilger, Bristol
5. Imre, A., Martinas, K. and Rebelo, L.P.N. (1998) Thermodynamics of Negative Pressures in Liquids, *J. Non-Equilib. Thermodyn.*, **23**, 351-375
6. Maris, H. and Balibar, S. (2000), Negative Pressures and Cavitation in Liquid Helium, *Phys. Today*, **53**, 29-34

7. Debenedetti, P.G. (1996) *Metastable Liquids: Concepts and Principles*, Princeton University Press, Princeton, N.J.

8. Clausius, R. (1864) *Abhandlungen über die mechanische Waermetheorie I.* Vieweg und Sohn, Braunschweig

9. Katchalsky, A., Curran, P. (1965) *Nonequilibrium Thermodynamics in Biophysics.* Harvard University Press, 1965

10. Truesdell, C. (1984) *Rational Thermodynamics*, Springer-Verlag, New York

11. Mallinckrodt A.J. and Leff, H. S. (1992) All about work, *Am. J. Phys.*, **60**, 356-365

12. Caratheodory, C. (1909) Untersuchungen über die Grundlagen der Thermodynamik. *Math. Ann.* Bd.**67**, 355-386

13. Born, M. (1921) Kritische Betrachtungen Zur traditionellen Darstellung der Thermodynamik. *Physik. Zeitsch.* **XXII**, 218-224, 249-254, 282-224.

14. Born, M. (1949) *Natural Philosophy of Cause and Chance*, Clarendon Press, Oxford

15. Landsberg, P. T. (1961) *Thermodynamics,* Interscience Publishers, New York London

16. Fowler, R. (1936) *Statistical Mechanics*, Cambridge University Press, Cambridge

17. Lieb, E.H. and Yngvason, J. (1998) A Guide To Entropy And The Second Law Of Thermodynamics, *Notices of the Amer. Math. Soc.*, **45**, 571-581.

18. Lieb, E.H. and Yngvason, J. (1998), The Physics And Mathematics Of The Second Law Of Thermodynamics, *Physics Reports*, **310**, 1-96.

LIMITING TENSIONS FOR LIQUIDS AND GLASSES FROM LABORATORY AND MD STUDIES

QING ZHENG, JENNY GREEN, JOHN KIEFFER, PETER H. POOLE,
JUN SHAO, GEORGE H. WOLF AND C. AUSTEN ANGELL
*Dept. of Chemistry and Biochemistry, Arizona State University
Tempe, Arizona 85287-1604, USA*

Abstract. We discuss first the systematic passage of liquids from states of positive pressure to states of isotropic tension, using spectroscopic characterization methods. We then describe some experimental methods for characterizing the range of temperature under which common liquids and glassformers can be studied in such negative pressure states. While rather large tensions, up to nearly 200MPa have been obtained in these laboratory studies, much larger tensions can be generated in systems with long range attractive forces. We characterize the limiting tensions for such cases using the methods of molecular dynamics. In practice, liquids fail by homogeneously nucleated cavitation at tensions far below the spinodal limits defined by equations of state. We anticipate a crossover from nucleated to spinodal failure as the glass transition temperature meets the spinodal. We lay the basis for the experimental testing of this notion, using laboratory data for the case of the model glassformer, o-terphenyl. In the case of SiO_2 glass we are able to characterize the failure mechanism.

1. Introduction

Of the thermodynamic states under which condensed matter has been studied, that of isotropic tension is certainly the least well explored. Although very large regions of PT space in which mechanically stable states of typical systems can be studied, fall into the "negative pressure" zone, almost nothing is known, other than by extrapolation, of the physical behavior of liquids in this domain. This dearth of experimental information is disadvantageous to our understanding of the liquid state since it is in this domain that a crossover from dominance of behavior by the repulsive part of the interaction potential, to dominance by the attractive part, should occur. Information from this crossover region, particularly at low temperatures where the vibrational and configurational components of the liquid properties can be separated (due to slow relaxational dynamics) [1], would certainly be helpful in unraveling the complexities of the "liquid problem".

The extent of the mechanically stable but stretched domain for water, as suggested by the Haar-Gallagher-Kerr (HGK) equation of state [2], is demonstrated in Fig. 1. We should note that Fig. 1 differs from previously published versions of this figure [3] by inclusion of a spinodal between low density supercooled water and the (strictly unphysical) negative pressure vapor phase. The origin of this branch of the spinodal is found in plots of P vs. V from the HGK equation. These show a second spinodal developing at temperatures below 277°C and pressures initially near zero. While this may be merely an artifact of the multiparameter equation extrapolation beyond the

33

A.R. Imre et al. (eds.), Liquids Under Negative Pressure, 33–46.

temperature range in which the parameters were established, we use it here to justify the inclusion of a low temperature tension limit, that must be present, on the low density amorphous form of water LDA in which most of the universe's water exists.

The HGK equation is not the only equation of state that yields such features, indeed the reentrant spinodal which forms the heavy boundary on the dark area of Fig. 1 seems to be a common feature of such experimentally based EOSs as are currently in use [2-5]. On the other hand, the existence of a spinodal (implying a line of first order transitions between two distinct states of water) is contradicted by studies of water by computer simulations using several different pairwise additive potentials [5]. These show that the continuity of the spinodal shown in Fig. 1 is interrupted at negative pressures. However they also show that a liquid-liquid equilibrium line (rimmed by spinodals) exists at significantly positive pressures, following a liquid-liquid critical point in the pressure range 50-200MPa. For laboratory water (and glass [6]) it is clear that there is a major anomaly, related to these considerations, lying in the low temperature liquid state [7] but unfortunately it cannot be studied directly because of preemptive crystallization. On the other hand there is *indirect* experimental support of a second critical point in this vicinity from studies of metastable melting of high pressure polymorphs of water [8].

Although Fig. 1 may be inaccurate in detail, it serves the purpose of illustrating the large P-T range in which it should, in principle, be possible to expand our knowledge base of liquid behavior. The major part of this range occurs in the negative pressure domain.

One of the problems of negative pressure studies is the lack of a simple means of measuring the pressure in the microscopic samples which must be used to avoid premature cavitation. The pioneering physical measurements of Henderson and Speedy [9] used sample containers which, being Bourdon pressure gauges themselves, provided the pressure of the measurements. However this clever strategy cannot easily be applied to the type of samples that are needed to avoid heterogeneous nucleation and achieve truly large tensions, near the homogeneous nucleation limits, where new physics can be expected. Rather an internal pressure gauge in the form of some molecular probe that can be calibrated at positive pressure and remains in its linear response domain even at large tensions, must be developed. While we can not provide an answer to this problem here, we can give a useful example of an internal pressure probe that can concommitantly serve as a sort of "visual' demonstration of existence of negative pressure by showing how the color of a liquid can change continuously with decreasing positive pressure through zero and across into the negative pressure domain. This is done in the next section.

2. Transition metal ion spectra under tension

Here we summarize the results of a study of the visible spectrum of the Co(II) ion in a concentrated aqueous solution of LiCl in which the cobalt ion distributes itself between two local sites, one tetrahedral corresponding to the anion $CoCl_4^{2-}$, and the other, octahedral, corresponding to the cation $Co(H_2O)^{2+}$. The former lends the solution a deep blue color while the latter renders it pink. The equilibrium is pressure-dependent, the octahedral site being favored by high pressure [10]. A sequence of spectra, as pressure is varied between 400 and 1 bar (40 and 0.1MPa) at 35°C, is shown in Fig. 2. By sealing the filled cell off under vacuum and then eliminating the remaining small vapor bubble by careful increase of temperature (see Fig. 3 insert), the same system can be studied during simultaneous temperature and pressure decrease in the manner of Berthelot [11]. In this case the initial pressure is the vapor pressure at the temperature where the bubble disappears, and is above 1 atm for T > 100°C. As the cell cools the pressure decreases and becomes negative. The spectrum at the temperature 35°C, just before cavitation, is shown in Fig. 3, along with the spectrum just after cavitation. The shift between the two is obvious and can only be accounted for the by recognizing that a true negative pressure is bearing on the sample.

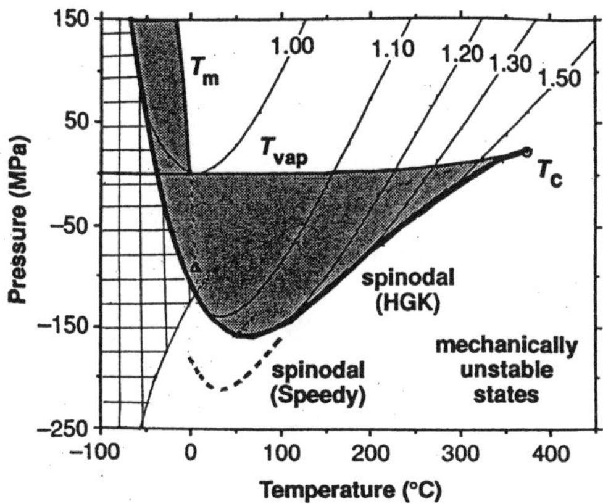

Figure 1. Phase diagram of water including metastable regions of the liquid state and showing isochores of different specific volume. Below 277°C the HGK equation generates a second spinodal branch the meaning of which is unclear. This line emerges from the dark area of Fig. 1 below 0°C, and we use it as a bound on tensile strength of the low temperature open network form of water (open-hatched area) which, however can only be obtained by extremely rapid liquid quenching, or by non-liquid formation routes [6,7]. The spinodal running between the two liquid regions implies a line of first order phase transitions between the two liquid forms which is not confirmed by pair potential MD studies except at positive pressures (above 100MPa) [5].

We can determine the magnitude of the tension by making a calibration curve using the positive pressure series of Fig. 2. The pressure indicated thereby for the stretched solution is −70 bar and the same value is obtained by extrapolating the isobaric equation of state of a solution of NaCl of the same concentration (LiCl data are not available). The tension reached is not large, but it suggests how internal spectroscopic pressure probes could be used for wider range studies. A ruby chip of course is the standard internal pressure probe for pressure studies, but this is most appropriate for very high pressures.

Needed for pressures near the spinodal in liquids is a probe with greater sensitivity, solubility in the medium being probed so that it can be carried into the inclusion samples to be described, but also a considerable degree of inertness to the medium so that changes in the latter, other than physical pressure, will not be influential. At the moment such a probe remains to be identified, though the UV edge of colloidal CdS, precipitated TlCl, or perhaps dissolved $BiCl_3$, are possibilities to be evaluated.

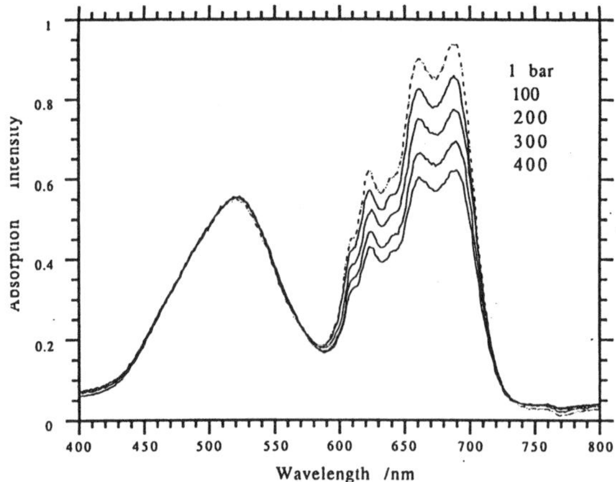

Figure 2. Visible range electronic spectrum of Co(II) ions in a concentrated aqueous lithium chloride solution as a function of pressure at the temperature 35°C. The longer wavelength, pressure dependent, component is due to the four-coordinated $CoCl_4^{2-}$ site, which is a high volume state.

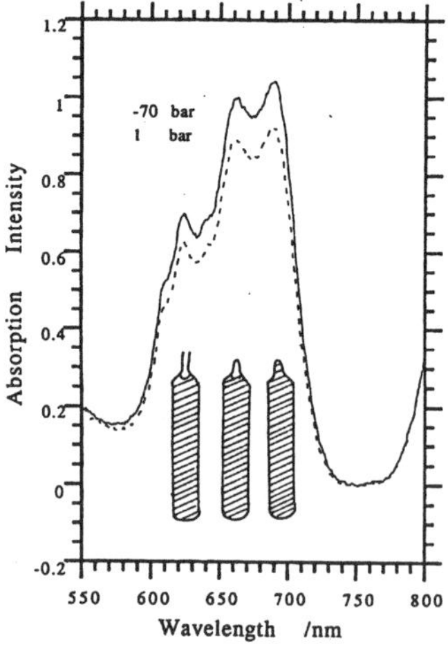

Figure 3. Visible spectrum of Co(II) ion in a concentrated aqueous chloride solution as a function of temperature at constant volume. The pressure at the lowest temperature before cavitation was found to be -70 bar, by positive pressure calibrations. Insert: three stages of creation of the solution filled Berthelot tube. Once stage 3 is reached by gentle warming, subsequent cooling will generate tension as temperature is decreased, because the bubble cannot be regenerated without crossing a nucleation energy barrier.

3. Stretching to the homogeneous cavity nucleation limit

Here we review the published works on inclusion studies that have permitted water to be stretched to the homogeneous cavitation limit. The possibility of such studies was first indicated by the work of Roedder who reported that water inclusions in minerals permitted the observation of melting in ice to be observed at +8°C [12]. By extrapolation of the melting point vs. pressure line, Roedder pointed out that his observation indicated a negative pressure of 800 atm (~80MPa), larger than any previously reported. The reason that such pressures can be realized is clearly associated with the microscopic dimensions of these samples and the exclusion of heterogeneous nuclei in the formation process. The large negative pressure range available was confirmed by Green et al [1] after they realized that the cavitated inclusions found in various crystalline hosts could be used as miniature Berthelot tubes. Using a long needle-like inclusion formed in a high Q quartz crystal, they showed that a pressure of some 1600 atm could be observed in an aqueous sample that had a dissolved solids equivalent of a 0.8M NaCl solution, according to its micro-Raman spectrum. The tensions were estimated by extrapolation of the equation of state data of ref. 13.

Stimulated by these observations, Zheng et al [14] then systematically investigated the behavior of pure water in inclusions of (uncavitated) density up to 1.7ml/g. The methods for synthesizing these variable density isochoric samples of high purity water are described in detail elsewhere [15].

The results of Zheng et al were proven reliable by comparison, in the positive pressure domain, with the best available data for superheating of water. The comparison with pulse heating boiling data of Skripov and co-workers [17] is shown in Fig. 5. The inclusion data are at least as precise as those of Skripov et al. It is to be noted that the two studies, the one involving heating to the limit, and the other, cooling to the limit (isochorically) together establish the superheating limit of water with high reliability. The accord with the theoretical results of Fisher [18] and of Blander and Katz [19] is to be emphasized.

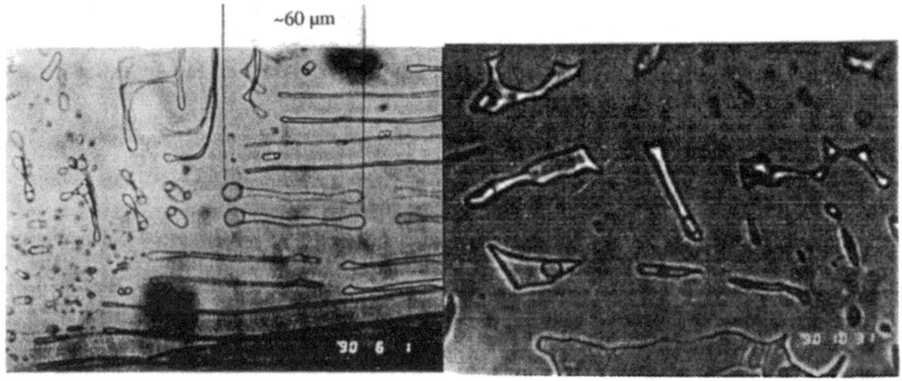

Figure 4. Cavitated inclusions of water in autoclave-healed calcite (left) and quartz (right) crystals. The inclusions often take the form of miniature ~60µm Berthelot tubes. The scale is indicated in the LH panel, which applies also to the other micrographs in this paper.

Our isochoric inclusion cooling method permits the study of the homogeneous cavitation limit to liquid stability to be followed deep down into the negative pressure region. Data obtained for this additional and novel range are displayed in Fig. 5. For the initial range of tensions down to about –40MPa pressure the data are highly reproducible. All inclusions cavitate at almost the same temperature. For greater tensions, however, lower initial densities were needed in the inclusion formation

38

process, and these involved annealing the quartz under autoclave conditions which were not suitable for producing optimally formed inclusions (because solubilities were not sufficient at the lower temperatures). In these cases, there were evidently heterogeneous cavitation sites left unhealed and a scatter of cavitation temperatures was observed. Naturally the lowest temperature process was the one that most closely approached the homogeneously nucleated limit, and only the lowest temperature, largest tension values are included in the figure.

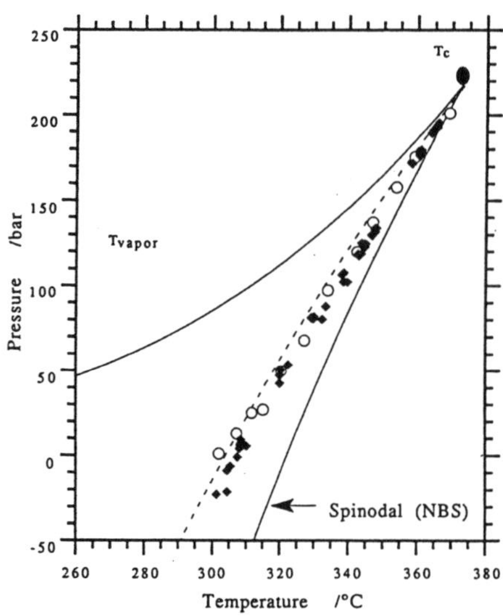

Figure 5. Comparison of superheating limits determined by isochoric cooling in the present work (filled diamonds) with data obtained by pulsed boiling technique in the Skripov laboratory, ref. 17 (open circles). The dotted line is the theoretical prediction of Fisher [18]. (Reproduced from ref. 14 by permission of AAAS).

Included in Fig. 6 are data for water inclusions in the alternative crystal hosts, calcite and fluorite. These are seen to be in good accord with the quartz inclusion data at both small and large tensions. Altogether the data provide a very extended set with which to test the theoretical predictions for the homogeneous nucleation of cavities in liquids under tension. The dashed line in Fig. 5 is the plot of the prediction of Fisher [18], which is almost indistinguishable from the later prediction of Blander and Katz [19]. The agreement is almost quantitative until the ability to increase the tension on the liquid by the ischoric cooling process reaches its maximum value, due to the change in sign of the isochore (see Fig. 1). The fact that samples, prepared at densities higher than that of the last experimental point on Fig. 5 (at −140MPa), do not cavitate at any temperature, may be taken as evidence that the extremum in the isochore predicted by the equation of state really exists at these large negative pressures. The largest tension point itself showed very interesting behavior. On repeated cycling between filled and cavitated states, it would sometimes cavitate and sometimes remain uncavitated to temperatures well below the recorded value. This means of course that at about the pressure of −140MPa the tension stopped increasing with decreasing temperature and presumably started to decrease. This identifies the temperature of the density maximum for water as 35 °C at an approximate negative pressure of –140MPa. We describe the pressure as approximate because the inclusions, for which these observations were made, were not of well-rounded form, like those on which the reliable and reproducible high temperature data were obtained. This is because of the low temperature at which the cracks in the quartz crystal had to be annealed (at the maximum autoclave pressure) in order to achieve the high density of this isochore. Therefore the possibility of relaxation of the cavity to less than its initial volume due to the elastic deformation of the surrounding quartz has to be allowed for. On the other hand, any reduction of the pressure would imply that the slope of the density maximum would have to be larger than predicted by the equation of state, rather than smaller as the results of computer simulations of water in various pair potential models [5,6,7] would imply. This is a question that will require further studies using either higher autoclave pressures or accelerated vesicle healing strategies (e.g. by raising the pH of the water slightly) to answer. The significance of support for the equation of state predictions offered by these data will be discussed elsewhere [20].

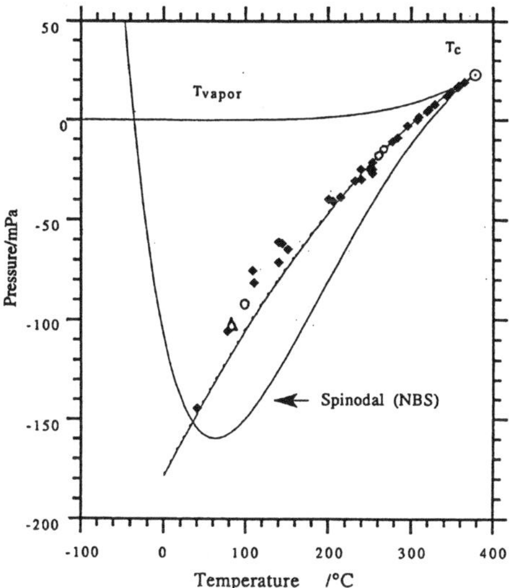

Figure 6. Limiting tensions for water by isochoric cooling using inclusions in different crystalline media. Filled diamonds are for quartz samples. Open circles for fluorite inclusions and open triangles for calcite inclusions. The line passing through the high T points is the prediction of the Fisher theory. (adapted from ref. 14)

The existence of water at tensions as high as 140MPa needs to be put in perspective. It is helpful to realise that, numerically, it is considerably larger (more negative) than the largest positive pressure at which bulk water can be found on the earth's surface. The latter pressure is 99MPa, the pressure of water at the deepest part of the Marianus trench (off the Phillipines coast)!

The reason for the existence of such large tension limits lies in the strength of the hydrogen bonds which must be broken in order to form a cavity that is large enough to act as a stable nucleus for cavitation. No common van der Waals bonded liquid, even

40

including examples with much higher critical temperatures than water (such as o-
terphenyl described in the next section), can sustain such a large tension.

4. Non-aqueous systems

An example of a van der Waals liquid with large T_c is provided by the case of o-
terphenyl (oTP), even the boiling point of which, 610°C, is in excess of T_c for water.
Samples of this liquid, to which the present isochoric cooling method for limiting
tension detection can be applied, were obtained by the strategy of healing the cracked
quartz crystals in the presence of surrounding liquid mixture (two phases) containing
about 10% water and 90% oTP. Vesicles formed under these condition proved to have
both liquid phases present, but in the opposite volume fractions, see Fig. 7. (It proved
impossible to form vesicles filled only with o-terphenyl, and had it been possible it
would not have been useful since the equation of state of oTP is unknown and so the
tension on cavitation would be unknown.

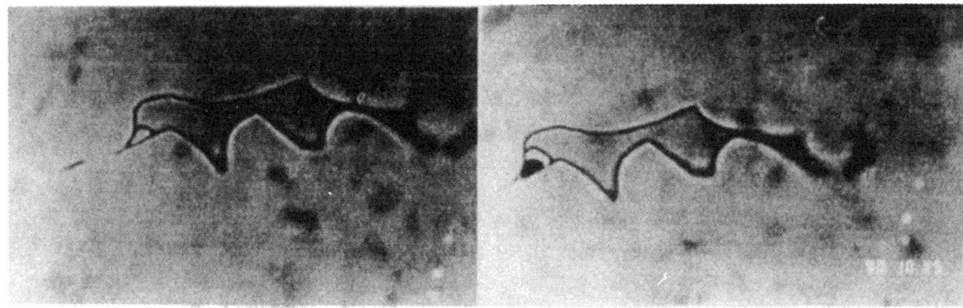

Figure 7. Sample of o-terphenyl in a water-filled inclusion in a quartz crystal supporting environment: left
panel: before cavitation; the oTP is the light-colored segment at the leftmost portion of the inclusion; right
panel shows vapor bubble in the oTP segment after cavitation. The limiting tension for o-terphenyl, which is
less than that for pure water, is obtained as a function of temperature by observing such samples prepared at
different densities.

In a cavity filled mostly with water, there are three major advantages. The first is
that water wets the surface of the quartz preferentially, guaranteeing that the cavitation
of oTP is not initiated at the inclusion wall. The second is that, because most of the
liquid filling the cavity is water, the pressure in the inclusion will be dominated by the
thermodynamics of the water. Thus the water equation of state acts as the pressure
gauge for the experiment. The third is that water has a higher tensile limit than oTP, so
that when cavitation finally occurs it always occurs in the oTP phase, see Fig. 7, right
panel. The tension at which the cavitation limit for oTP is reached can thus be obtained
to good approximation from the isochore of water established by the bubble
disappearance temperature (on heating) for the two phase vesicles. As in the case of
pure water, all the vesicles cavitated at essentially the same temperature for low density
isochores, and some scatter was observed for the higher density samples (which
cavitated at the lower temperatures). Again only the lowest cavitation temperatures are
recorded.

The homogeneous nucleation limited tension maximum for o-TP at different densities is shown in Fig. 8 by the points through which a solid line with dashed extension is drawn. The homogeneous nucleation limit for water is shown as a dotted line and the HGK spinodal for water is shown as a solid line. The limit on oTP tension measurements for oTP is set by the circumstance that water has a maximum tension, due to its density maximum. However this limitation could be lifted if a concentrated saline solution (NaCl for instance, since its equation of state is known) were used as the SiO_2-wetting component. Then data could in principle be obtained into the doubly metastable range where oTP is metastable with respect to both cavitation and crystallization.

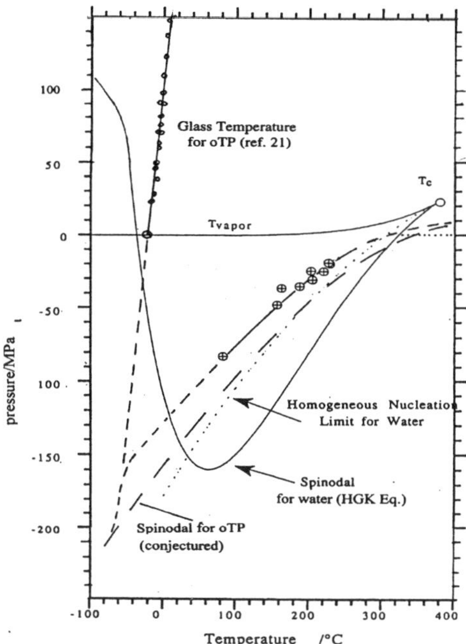

Figure 8. The homogeneous nucleation tension maximum for oTP at different densities. The homogeneous nucleation limit for water is shown as a dotted line and the spinodal for water is a solid line. Also shown is a conjectured spinodal for oTP (long-dashed line). And the line of glass transition temperatures measured at positive pressures in ref. 21. Where the cavitation temperature intersects the T_g line (shown extrapolated as a dashed line) there should be an abrupt increase in the observable limiting tension as the mechanism of rupture changes from nucleated to spinodal.

While this extension of the present study must remain for the future, it is interesting to anticipate its findings by using a reasonable extrapolation of the Fig. 8 findings, and including an extrapolation of the known variations with pressure of the glass transition temperature of oTP [21]. The extrapolations are aided by the computer simulation studies of Sastry [22] for the binary Lennard-Jones system of Kob and Andersen [23], and the mean field calculations of Debenedetti et al [24]. These studies showed that the theoretical limits of stretching and supercooling (i.e. the spinodal and Kauzmann temperature loci, either meet at their zero temperature limits [24] or intersect somewhat before 0K is reached [22]. While these events involve unrealizable quantities, T_s and

42

T_K, the intersection of the measurable limits, glass transition temperature T_g and homogeneously nucleated cavitation T_c, should be realizable, and should have interesting consequences. For instance, when molecular motion becomes arrested at the glass transition then the diffusion involved in forming a cavity nucleus should be precluded. The tensile failure could only occur via a true mechanical instability and the cavitation phenomenon will crossover to a tensile crack failure mechanism, which should require a larger tension. This failure mechanism was also considered by Fisher in the same paper in which the successful theory of cavitation was developed [18]. It would certainly be of interest to observe the crossover phenomenon, and it offers a challenge to future workers.

5. Increasing the tension limit with coulomb interactions

To obtain liquid tensile stress limits greater than those of water, it is necessary to introduce attractive forces stronger than those of hydrogen bonds. The obvious case of ionic solutions was already indicated by our initial study [3] in which Raman spectroscopy of, and ice melting point data in, the inclusion liquid indicated a saline content equivalent to 0.8M NaCl. The tensile limit was identified in that study as - 180MPa. We have obtained comparable tensions in inclusions prepared with high concentrations of e.g. 26 wt % NaCl for which equation of state data are available [13]. Fig. 9 shows examples of inclusions of this solution after it has reached equilibrium at ambient by both cavitating and also precipitating a halite crystal. Insufficient measurements on this system were made to justify a diagrammatic display, but tensions as high as 180MPa were estimated. The isochores are very similar for 19 and 26 wt % NaCl[13]. The tensions are not higher than in dilute solutions and the cavitation is probably initiated by halite crystal nucleation.

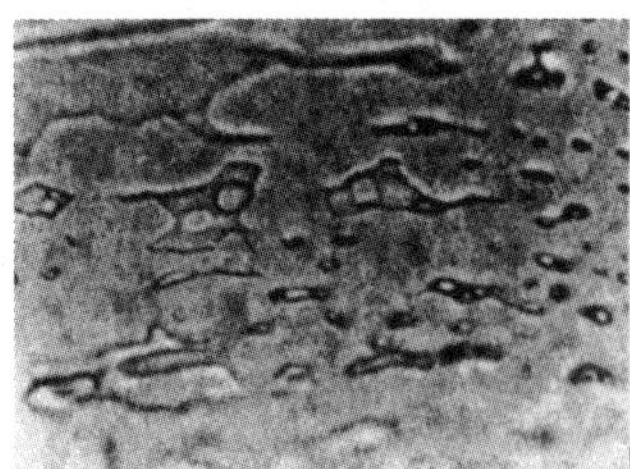

Figure 9. Room temperature state of inclusions prepared with 26 wt % NaCl in H_2O, showing halite crystals and vapor bubbles.

To reach higher tensions, we turned to computer simulation studies [25]. The "standard salt", KCl, was studied in the Tosi-Fumi potential [26], using the program "ion" [27] with a periodic box containing a total of 400 ions. The spinodal limits, based on the extremes in the P vs. V relation, are shown in Fig. 10a and 10a (insert). In computer experiments with small primary boxes, the unstable region of the van der Waals loop is always observed. Interestingly, spinodal limits more negative than in aqueous solutions were only encountered at temperatures far below the melting point.

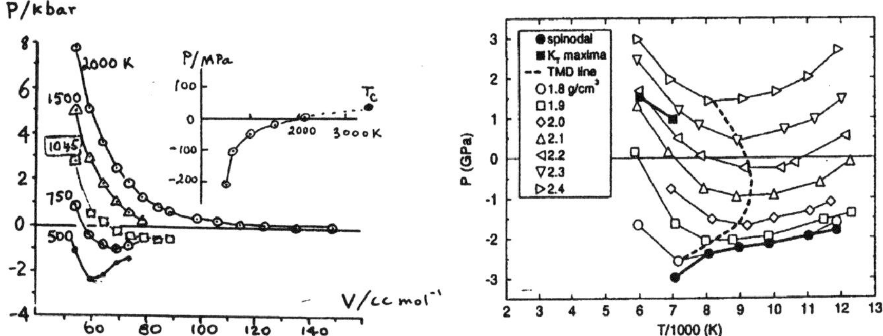

Figure 10(a). P-V plots showing extrema, i.e. compressibility divergences, and (insert) the spinodal limit of mechanical stability for molten KCl, using potential function of ref. 27. The rapid increase between 750 and 500K may be associated with vitrification on the time scale of these computer "experiments".
(b) The spinodal for liquid SiO_2 in the rigid ion model WAC potential [26] (solid points) compared with the isochores for this system at different densities. The dashed line is the line of density maxima for this potential. Note that it passes through a maximum temperature and, as for water, never intersects the spinodal. (From ref. 28, by permission of Amer. Phys. Soc.)

For the case of liquid SiO_2, data are available in the work of Poole et al. [28] on the rigid ion model in the WAC potential [26]. They are shown in Fig. 10(b) along with the variation of the pressure with temperature along isochores of different density. It is seen that the limiting tension is considerably in excess of that of KCl, notwithstanding the higher temperatures of the simulation. In the case of SiO_2, however, we have both liquid and solid (glass) data, and the expectation of Fig. 8 for the effect of lowering temperature through the glass temperature can be checked, at least qualitatively. This will be done in the final section of this paper.

6. Fractal rupture of glassy states, under isotropic tension

Finally we reproduce some results of earlier studies of SiO_2 in the glassy state [29] using the same potential function [26]. Although the cavitation of SiO_2 was recorded (as an extremum in the pressure volume function) even earlier [30] it was only studied in detail some time later, by Kieffer [29]. The variation of pressure with volume, relative to the zero pressure volume, is shown in Fig. 11. Note that, at ambient temperature, the glass sustains a volume extension of almost 30% before the tensile limit is reached. The value of the tension at rupture is much higher than any seen in Fig. 10, as we anticipated for oTP in discussing Fig. 8. The enormous value, -7GPa (the negative equivalent of the pressure at a depth of ~200 km beneath the earth's surface) is presumably also approached at the center of a Prince Rupert's drop of soda-lime glass. (It is the consequent compression of the surface that is responsible for the ability of Prince Rupert's drops to withstand the hammering on a steel forge for which they are famous).

The patterns of bonds for the cavitated structure is reproduced in Fig. 12 from ref. 29. The fractal dimension was found to change continuously with the final relative volume of the cavitated glass. It is notable how the failure of this glassy structure is uniform rather than being concentrated into a single crack, but we are unable to say under what conditions the crack failure, which would release the stress with much less total work, would be observed.

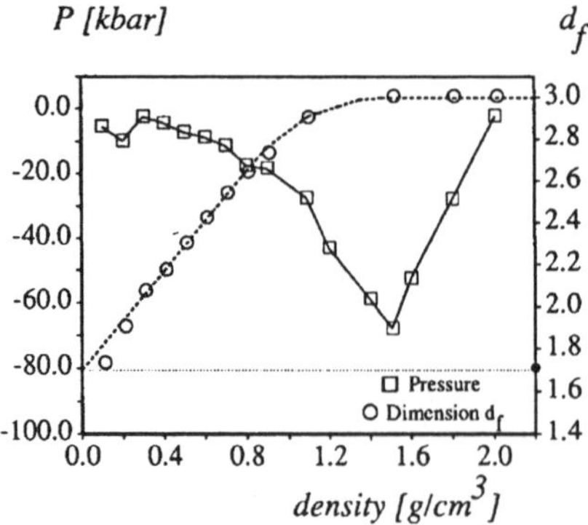

Figure 11. Increase of tension with decreasing density in ambient temperature glassy SiO_2 in the rigid ion model, showing rupture at -7 GPa. The glassy tension limit is larger by a factor of 2 than any seen in the liquid state (Fig 10(b)). The fractal dimension (right hand scale) remains 3 until rupture and then decreases as the cavity structure seen in Fig. 12 is generated. (from ref. 29, by permission).

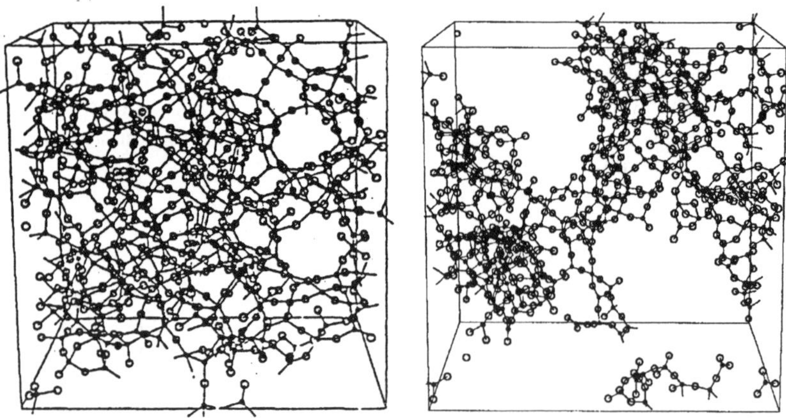

Figure 12. Silica structure after rupture according to MD studies of the rigid ion model of SiO_2. (from ref. 29, by permission of Elsevier).

7. Concluding remarks

The samples we have considered in this work, while is some cases microscopic in dimensions, are in all cases tens of thousands of molecules in cross section, hence are expected to behave as macroscopic samples of the liquids in question. We should, in concluding this overview, take note of the existence of these same tensile limit phenomena in the case of systems in which the dimensions of the liquids under study are of much smaller dimensions, indeed as small as a few molecules across. We think here of the studies of hysteritic adsorption isotherms for liquids in nanoporous materials, reported by Machin and co-workers [31-33] in recent years which they convincingly interpret in terms of limiting tensions in the pore fluids. In these systems the fluctuations that give rise to nucleation of cavities must be much less probable hence, as with crystal nucleation in small systems, the nucleation events must become much less probable. In such case the rupture should be found to occur much closer to the spinodal than in our samples, hence to the extent that the equation of state for bulk water has any relevance to such samples, the tensile limit may become even larger than we have been able to record. Indeed Machin and coworkers report tensile limits for water that considerably exceed our own.

It seems unlikely that, in the circumstances of extreme confinement, and with the competition of pore surface interactions, the open network that characterizes water and allows it to have a tension maximum, could continue to dominate the system's behavior. This aspect of these nanoscopic systems' behavior will be interesting to follow up in the same way as the influence of network breaking, density maximum destroying, components added to our samples will be interesting to follow up.

8. Acknowledgements

The work reported here could not have been performed without the consistent support of the National Science Foundation, Solid State Chemistry program, currently under Grant No. DMR0082535.

9. References

[1]   See articles in Science (special issue) 267 (1995) 1924-1953.

[2]   L. Haar, J. Gallagher, G. S. Kell, *National Bureau of Standards-National Research Council Steam Tables*, McGraw-Hill: New York (1985).

[3]   J. L. Green, D. J. Durben, G. H. Wolf and C. A. Angell, *Science* **249** (1990) 649-652.

[4]   S. B. Kiselev, J. M. H. Levelt Sengers and Q. Zheng, in *Physical Chemistry of Aqueous Solutions: Meeting the Needs of Industry*, eds. H. J. White Jr., J. V. Sengers, D. B. Neumann and J. C. Bellows, Begell House: New York (1995) 378.

[5]   (a) P. H. Poole, F. Sciortino, U. Essmann and H.E. Stanley, *Nature* (London), **360**(6402) (1992) 324-328.
       (b) S. Harrington, R. Zhang, P. H. Poole, F. Sciortino and H. E. Stanley, *Phys. Rev. Lett.* **78**(12) (1997) 2409-2412.
       (c) M. Yamada, S. Mossa, H. E. Stanley and F. Sciortino, http://xxx.lanl.gov/pdf/cond-mat/0202094.

[6]   (a) P. Brüeggeller and E. Mayer, *Nature* **288b** (1980) 569; **298** (1982) 715.
       (b) J. Dubochet and A. W. McDowall, *J. Microsc.* **124** (1981) RP3.
       (c) G.P. Johari, A. Hallbrucker and E. Mayer, *Nature* **330** (1987) 552-553.
       (d) O. Mishima, *J. Chem. Phys.* **100** (1994) 5910.

[7]   C. A. Angell, *Ann. Rev. Phys. Chem.* **34** (1983) 593-630.

[8]   O. Mishima and H. E. Stanley, *Nature* **396** (1998) 329.

[9]   (a) S. J. Henderson and R. J. Speedy, *J. Phys. E.* **13**(7) (1980) 778-782.
       (b) S. J. Henderson and R. J Speedy, *J. Phys. Chem.* **91** (1987) 3069-3072.

[10]   V. E. Rodgers and C. A. Angell, *J. Chem. Ed.* **60** (1983) 602.

[11]   M. Berthelot, *Ann. Chem. Ser. 3* **30** (1850) 232.

46

[12]   E. Roedder, *Science* **155** (1967) 1413.
[13]   G. C. Kennedy, *Econ. Geol.* **45** (1950) 629; G. J. Wasserburg, H. C. Heard, R. C. Newton, *Am. J. Sci.* **260** (1962) 501; J. L. Haas, Jr., *U.S. Geol. Surv. Bull. 1421* (1976), data in part C by R. W. Potter II and D. L. Brown.
[14]   Q. Zheng, D. J. Durben, G. H. Wolf and C. A. Angell, *Science* **254** (1991) 829.
[15]   S. M. Sternar and R. J. Bodnar, *Geochim. Cosmochim. Acta* **48** (1984) 2659.
[16]   Q. Zheng, Ph. D. Thesis, Arizona State University (1991).
[17]   V. P. Skripov, in *Water and Steam*, eds. J. Straub and K Scheffler, Pergamon: Elmsford, NY (1980).
[18]   J. C. Fisher, *J. Appl. Phys.* **19** (1948) 1063.
[19]   M. Blander and J. Katz, *Am. Inst. Chem. Eng. J.* **21** (1975) 833.
[20]   C. A. Angell, *J. Phys. Chem.* (feature article, to be published).
[21]   T. Atake and C. A. Angell, *J. Phys. Chem.* **83** (1979) 3218.
[22]   S. Sastry, *Phys. Rev. Lett.* **85** (2000) 590-593.
[23]   W. Kob and H. C. Andersen, *Phys. Rev. E.* **51** (1995) 4626-4641.
[24]   P. G. Debenedetti, F. H. Stillinger, T. M. Truskett and C. J. Roberts, *J. Phys. Chem. B* **103** (1999) 390.
[25]   J. Shao and C. A. Angell (unpublished work).
[26]   L. V. Woodcock, C. A. Angell and P. A. Cheeseman, *J. Chem. Phys.* **65** (1976) 1565.
[27]   J. Kieffer and C. A. Angell, *J. Chem. Phys.* **90** (1989) 4982-4991.
[28]   P. H. Poole, M. Hemmati and C. A. Angell, *Phys. Rev. Lett.* **79**(12) (1997) 2281-2284.
[29]   J. Kieffer and C. A. Angell, *J. Non-Cryst. Sol.* **106** (1988) 336-342.
[30]   C. A. Angell, P. A. Cheeseman, C. C. Phifer, *Mat. Res. Soc. Symp. Proc.* **63** (1986) 85-94.
[31]   W. D. Machin and J. T. Stuckless, *J. Colloid Interface Sci* **108** (1985) 46-49.
[32]   W. D. Machin and P. D. Golding, *J. Chem. Soc. Faraday Trans. 1* **83** (1987) 1203-1212.
[33]   W. D. Machin and J. T. Stuckless, *J. Chem. Soc. Faraday Trans. 1* **81** (1985) 597-600.

INVESTIGATION OF THE THERMODYNAMIC PROPERTIES OF A POLARIZABLE WATER MODEL IN A WIDE RANGE OF PRESSURES AROUND THE TEMPERATURE OF MAXIMUM DENSITY

P. JEDLOVSZKY
Department of Colloid Chemistry, Eötvös Loránd University
H-1117 Budapest, Pázmány Péter stny. 1/a, Hungary

R. VALLAURI
INFM e Dipartimento di Fisica, Università degli Studi di Trento
14 via Sommarive, I-38050 Povo (Trento), Italy

1. Introduction

The understanding of the molecular level origin of the anomalous thermodynamic behaviour of liquid and supercooled water has been the subject of scientific research for more than a century [1-4]. Such investigations led to the 'iceberg' model of Némethy and Scheraga [5-7], which assumes the existence of tetrahedrally ordered clusters ('icebergs') in a disordered, normal liquid-like medium. Research in this field has been largely accelerated by the spreading of fast computers, which enabled scientists to study this problem also by computer simulation methods. Early simulation studies have contradicted the iceberg theory, showing that water molecules are forming gel-like space-filling percolating hydrogen bond network in the liquid phase [8-12].

It has been known from scattering experiments [13-17] that the first co-ordination shell of the water molecules contains considerably more neighbours than what can be expected by assuming tetrahedral co-ordination (i.e., 4.4 instead of 4). Therefore, as an alternative explanation, it has been assumed that some of the water molecules are located in the cavities of the tetrahedral network built up by the other molecules. Such interstitial neighbours have been identified in 1993 by Kusalik and Svishchev [18] in SPC/E [19] water, and their existence has subsequently been demonstrated by computer simulations with various water models several times [20-22]. *Ab initio* calculations have shown that interstitial-like neighbours are already present in low-lying minimum conformations of water clusters built up by 6-8 molecules [23,24]. The presence of the interstitial neighbours implies, as shown by Kusalik and Svishchev [18], that there are two different preferred $O\cdots O$ separations of contacting water pairs. Thus, the preferred distance of the interstitial neighbours, located in specific non-tetrahedral directions around the central molecule, is about 3.5Å, considerably larger than the distance of 2.7Å preferred by the tetrahedrally aligned hydrogen bonded neighbours. In 1999 it has been shown by Jagla [25] that even the simplest system in which there are two competing equilibrium nearest neighbour separations (i.e., a system of spherical

47

A.R. Imre et al. (eds.), Liquids Under Negative Pressure, 47–57.

particles interacting through a potential consisting of a hard sphere repulsion and a linear repulsive shoulder) exhibits most of the anomalous properties of water. Recently we have demonstrated [22] that upon warming an increased fraction of water molecules leaves the tetrahedral network and becomes interstitial. Since interstitial molecules are located in the cavities of the tetrahedral hydrogen bonded network, a more compact structure can be realised this way, which can, up to a certain point, compensate the effect of the increasing thermal motion. The appearance of the density maximum is resulted from the balance of these two opposite effects [22].

A lot of effort has also been expended in the past decades to the understanding of the thermodynamic background of the anomalous properties of water, as well. It became clear that the important features of the phase diagram of water, which can explain the anomalous behaviour, are located in extreme conditions, i.e., at negative pressures and at deeply supercooled states that are extremely difficult to access by experimental methods. This fact again enhances the importance of the computer simulation methods in such investigations.

A possible phase diagram has been proposed by the 'Stability Limit Conjecture' of Speedy and Angell [26-29]. It assumes that the liquid spinodal line, which is originated at the critical point and goes through lower pressure points with decreasing temperature, changes slope at a certain point in the negative pressure regime and goes to higher pressures upon further decrease of the temperature. Thus, it becomes re-entrant, i.e., enters back to positive pressures in the deeply supercooled region. An alternative phase diagram, containing a monotonous liquid spinodal line has been suggested by Stanley and co-workers [30-33]. In this phase diagram a first order phase transition is assumed between two liquid phases of water in the supercooled region. The transition line ends at a second critical point. The two distinct supercooled liquid phases have been identified with the experimentally known low density amorphous ice (LDA) [34] and high density amorphous ice (HDA) [35]. Thermodynamically stable liquid water represents the supercritical state of this phase co-existence.

The temperature of maximum density (TMD) line of water is known to be closely related to the liquid spinodal. It has been shown by thermodynamic arguments that the TMD and spinodal lines can basically have two different arrangements in a thermodynamically consistent phase diagram. In the first type of phase diagram the TMD line is monotonous in the p-T plane, and terminates at an intersection with the spinodal line [36]. At the intersection the spinodal line changes slope and becomes re-entrant at low temperatures [27,37]. In the second case the TMD line changes slope at a certain pressure, and further decrease of the pressure leads to decreasing TMD. In this type of phase diagram the TMD line never intersects the spinodal [37], which is monotonous in the entire phase diagram. Experimental evidences show that that if the TMD line indeed changed slope at a certain pressure, this must happen below −200 bar [38]. The first type of phase diagram supports the Stability Limit Conjecture, whereas the second one is consistent with the assumption of the liquid-liquid critical point as well as with the singularity free explanation of the water anomalies.

All these findings show the importance of the investigation of the properties of water under extreme conditions, in particular under negative pressures and at deeply supercooled states. Due to the experimental difficulties in accessing such states the computer simulation studies are of great importance. Several computer simulation studies have targeted the properties of water at negative pressures [30-33,37,39]. However, these studies have described water by simple, pairwise additive potential models, which often proved to be inaccurate in reproducing anomalous properties of water. Thus, for instance, at atmospheric pressure the TMD of water is overestimated by about 50K by the ST2 [40] model [30,37]. On the other hand, the TIP4P [41] and SPC/E [19] models result in the atmospheric TMD values of 255K [42] and of 235K-245K [43,44], respectively. Moreover, the TIP3P [41] and SPC [45] models are shown not to have TMD at atmospheric pressure between 223K and 373K [42].

A possible way of improving the accuracy of the potential models is the explicit inclusion of the polarizability of the molecules. In this way the effect of the local electric field on the molecules can be taken into account, and the molecules can be adopted to the changes of this field during the simulation. We have recently shown that a representative of these polarizable water models, the Brodholt-Sampoli-Vallauri (BSV) potential [46] is indeed able to reproduce the TMD of water within the accuracy of 5K not only at atmospheric pressure [47] but also in the entire range of its existence in thermodynamically stable liquid water [48]. Several other experimentally measured properties of water have also been reproduced well by the BSV model in a wide range of pressures around the TMD [22,47-49]. Therefore, we have extended our investigations on the properties of BSV water also to the range of negative pressures. In a recent paper we have demonstrated that the TMD line of BSV water changes slope at −32 bar, and thus the phase diagram of this model supports the assumption of a liquid-liquid critical point or a singularity free explanation of the water anomalies, and contradicts the Stability Limit Conjecture [50]. In the present study the pressure and temperature dependence of the thermodynamic properties of BSV water is investigated in the vicinity of the TMD in a broad range of pressures covering both negative and positive pressure states.

2. Calculation Details

Monte Carlo (MC) simulations of liquid water have been performed on the (N,p,T) ensemble with 256 water molecules at four pressures, i.e., at −100 bar, 1 bar, 100 bar and 200 bar, and at six temperatures, i.e., 268K, 273K, 278K, 283K, 288K and 293K. The interaction of the molecules has been described by the polarizable BSV potential [46]. The calculation of the total energy of the system has been described in detail elsewhere [49]. The BSV model describes the polarisation of the molecules by induced point dipoles. The determination of the dipole induced on the ith molecule μ_i requires also the dipoles induced on the other molecules. Therefore, the entire set of induced dipole moments can only be calculated by an iterative procedure, which slows down the simulation considerably, and sets a practical limit of extending the range of thermodynamic states investigated as well as of improving the precision of the results.

In the MC simulations every 256 particle displacement steps have been followed by a volume change attempt. In a particle displacement step a randomly chosen molecule has been translated randomly by no more than 0.1Å and rotated around a randomly selected space fixed axis by no more than $10°$. In a volume change step the volume of the simulation box has been attempted to change isotropically by no more than $200Å^3$.

50

The ratio of the accepted and tried moves have resulted in about 1:2 and 1:10 for particle displacement and volume change steps, respectively. As the attempted changes represent rather small perturbations of the charge distribution of the system, the iteration of the induced point dipoles, performed in every attempted move, has converged rapidly, usually after one iteration step. Runs have started from configurations resulted in previous simulations under slightly different thermodynamic conditions. Starting configurations have been equilibrated by 3-5 million MC steps. Thermodynamic properties have been averaged over 40000 equilibrium configurations, separated by 256 particle displacement steps and one volume change step each. The total computing time required by the production phase of each simulation was 6 weeks in a single R10000 SGI processor.

3. Results and Discussion

3.1. DENSITY AND DERIVATIVES

The temperature dependence of the density of BSV water ρ is shown and compared with experimental data in Figure 1 along the four isobars simulated. The data regarded as 'experimental' are resulted from the Saul-Wagner equation-of-state [51]. The pressure dependence of the density along five simulated isotherms are presented in Figure 2. In order to emphasise the comparison of the *change* of the density with the thermodynamic conditions the densities are shown in reduced units in these figures, i.e., $\rho^*=\rho/\rho_0$ is plotted where ρ_0 is the density corresponding to the TMD at atmospheric pressure. As is seen from Fig. 1, the simulated $\rho^*(T)$ data can be well fitted by a third order polynomial at each pressure. The simulated $\rho^*(T)$ curves reproduce well the experimental TMD values, however, they change much faster with the temperature than the experimental data. Since the thermal expansion coefficient α is related to the derivative of $\rho(T)$ as

$$\alpha = -\frac{1}{\rho}\left(\frac{\partial \rho}{\partial T}\right)_p = -\frac{1}{\rho^*}\left(\frac{\partial \rho^*}{\partial T}\right)_p , \tag{1}$$

the above finding indicates that the magnitude of the thermal expansion coefficient of the BSV model is considerably, about an order of magnitude larger than that of real water. This failure, the origin of which is not fully understood yet, seems to be a common problem of the polarizable water models [47,49,52-55], and, although in a smaller extent, affects nonpolarizable models, as well [42,56]. Due to the large difference of the simulated and experimental α values, the $\rho^*(p)$ isotherms of the BSV model are shifted to lower ρ^* values relative to the experimental curves at all temperatures apart from the TMD. Hence, for clarity, the comparison of the experimental and simulated $\rho^*(p)$ curve is only given at 278K in Fig. 2. As is seen, at this temperature the BSV model can reproduce quite well the experimental $\rho^*(p)$ function. Apart from the shift discussed above, the experimental curves are reproduced with the same quality at the other isotherms, as well.

The simulated $\rho^*(p)$ data can be fitted well by linear functions at all the isotherms studied. The steepness of these lines is related to the isothermal compressibility κ of the model through the relation

$$\kappa = \frac{1}{\rho}\left(\frac{\partial \rho}{\partial p}\right)_T = \frac{1}{\rho^*}\left(\frac{\partial \rho^*}{\partial p}\right)_T, \tag{2}$$

The fact that the model describes well the steepness of the $\rho^*(p)$ curves implies that it is able to reproduce the isothermal compressibility well. The isothermal compressibility can also be calculated in each simulation directly, from the fluctuation of the volume of the system V as

$$\kappa = \frac{\left\langle V^2\right\rangle - \left\langle V\right\rangle^2}{k_B T \left\langle V\right\rangle}, \tag{3}$$

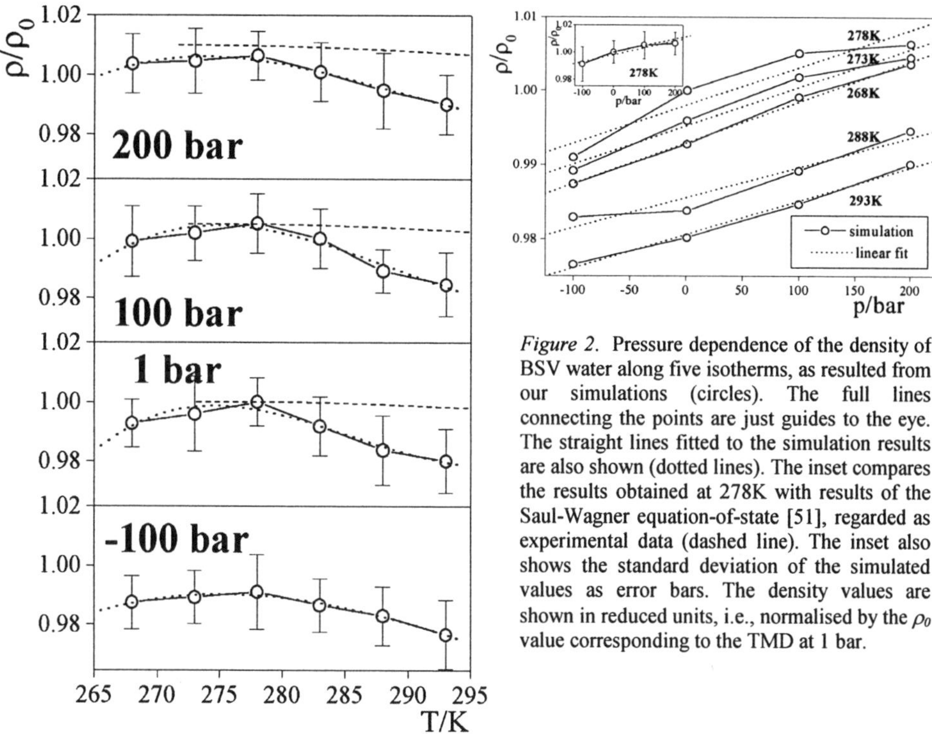

Figure 2. Pressure dependence of the density of BSV water along five isotherms, as resulted from our simulations (circles). The full lines connecting the points are just guides to the eye. The straight lines fitted to the simulation results are also shown (dotted lines). The inset compares the results obtained at 278K with results of the Saul-Wagner equation-of-state [51], regarded as experimental data (dashed line). The inset also shows the standard deviation of the simulated values as error bars. The density values are shown in reduced units, i.e., normalised by the ρ_0 value corresponding to the TMD at 1 bar.

Figure 1. Temperature dependence of the density of BSV water along four isobars, as resulted from our simulations (circles). The full lines connecting the points are just guides to the eye. Standard deviations are shown as error bars. The third order polynomials fitted to the simulation results (dotted lines) and experimental curves (dashed lines) are also shown. Results of the Saul-Wagner equation-of-state [51] are regarded as experimental data. The density values are shown in reduced units, i.e., normalised by the ρ_0 value corresponding to the TMD at 1 bar.

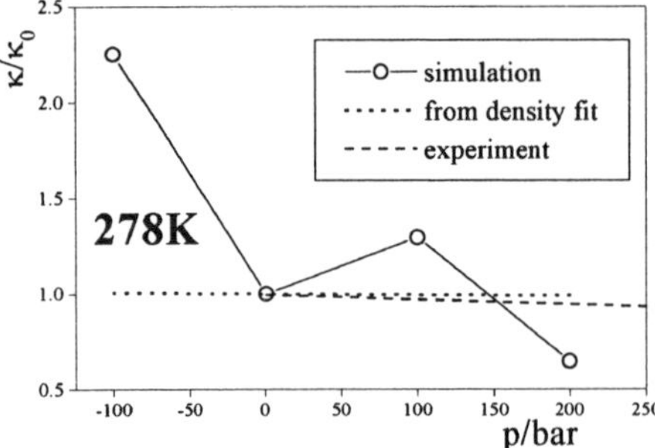

Figure 3. Pressure dependence of the isothermal compressibility of BSV water at 278K, as obtained from the derivation of the linear fit of the simulated $\rho^*(p)$ data (see eq. 2) (dotted line), and directly from the simulations using the volume fluctuation formula (eq. 3) (circles). The full line connecting the points is just a guide to the eye. Results of the Saul-Wagner equation-of-state [51] are shown for comparison as experimental data (dashed line). The isothermal compressibility values are shown in reduced units, i.e., normalised by the κ_0 value corresponding to the TMD at 1 bar.

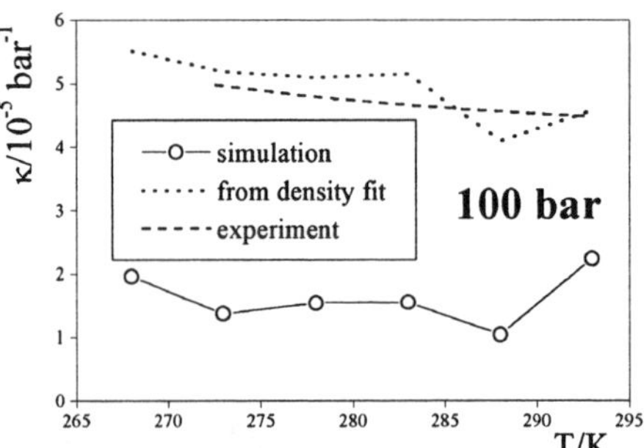

Figure 4. Temperature dependence of the isothermal compressibility of BSV water at 100 bar, as obtained from the derivation of the linear fit of the simulated $\rho^*(p)$ data (see eq. 2) (dotted line), and directly from the simulations using the volume fluctuation formula (eq. 3) (circles). The full line connecting the points is just a guide to the eye. Results of the Saul-Wagner equation-of-state [51] are shown for comparison as experimental data (dashed line).

where k_B is the Boltzmann constant, and $<...>$ denotes ensemble averaging. However, the direct calculation of κ is usually rather inaccurate, as it needs not only the volume of the system but also its fluctuation to be equilibrated, which requires extremely long simulation. However, in the present study the length of the simulations performed have been limited by the large computing cost of the calculation of the induced point dipoles. The pressure dependence of the isothermal compressibility of the BSV model is shown in Figure 3 as calculated both from the linear fit of the $\rho^*(p)$ data and from the volume fluctuation of the system along the T=278K isotherm. The simulation results are also compared with experimental data. As is seen, the direct calculation of κ^* indeed leads to rather inaccurate results, as the $\kappa^*(p)$ data obtained in this way are scattered with a large fluctuation around both the experimental line and the curve obtained from the fit of the simulated $\rho^*(p)$ data. Nevertheless, it reproduces the experimental line in a qualitative way, indicating that the isothermal compressibility of the system is increasing when decreasing the pressure of the system and go to the negative pressure regime. On the other hand, the $\kappa^*(p)$ function obtained from the fit of the $\rho^*(p)$ data agrees very well with the experimental curve. Similar conclusions can be drawn from the comparison of the $\kappa(T)$ functions, shown in Figure 4 along the p=100 bar line. In order to demonstrate that the BSV model can well reproduce not only the temperature and pressure dependence of κ but also its absolute values at different thermodynamic states, we have plotted the absolute values of κ rather then using reduced units here. As is seen, the results obtained from the fit of the $\rho^*(p)$ function are indeed in an excellent agreement with the experimental data, whereas the direct calculation of κ leads again to rather inaccurate results. The general shape of the experimental $\kappa(T)$ function is, however, reproduced in both ways, as the $\kappa(T)$ curve of the BSV model is found to go through a minimum at 288K by both methods. This is a remarkable result, since the fact that the isothermal compressibility of water exhibits a minimum as a function of the temperature is one of its important anomalies. This anomaly is clearly reproduced by the BSV model.

3.2. ENTHALPY AND HEAT CAPACITY

The temperature dependence of the enthalpy H of BSV water is shown and compared with the experimental data in Figure 5 along the four isobars investigated. As is seen, the obtained $H^*(T)$ data can well be fitted by straight lines at each pressure studied. The results are in a fairly good agreement with the experimental data. The temperature dependence of the enthalpy along a given isobar is related to the constant pressure heat capacity of the system c_p as

$$c_p = \left(\frac{\partial H}{\partial T}\right)_p = H_0\left(\frac{\partial H^*}{\partial T}\right)_p . \tag{4}$$

Thus, the heat capacity of BSV water can be calculated from the steepness of the lines fitted to the obtained $H^*(T)$ data. The constant pressure heat capacity can also be calculated directly in a simulation, from the fluctuation of the enthalpy of the system:

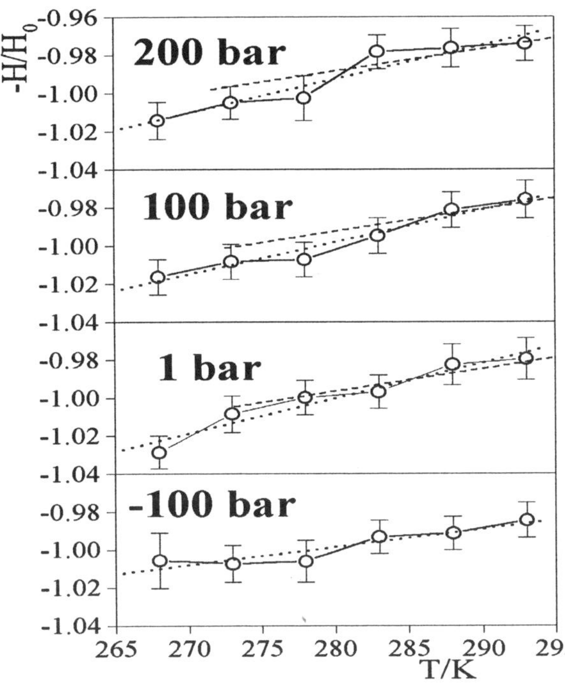

Figure 5. Temperature dependence of the enthalpy of BSV water along four isobars, as resulted from our simulations (circles). The full lines connecting the points are just guides to the eye. Standard deviations are shown as error bars. The straight lines fitted to the simulation results (dotted lines) and results of the Saul-Wagner equation-of-state [51], regarded as experimental data (dashed lines) are also shown. The enthalpy values are shown in reduced units, i.e., normalised by the H_0 value corresponding to the TMD at 1 bar.

54

$$c_p = \frac{\langle H^2 \rangle - \langle H \rangle^2}{N k_B T}.$$

(5)

However, similarly to the isothermal compressibility, the calculation of the heat capacity with this fluctuation formula is also supposed to be rather inaccurate, although the enthalpy of the system changes more rapidly than the volume in a simulation, and hence the fluctuation of the enthalpy can be equilibrated considerably faster than that of the volume. The pressure dependence of the heat capacity along the T=283K isotherm is shown in Figure 6, whereas the temperature dependence of c_p along two isobars is plotted in Figure 7. The heat capacity values obtained in both ways are shown and compared with experimental data in these figures. As is seen, the linear fit of the $H^*(T)$ data indeed results in c_p values that are in a better agreement with the experimental data than those resulted from the enthalpy fluctuation. However, the later method is now far more accurate than in the case of the isothermal compressibility. Thus, the κ values obtained from the fluctuation formula deviate by 70-80% from the experimental data, whereas here this deviation is only about 20-30%. As is seen from Fig. 6, the two methods predict rather different c_p values at –100 bar. Considering the fact that the c_p function is rather close to constant at positive pressures, and this trend is more or less reproduced by the results obtained in both ways at the $p>0$ region, we can consider the prediction obtained by using the fluctuation formula as the more reliable one.

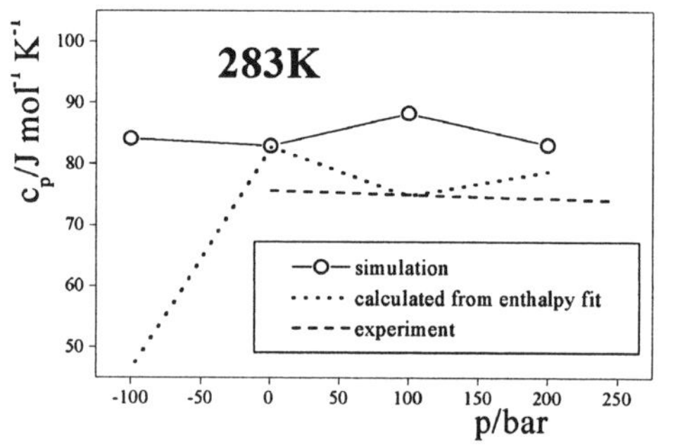

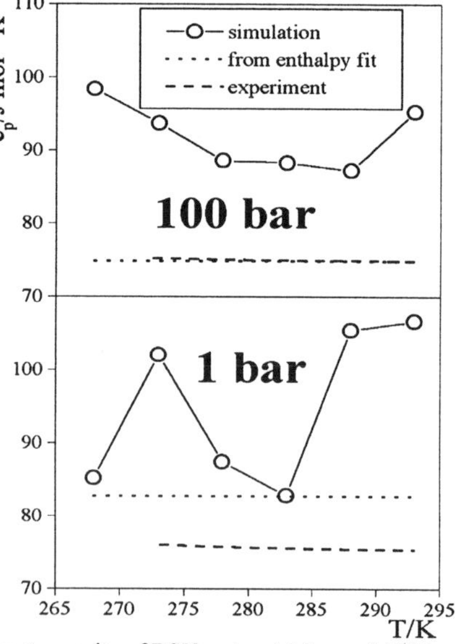

Figure 6. Pressure dependence of the constant pressure heat capacity of BSV water at 283K, as obtained from the derivation of the linear fit of the simulated $H^*(T)$ data (see eq. 4) (dotted line), and directly from the simulations using the enthalpy fluctuation formula (eq. 5) (circles). The full line connecting the points is just a guide to the eye. Results of the Saul-Wagner equation-of-state [51] are shown for comparison as experimental data (dashed line).

Figure 7. Temperature dependence of the constant pressure heat capacity of BSV water at 1 bar and 100 bar, as obtained from the derivation of the linear fit of the simulated $H^*(T)$ data (see eq. 4) (dotted lines), and directly from the simulations using the enthalpy fluctuation formula (eq. 5) (circles). The full lines connecting the points are just guides to the eye. Results of the Saul-Wagner equation-of-state [51] are shown for comparison as experimental data (dashed line).

Since the simulated $H^*(T)$ data have been fitted by straight lines, the $c_p(T)$ functions obtained from eq. 4 are constants. Hence, the temperature dependence of the heat capacity of the BSV model can only be analysed on the basis of the c_p data calculated directly in the simulations, using eq. 5. As is seen, the obtained $c_p(T)$ curves go through a minimum at both pressures, reproducing another of the anomalies of water. Even the pressure dependence of the temperature corresponding to the minimum heat capacity is reproduced qualitatively by the model, as at higher pressures this minimum appears at higher temperature values.

4. Conclusions

In this paper the temperature and pressure dependence of the thermodynamic properties of the polarizable BSV water model have been analysed in detail in a broad pressure range, covering both negative and positive pressures, in the vicinity of the temperature of maximum density. It is found that, apart from the steepness of the change of the density with temperature (i.e., the thermal expansion coefficient), the BSV model reproduces very well the temperature and pressure dependence of the thermodynamic properties of water. Thus, apart from the density anomaly two more anomalous properties are found to be reproduced by the model, namely the fact that both the constant pressure heat capacity and the isothermal compressibility goes through a minimum as a function of the temperature at a given pressure. The present calculations predict that the isothermal compressibility of water increases whereas the constant pressure heat capacity decreases when decreasing the pressure of the system and go to the region of negative pressures. The present results have also demonstrated that thermodynamic response functions can be calculated with a much higher accuracy by fitting the temperature or pressure dependence of the corresponding thermodynamic function (i.e., pressure or density) than from the fluctuation formulae, and hence such curve fitting procedure should be used for this purpose whenever possible.

Acknowledgement

P. J. is a Magyary Zoltán fellow of the Foundation for Hungarian Research and Higher Education, Ministry of Education, Hungary, which is gratefully acknowledged.

References

1. Röntgen, W. C. (1892) Ueber die Constitution des flüssigen Wassers, *Ann. Phys.* **45**, 91-97.
2. Bernal, J. D. and Fowler, R. H. (1933) A Theory of Water and Ionic Solution, with Particular Reference to Hydrogen and Hydroxyl Ions, *J. Chem. Phys.* **1**, 515-548.
3. Eisenberg, D. and Kauzmann, W. (1969) *The Structure and Properties of Water*, Oxford University Press, New York, and references therein
4. F. Franks (ed.) (1972) *Water: A Comprehensive Treatise,* Plenum, New York, and references therein.
5. Némethy, G. and Scheraga, H. A. (1962) Structure of Water and Hydrophobic Bonding in Proteins. I. A Model for the Thermodynamic Properties of Liquid Water, *J. Chem. Phys.* **36**, 3382-3400.

56

6.   Némethy, G. and Scheraga, H. A. (1962) Structure of Water and Hydrophobic Bonding in Proteins. II. Model for the Thermodynamic Properties of Aqueous Solutions of Hydrocarbons, *J. Chem. Phys.* **36**, 3401-3417.

7.   Némethy, G. and Scheraga, H. A. (1964) Structure of Water and Hydrophobic Bonding in Proteins. IV. The Thermodynamic Properties of Liquid Deuterium Oxide, *J. Chem. Phys.* **41**, 680-689.

8.   Geiger, A., Stillinger, F. H., and Rahman, A. (1979) Aspects of the percolation process for hydrogen-bond networks in water, *J. Chem. Phys.* **70**, 4185-4193.

9.   Stanley, H. E. and Teixeira, J. (1980) Interpretation of the unusual behavior of H_2O and D_2O at low temperatures: Tests of a percolation model, *J. Chem. Phys.* **73**, 3404- 3422.

10. Mezei, M. and Beveridge, D. L. (1981) Theoretical studies of hydrogen bonding in liquid water and dilute aqueous solutions, *J. Chem. Phys.* **74**, 622-632.

11. Geiger, A. and Stanley, H. E. (1982) Low-Density "Patches" in the Hydrogen-Bond Network of Liquid Water: Evidence from Molecular-Dynamics Computer Simulations, *Phys. Rev. Lett.* **49**, 1749-1752.

12. Stanley, H. E., Blumberg, R. L., and Geiger, A. (1983) Gelation models of hydrogen bond networks in liquid water, *Phys. Rev. B* **28**, 1626-1629.

13. Narten, A. H. and Levy, H. A. (1969) Observed Diffraction Pattern and Proposed Models of Liquid Water, *Science* **165**, 447-454.

14. Walford, G. and Dore, J. C. (1977) Neutron-diffraction studies of the structure of water II. Temperature variation effects for heavy water, *Mol. Phys.* **34**, 21-32.

15. Pálinkás, G., Kálmán, E., and Kovács, P. (1977) Liquid water II. Experimental atom pair-correlation functions of D_2O, *Mol. Phys.* **34**, 525-537.

16. Soper, A. K. and Phillips, M. G. (1986) A new determination of the structure of water at 25°C, *Chem. Phys.* **107**, 47-60.

17. Soper, A. K., Bruni, F., and Ricci, M. A. (1997) Site-site pair correlation functions of water from 25 to 400 °C: Revised analysis of new and old diffraction data, *J. Chem. Phys.* **106**, 247-254.

18. Svishchev, I. M. and Kusalik, P. G., (1993) Structure in liquid water: A study of spatial distribution functions, *J. Chem. Phys.* **99**, 3049-3058.

19. Berendsen, H. J. C., Grigera, J. R., and Straatsma, T. P. (1987) The Missing Term in Effective Pair Potentials, *J. Phys. Chem.* **91**, 6269-6271.

20. Jedlovszky, P., Bakó, I., Pálinkás, G., Radnai, T., and Soper, A. K. (1996) Investigation of the uniqueness of the reverse Monte Carlo method: Studies on liquid water, *J. Chem. Phys.* **105**, 245-254.

21. Yeh, Y. L. and Mou, C. Y. (1999) Orientational Relaxation Dynamics of Liquid Water Studied by Molecular Dynamics Simulation, *J. Phys. Chem. B* **103**, 3699- 3705.

22. Jedlovszky, P., Mezei, M., and Vallauri, R. (2000) A molecular level explanation of the density maximum of liquid water from computer simulations with a polarizable potential model, *Chem. Phys. Lett.* **318**, 155-160.

23. Kryachko, E. S. (1997) Norbornane-type water heptamer, *Chem. Phys. Lett.* **272**, 132-138.

24. Kryachko, E. S. (1998) Water Cluster Approach To Study Hydrogen-Bonded Pattern in Liquid Water: Ab Initio Orientational Defects in Water Hexamers and Octamers, *Int. J. Quant. Chem.* **70**, 831-853.

25. Jagla, E. A. (1999) Core-softened potentials and the anomalous properties of water, *J. Chem. Phys.* **111**, 8980-8986.

26. Speedy, R. J. and Angell, C. A. (1976) Isothermal compressibility of supercooled water and evidence for a thermodynamic singularity at -45°C, *J. Chem. Phys.* **65**, 851-858.

27. Speedy, R. J. (1982) Stability-Limit Conjecture. An Interpretation of the Properties of Water, *J. Phys. Chem.* **86**, 982-991.

28. Speedy, R. J. (1982) Limiting Forms of the Thermodynamic Divergences at the Conjectured Stability Limits in Superheated and Supercooled Water, *J. Phys. Chem.* **86**, 3002-3005.

29. Speedy, R. J. (1987) Thermodynamic Properties of Supercooled Water at 1 atm, *J. Phys. Chem.* **91**, 3354-3358.

30. Poole, P. H., Sciortino, F., Essmann, U., and Stanley, H. E. (1992) Phase behaviour of metastable water, *Nature* **360**, 324-328.

31. Stanley, H. E., Angell, C. A., Essmann, U., Hemmati, M., Poole, P. H., and Sciortino, F. (1994) Is there a second critical point in liquid water?, *Physica A* **205**, 122-139.

32. Poole, P. H., Essmann, U., Sciortino, F., and Stanley, H. E. (1993) Phase diagram for amorphous solid water, *Phys. Rev. E* **48**, 4605-4610.

33. Sciortino, F., Poole, P. H., Essmann, U., and Stanley, H. E. (1997) Line of compressibility maxima in the phase diagram of supercooled water, *Phys. Rev. E* **55**, 727-737.

34. Burton, E. F. and Oliver, W. F. (1935) X-Ray Diffraction Patterns of Ice, *Nature* **135**, 505-506.

35. Mishima, O., Calvert, L. D., and Whalley, E. (1984) 'Melting ice' I at 77 K and 10 kbar: a new method of making amorphous solids, *Nature* **310**, 393-395.

36. Debenedetti, P. G., Raghavan, V. S., and Borrick, S. S. (1991) Spinodal Curve of Some Supercooled Liquids, *J. Phys. Chem.* **95**, 4540-4551.

37. Poole, P. H., Sciortino, F., Essmann, U., and Stanley, H. E. (1993) Spinodal of liquid water, *Phys. Rev. E* **48**, 3799-3817.

38. Veiga, H. I. M., Rebelo, L. P. N., Nunes de Ponte, M., and Szydlowski, J. (2001) Water and Gallium at Absolute Negative Pressures. Loci of Maximum Density and of Melting., *Int. J. Thermophys.* **22**, 1159-1174.

39. Ruocco, G., Sampoli, M., Torcini, A., and Vallauri, R. (1993) Molecular dynamics results for streched water, *J. Chem. Phys.* **99**, 8095-8104.

40. Stillinger, F. H. and Rahman, A. (1974) Improved simulation of liquid water by molecular dynamics, *J. Chem. Phys.* **60**, 1545-1557.

41. Jorgensen, W. L., Chandrasekhar, J., Madura, J. D., Impey, R. W., and Klein, M. L. (1983) Comparison of simple potential functions for simulating liquid water, *J. Chem. Phys.* **79**, 926-935.

42. Jorgensen, W. L. and Jenson, C. (1998) Temperature Dependence of TIP3P, SPC and TIP4P Water from NPT Monte Carlo Simulations: Seeking Temperatures of Maximum Density, *J. Comp. Chem.* **19**, 1179-1186.

43. Báez, L. A. and Clancy, P. (1994) Existence of a density maximum in extended simple point charge water *J. Chem. Phys.* **101**, 9837-9840.

44. Bagchi, K., Balasubramanian, S., and Klein, M. L. (1997) The effects of pressure on structural and dynamical properties of associated liquids: Molecular dynamics calculations for the extended simple point charge model of water, *J. Chem. Phys.* **107**, 8561-8567.

45. Berendsen, H. J. C., Postma, J. P. M., van Guntseren, W. F., and Hermans, J. (1981) Interaction models for water in relation to protein hydration., in B. Pullman (ed.), *Intermolecular Forces*, Reidel, Dordrecht, pp. 331-342.

46. Brodholt, J. P., Sampoli, M., and Vallauri, R. (1995) Parametrizing a polarizable intermolecular potential for water, *Mol. Phys.* **86**, 149-158.

47. Jedlovszky, P. and Vallauri, R. (1999) Temperature dependence of the thermodynamic properties of a polarizable potential model of water, *Mol. Phys.* **97**, 1157-1163.

48. Balucani, U., Brodholt, J. P., Jedlovszky, P., and Vallauri, R. (2000) Viscosity of liquid water from computer simulations with a polarizable potential model, *Phys. Rev. E* **62**, 2971-2973.

49. Jedlovszky, P. and Vallauri, R. (2001) Thermodynamic and structural properties of liquid water around the temperature of maximum density in a wide range of pressures: A computer simulation study with a polarizable potential model, *J. Chem. Phys.* **115**, 3750-3762.

50. Jedlovszky, P. and Vallauri, Investigation of the temperature of maximum density line of a polarizable water model, R. *Phys. Rev. Lett.*, submitted for publication.

51. Saul, A. and Wagner, W. (1989) A fundamental equation for water covering the range from the melting line to 1273 K at pressures up to 25,000 MPa , *J. Phys. Chem. Ref. Data* **18**, 1537-1564.

52. Svishchev, I. M., Kusalik, P. G., Wang, J., and Boyd, R. J. (1996) Polarizable point-charge model for water: Results under normal and extreme conditions, *J. Chem. Phys.* **105**, 4742-4750.

53. Kiyohara, K., Gubbins, K. E., and Panagiotopoulos, A. Z. (1998) Phase coexistence properties of polarizable water models, *Mol. Phys.* **94**, 803-808.

54. Jedlovszky, P. and Richardi, J. (1999) Comparison of different water models from ambient to supercritical conditions: A Monte Carlo simulation and molecular Ornstein-Zernike study, *J. Chem. Phys.* **110**, 8019-8031.

55. Jedlovszky, P., Vallauri, R., and Richardi, J. (2000) The change of th estructural and thermodynamic properties of water from ambient to supercritical conditions as seen by computer simulations, *J. Phys.: Condens. Matter* **12**, A115-A122.

56. Mahoney, M. W. and Jorgensen, W. L. (2000) A five-site model for liquid water and the reproduction of the density anomaly by rigid, nonpolarizable potential functions, *J. Chem. Phys.* **112**, 8910-8922.

WATER AT POSITIVE AND NEGATIVE PRESSURES

H. E. STANLEY,[1] M. C. BARBOSA,[1,2] S. MOSSA,[1] P. A. NETZ,[3]
F. SCIORTINO,[4] F. W. STARR,[5] AND M. YAMADA[1]
[1] *Center for Polymer Studies and Department of Physics, Boston University*
Boston, MA 02215 USA
[2] *Instituto de Física, Universidade Federal do Rio Grande do Sul, Caixa Postal 15051*
91501-970, Porto Alegre, RS, Brazil
[3] *Departamento de Química, Universidade Luterana do Brasil*
92420-280, Canoas, RS, Brazil
[4] *Dipartimento di Fisica Università di Roma La Sapienza, Istituto Nazionale di Fisica*
della Materia, and INFM Center for Statistical Mechanics and Complexity,
Piazzale Aldo Moro 2, 00185 Roma, Italy
[5] *Polymers Division and Center for Theoretical and Computational Materials Science*
National Institute of Standards and Technology, Gaithersburg, MD 20899 USA

Abstract. We review recent results of molecular dynamics simulations of two models of liquid water, the extended simple point charge (SPC/E) and the Mahoney-Jorgensen transferable intermolecular potential with five points (TIP5P), which is closer to real water than previously-proposed classical pairwise additive potentials. Simulations of the TIP5P model for a wide range of deeply supercooled states, including both positive and negative pressures, reveal (i) the existence of a non-monotonic "nose-shaped" temperature of maximum density (TMD) line and a non-reentrant spinodal, (ii) the presence of a low temperature phase transition. The TMD that changes slope from negative to positive as P decreases and, notably, the point of crossover between the two behaviors is located at ambient pressure (temperature $\approx 4\,°\text{C}$, and density ≈ 1 g/cm^3). Simulations on the dynamics of the SPC/E model reveal (iii) the dynamics at negative pressure shows a minimum in the diffusion constant D when the density is decreased at constant temperature, complementary to the known maximum of D at higher pressures, and (iv) the loci of minima of D relative to the spinodal shows that they are inside the thermodynamically metastable regions of the phase-diagram. These dynamical results reflect the initial enhancement and subsequent breakdown of the tetrahedral structure and of the hydrogen bond network as the density decreases.

A.R. Imre et al. (eds.), Liquids Under Negative Pressure, 59–67.
© 2002 *Kluwer Academic Publishers. Printed in the Netherlands.*

60

1. Introduction

Water is an important liquid in nature, and is also fundamental in chemical and technological applications. Although the individual water molecule has a simple chemical structure, water is considered a complex fluid because of its anomalous behavior [1, 2, 3, 4, 5, 6]. It expands on freezing and, at a pressure of 1 atm, the density has a maximum at 4°C. Additionally, there is a minimum of the isothermal compressibility at 46°C and a minimum of the isobaric heat capacity at 35°C [7]. These anomalies are linked with the microscopic structure of liquid water, which can be regarded as a transient gel–a highly associated liquid with strongly directional hydrogen bonds [8, 9]. Each water molecule acts as both a donor and an acceptor of bonds, generating a structure that is locally ordered, similar to that of ice, but maintaining the long-range disorder typical of liquids. Despite the extensive work that has been done on water, many aspects of its behavior remain unexplained.

Several scenarios have been proposed to account for the anomalous behavior of the thermodynamic response functions on cooling, each predicting a different behavior for the liquid spinodal, the line of the limit of stability separating the region where liquid water is metastable from the region where the liquid is unstable. (i) According the stability-limit conjecture [10, 11], the pressure of the spinodal line should decrease on cooling, become negative, and increase again after passing through a minimum. It reenters the positive pressure region of the phase diagram at a very low temperature, thereby giving rise to a line of singularities in the positive pressure region, and consequently the increase in the thermodynamic response functions on cooling in the anomalous region is due to the proximity of this reentrant spinodal. (ii) The critical point hypothesis [12, 13, 14, 15, 16, 17], proposes a new critical point at the terminus of a first-order phase transition line separating two liquid phases of different density. The anomalous increases of the response functions, compressibility, specific heat, and volume expansivity, is interpreted in terms of this critical point. (iii) The singularity-free hypothesis [9, 18, 19] proposes that actually there is no divergence close to the anomalous region; the response functions grow on lowering temperature but remain finite, attaining maximum values.

Water properties and anomalies can be strongly influenced by the physical or chemical properties of the medium [1, 3, 4, 5, 20, 21]. The effect not only of applied pressure, but also of negative pressure ("stretching") is remarkable. The study of the behavior of this fluid under negative pressures is relevant not only from the academic point of view, but also for realistic systems. For example, negative pressures are observed [22], and seem to play an important role in the mechanism of water transport in plants. Therefore, properties that modify the structure of water, especially if this modification is similar to the effect of stretching (as is the case in some hydrogels [20]), also influence its dynamical behavior.

Dynamic properties, such as the diffusion constant, have been studied in detail for water systems at atmospheric and at high positive pressures, both experimentally [23, 24] as well as by computer simulations [25, 26, 27, 28, 29, 30, 31]. The increase of pressure increases the presence of defects and of interstitial water molecules in the network [26]. They disrupt the tetrahedral local structure, weakening the hydrogen bonds, and thus increasing the diffusion constant [30, 31]. However, a further increase in the pressure leads to steric effects which works in the direction of lowering the mobility. The interplay of these factors leads to a maximum in the diffusion constant [30, 31] at some high density $\rho_{\max}$. Above this density (or corresponding pressure), the diffusion of water is in some sense like that of a normal liquid, controlled by hindrance, with the hydrogen bonds playing a secondary role. However, the behavior at very low ρ is less well understood.

2. Location of the Spinodal at Positive and Negative Pressures

Relatively few experimental works [32, 33] and simulations [12, 16, 29, 30, 35, 34] have been performed on "stretched" water. In this negative pressure region of the phase diagram the system is metastable, and becomes unstable beyond the spinodal line, so locating the spinodal we can ensure that our simulated state points lie in the metastable and not in the unstable region. Moreover, the shape of the spinodal can test the stability-limit conjecture against the critical point hypothesis and the singularity-free interpretation, so we first discuss the density and pressure of the spinodal, which we denote $\rho_{\mathrm{sp}}(T)$ and $P_{\mathrm{sp}}(T)$, respectively.

Yamada and her coworkers [36] simulated a system of $N = 343$ molecules interacting with the TIP5P potential [37]. TIP5P is a five-site, rigid, non-polarizable water model, not unlike the ST2 model [38]. The TIP5P potential accurately reproduces the density anomaly at 1 atm and exhibits excellent structural properties when compared with experimental data [37, 39]. The TMD shows the correct pressure dependence, shifting to lower temperatures as pressure is increased. Under ambient conditions, the diffusion constant is close to the experimental value, with reasonable temperature and pressure dependence away from ambient conditions [37]. Equilibration runs were performed at constant T. After thermalization at $T = 320$ K the thermostat temperature was set to the temperature of interest. The system evolved for a time longer than the structural relaxation time τ_α, defined as the time at which $F_s(Q_0, \tau_\alpha) = 1/e$, where $F_s(Q_0, t)$ is the self-intermediate scattering function evaluated at $Q_0 = 18 \ \mathrm{nm}^{-1}$, the location of the first peak of the static structure factor. In the time τ_α, each molecule diffuses on average a distance of the order of the nearest neighbor distance. We use the final configuration of the equilibration run to start a production run of length greater than several τ_α and then analyze the calculated trajectory.

Figure 2 shows results for pressure along isotherms. At lower temperatures an inflection develops, which becomes a "flat" isotherm at the lowest temperature, $T = 215$ K. The presence of a flat region indicates that a phase separation takes place; the critical temperature is $T_{C'} = 217 \pm 3$ K, the critical pressure is $P_{C'} = 340 \pm 20$ MPa, and the critical density $\rho_{C'} = 1.13 \pm 0.04 \ \mathrm{g/cm}^3$.

Figure 2(a) plots the pressure along isochores. The curves show minima as a function of temperature; the locus of the minima is the TMD line, since $(\partial P/\partial T)_V = \alpha_P/K_T$. Note that the pressure exhibits a minimum if the density passes through a maximum ($\alpha_p = 0$). It is clear that, as in the case of ST2 water, TIP5P water has a TMD that changes slope from negative to positive as P decreases. Notably,

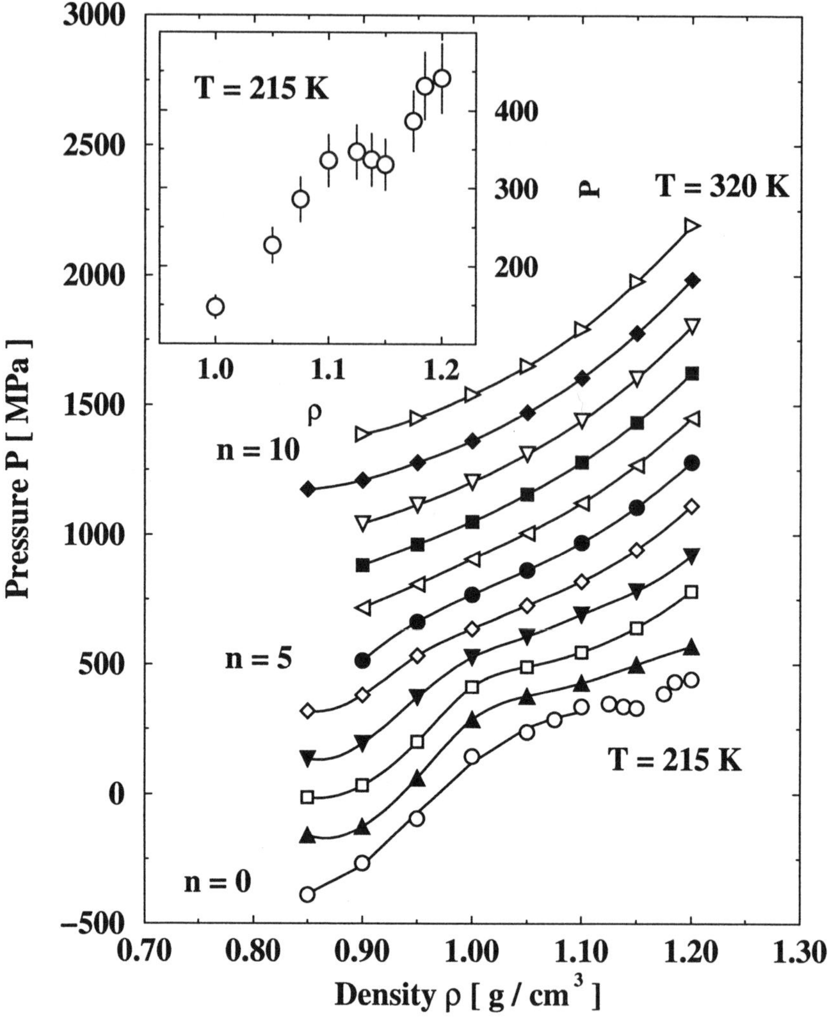

Figure 1. Dependence on density of the pressure at all temperatures investigated ($T = 215, 220, 230, 240, 250, 260, 270, 280, 290, 300, 320$ K, from bottom to top). Each curve has been shifted by $n \times 150$ MPa to avoid overlaps. An inflection appears as T is decreased, transforming into a "flat" coexistence region at $T = 215$ K, indicating the presence of a liquid-liquid transition. Inset: A detailed view of the $T = 215$ K isotherm. Adapted from [36].

the point of crossover between the two behaviors is located at ambient pressure, $T \approx 4\,°\mathrm{C}$, and $\rho \approx 1$ g/cm³.

Also plotted the spinodal line, obtained by fitting the isotherms (for $T \geq 300K$) of Fig. 2 to the form $P(T, \rho) = P_s(T) + A\left[\rho - \rho_s(T)\right]^2$, where $P_s(T)$ and $\rho_s(T)$ denote the pressure and density of the spinodal line. This functional form is the

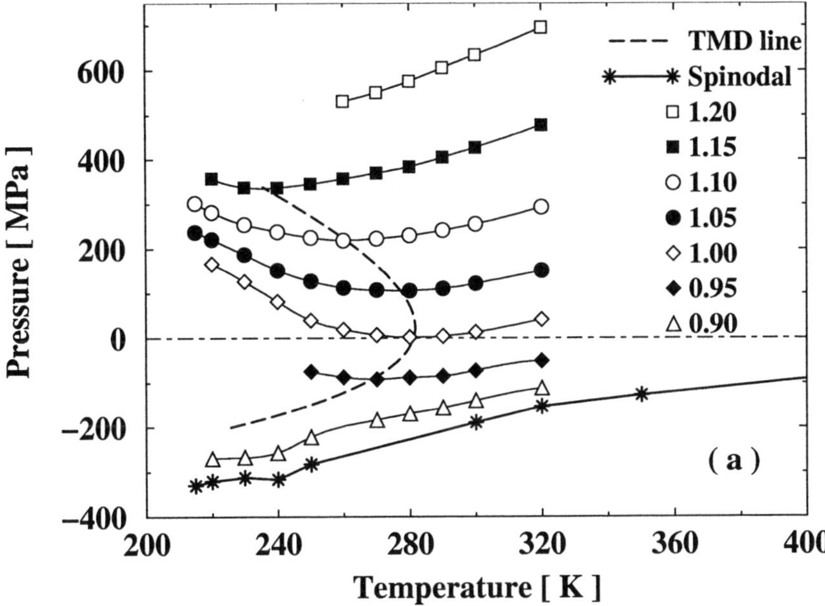

Figure 2. Pressure along seven isochores; the minima correspond to the temperature of maximum density line (dashed line). Note the "nose" of the TMD line at $T = 4\ °C$. Stars denote the liquid spinodal line, which is not reentrant, and terminates at the liquid-gas critical point. Adapted from [36].

mean field prediction for $P(\rho)$ close to a spinodal line. For $T \leq 250K$, $P_s(T)$ is calculated by estimating the location of the minimum of $P(\rho)$. The results in Fig.2 show that the liquid spinodal line is not reentrant and does not intersect the TMD line.

3. Dynamic Properties

We next discuss results on the dynamics of stretched water recently obtained by Netz and his collaborators [40]. While there are a large number of intermolecular potential functions used to simulate water, each of which gives slightly different results, the overall thermodynamics picture obtained from these models is generally very similar. Since dynamic properties are particularly sensitive to the potential choice, the extended simple point charge (SPC/E) potential is used since it reproduces both the maximum in diffusivity under pressure as well as the power-law behavior of dynamics properties on cooling. For understanding the properties of water at negative pressure, simulations are particularly important since experiments are very difficult to perform in this region.

The effect of extreme conditions on the flow of the liquid is assessed by calculating the diffusion constant D, defined by the asymptotic value of the slope

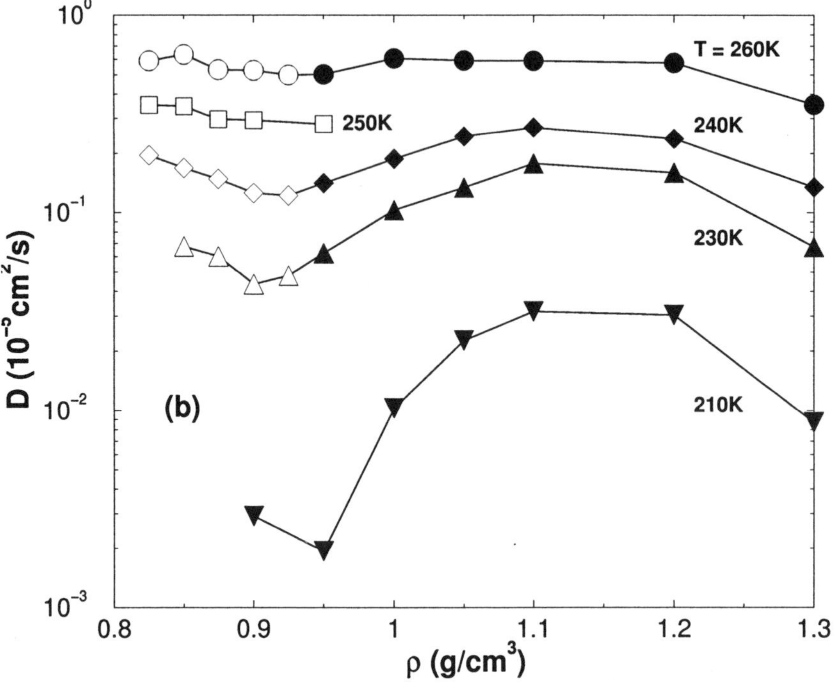

Figure 3. Dependence of the diffusion constant D on ρ along isotherms (for $\rho \leq 1.0$ g/cm^3). Open symbols are the new simulations we report, and filled symbols are from Ref. [30]. Adapted from [40].

of the mean square displacement versus time. We show D along isotherms in Fig. 3. For $T \leq 260$ K, D has a minimum value at $\rho \approx 0.9$ g/cm^3, which becomes more pronounced at lower T. This behavior can be understood considering the structural changes that occur with decreasing density. At low T, the decreased density enhances the local tetrahedral ordering, which leads to a decrease in D. Further decreases in density reduces the stability of the tetrahedral structure and causes an increase of D.

The location of the minimum is near the ice Ih density ≈ 0.915 g/cm^3, which is the density where the perfect tetrahedral order occurs. The behavior of the minimum of D, $D_{\mathrm{min}}(T)$, complements the known behavior of $D_{\mathrm{max}}(T)$ for the same model [30, 31, 35], where a maximum occurs due to breaking hydrogen bonds at high pressure; the density of the $D_{\mathrm{min}}(T)$ increases slightly with increasing T, while the density of $D_{\mathrm{max}}(T)$ decreases with increasing T [31]. This is expected, since the range of densities where anomalous behavior occurs expands with decreasing T. We show the loci of $D_{\mathrm{min}}(T)$ and $D_{\mathrm{max}}(T)$, along with the spinodal and locus of density maxima in Fig. 4.

Below the spinodal, D also increases, since the mobility of the gas is larger than

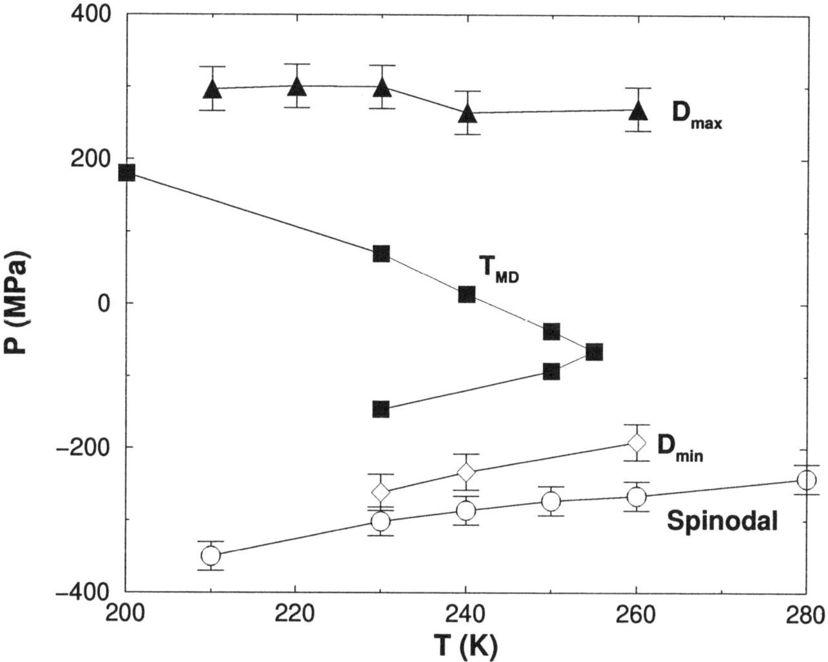

Figure 4. Relation of the loci of maxima and minima of D with T_{MD} and the spinodal. Open symbols are from the present work, and filled symbols are from Ref. [30]. Adapted from [40].

that of the liquid. However, the simulations clearly show that D_{min} for the liquid occurs prior to the onset of cavitation, and so the location of D_{min} we estimate is not affected by phase separation. Recently, Ref. [35] estimated the location of D_{min} for the same model along several isotherms, and associated D_{min} with a maximum in orientational order.

4. Conclusions

Water exhibits a very complex structure and its properties and anomalies are strongly influenced by variations of pressure. For high densities ($\rho > \rho_{max}$), water behaves as a normal liquid and the decrease of D with increasing pressure is governed by steric effects. For $\rho_{min} < \rho < \rho_{max}$, as the pressure is decreased, the presence of defects and interstitial water decrease, the tetrahedral structure dominates, with stronger hydrogen bonds. This process reaches its maximum at $\rho = \rho_{min} \approx \rho_{ice}$. Further stretching destabilizes the hydrogen bond network, leading to an increase in mobility. The locus of D_{min} roughly tracks the spinodal, not surprising since the same breakdown of tetrahedral order that gives rise to D_{min} also facilitates cavitation.

Acknowledgements

We thank D. R. Baker, S. V. Buldyrev, P. Debenedetti, G. Franzese, W. Kob, E. La Nave, M. Marquez and C. Rebbi for useful discussions, and NSF Grant CHE-0096892, the Conselho Nacional de Desenvolvimento Cientifico e Technologico (CNPq), the Fundacao de Amparo a Pesquisa do Rio Grande do Sul (Fapergs) for support. MY thanks NSF Grant GER-9452651 for support as a Graduate Research Trainee at the Boston University Center for Computational Science, FS thanks MURST COFIN 2000 and INFM Iniziativa Calcolo Parallelo, and FWS thanks the National Research Council.

References

1. For elementary introductions to recent work on liquid water, the reader may wish to consult P. Ball, *Life's Matrix: A Biography of Water* (Farrar Straus and Giroux, New York, 2000) or P. G. Debenedetti and H. E. Stanley, "The Novel Physics of Water at Low Temperatures", *Physics Today* (submitted).
2. V. Brazhkin. S. V. Buldyrev, V. N. Ryzhov, and H. E. Stanley [eds], *New Kinds of Phase Transitions: Transformations in Disordered Substances* Proc. NATO Advanced Research Workshop, Volga River (Kluwer, Dordrecht, 2002).
3. O. Mishima and H. E. Stanley, *Nature* **396**, 329 (1998).
4. M.-C. Bellissent-Funel, ed., *Hydration Processes in Biology: Theoretical and Experimental Approaches* (IOS Press, Amsterdam, 1999).
5. H. E. Stanley, S. V. Buldyrev, N. Giovambattista, E. La Nave, A Scala, F. Sciortino, and F. W. Starr, [Proc. IUPAP Statphys21, Cancun] *Physica A* **306**, 230–242 (2002).
6. S. V. Buldyrev, G. Franzese, N. Giovambattista, G. Malescio, M. R. Sadr-Lahijany, A. Scala, A. Skibinsky, and H. E. Stanley [Proc. International Conf. on Scattering Studies of Mesoscopic Scale Structure and Dynamics in Soft Matter] *Physica A* **304**, 23-42 (2002).
7. R. C. Dougherty and L. N. Howard, *J. Chem. Phys.* **109**, 7379 (1998).
8. A. Geiger, F. H. Stillinger, and A. Rahman, *J. Chem. Phys.* **70**, 4185 (1979).
9. H. E. Stanley and J. Teixeira, *J. Chem. Phys.* **73**, 3404 (1980).
10. R. J. Speedy, *J. Chem. Phys.* **86**, 982 (1982); *Ibid* **86**, 3002 (1992).
11. R. J. Speedy, *J. Chem. Phys.* **91**, 3354 (1987).
12. P. H. Poole, F. Sciortino, U. Essmann, and H. E. Stanley, *Nature* **360**, 324 (1992); *Phys. Rev. E* **48**, 3799 (1993); F. Sciortino, P. H. Poole, U. Essmann, and H. E. Stanley, Ibid. **55**, 727 (1997); S. Harrington, R. Zhang, P. H. Poole, F. Sciortino, and H. E. Stanley, *Phys. Rev. Lett.* **78**, 2409 (1997).
13. O. Mishima, *J. Chem. Phys.* **100**, 5910 (1994).
14. P. H. Poole, F. Sciortino, T. Grande, H. E. Stanley and C. A. Angell, *Phys. Rev. Lett.* **73**, 1632 (1994); C. F. Tejero and M. Baus, *Phys. Rev. E* **57**, 4821 (1998); T. M. Truskett, P. G. Debenedetti, S. Sastry, and S. Torquato, *J. Chem. Phys.* **111** 2647 (1999).
15. M.-C. Bellissent-Funel, *Europhys. Lett.* **42**, 161 (1998); O. Mishima and H. E. Stanley, Nature **392**, 192 (1998).
16. H. Tanaka, *J. Chem. Phys.* **105**, 5099 (1996).
17. A. Scala, F. W. Starr, E. La Nave, H. E. Stanley and F. Sciortino, *Phys. Rev. E* **62**, 8016 (2000).
18. S. Sastry, P. G. Debenedetti, F. Sciortino, and H. E. Stanley, *Phys. Rev. E* **53**, 6144 (1996).
19. L. P. N. Rebelo, P. G. Debenedetti, and S. Sastry, *J. Chem. Phys.* **109**, 626 (1998).
20. P. A. Netz and Th. Dorfmüller, *J. Phys. Chem. B* **102**, 4875 (1998).
21. K. Koga, X. C. Zeng, and H. Tanaka, *Chem. Phys. Lett.* **285**, 278 (1998).
22. W. T. Pockman, J. S. Sperry, and J. W. O'Leary, *Nature* **378**, 715 (1995).
23. J. Jonas, T. DeFries, and D. J. Wilbur, *J. Chem. Phys.* **65**, 582 (1976).

24. F. X. Prielmeier, E. W. Lang, R. J. Speedy, and H.-D. Lüdemann, *Phys. Rev. Lett.* **59**, 1128 (1987); *Ber. Bunsenges. Phys. Chem.* **92**, 1111 (1988).
25. M. Rami Reddy and M. Berkovitz, *J. Chem. Phys.* **87**, 6682 (1987).
26. F. Sciortino, A. Geiger, and H. E. Stanley, *Nature* **354**, 218 (1991); *Ibid.*, *J. Chem. Phys.* **96**, 3857 (1992).
27. N. Giovambattista, F. W. Starr, F. Sciortino, S. V. Buldyrev, and H. E. Stanley, *Phys. Rev. E* **65**, 041502-1 – 041502-6 (2002) cond-mat/0201028.
28. E. La Nave, A. Scala, F. W. Starr, H. E. Stanley and F. Sciortino, *Phys. Rev. E* **64**, 036102-1 – 036102-10 (2001); E. La Nave, H. E. Stanley and F. Sciortino, *Phys. Rev. Letters* **88**, 035501-1 to 035501-4 (2002) cond-mat/0108546.
29. P. Gallo, F. Sciortino, P. Tartaglia, and S.-H. Chen, Phys. Rev. Lett. **76**, 2730 (1996).
30. F. W. Starr, F. Sciortino, and H. E. Stanley, *Phys. Rev. E* **60**, 6757 (1999); F. W. Starr, S. T. Harrington, F. Sciortino, and H. E. Stanley, *Phys. Rev. Lett.*, **82**, 3629, (1999).
31. A. Scala, F. W. Starr, E. La Nave, F. Sciortino and H. E. Stanley, *Nature* **406**, 166 (2000).
32. S. J. Henderson and R. J. Speedy, *J. Phys. E: Scientific Instrumentation* **13**, 778 (1980).
33. J. L. Green, D. J. Durben, G. H. Wolf, and C. A. Angell, *Science* **249**, R649 (1990).
34. I. I. Vaisman, L. Perera, and M. L. Berkovitz, *J. Chem. Phys.* **98**, 9859 (1993).
35. J. R. Errington and P. G. Debenedetti, *Nature* **409**, 318 (2001).
36. M. Yamada, S. Mossa, H. E. Stanley, F. Sciortino, *Phys. Rev. Letters* **88**, 195701 (2002); cond-mat/0202094
37. M. W. Mahoney and W. L. Jorgensen, J. Chem. Phys. **112**, 8910 (2000); *Ibid.* **114**, 363 (2001).
38. F. H. Stillinger and A. Rahman, *J. Chem. Phys.* **60**, 1545 (1974).
39. J. M. Sorenson, G. Hura, R. M. Glaeser, and T. Head-Gordon, *J. Chem. Phys.* **113**, 9149 (2000).
40. P. A. Netz, F. W. Starr, H. E. Stanley, and M. C. Barbosa, *J. Chem. Phys.* **115**, 344–348 (2001); cond-mat/0102196; P. A. Netz, F. W. Starr, H. E. Stanley, and M. C. Barbosa, cond-mat/0201130; P. A. Netz, F. Starr, M. C. Barbosa, H. E. Stanley, cond-mat/0201138.

THE THERMOPHYSICAL PROPERTIES OF LIQUIDS ON THE MELTING LINE AT NEGATIVE PRESSURES

M. Z. FAIZULLIN and V. P. SKRIPOV
Institute of Thermophysics, Ural Branch of RAS
Pervomaiskaya Str. 91, GSP-828, Ekaterinburg, 620219, Russia

1. Introduction

The liquid-solid equilibrium line has a real extension beyond the triple point into the negative pressure region. For "normally" melting substances such an extension leads to a limiting pressure for the equilibrium curve: $p(T{\to}0)= - p_*$. Figure 1 shows a simple temperature-pressure diagram of the states for argon as an example. The extension AE of the melting line BA beyond the triple point corresponds to the Simon equation [1]:

$$\frac{p}{p_*}+1=\left(\frac{T}{T_0}\right)^c ,$$ (1)

where $p_* = p(T=0)$, $c > 1$ are individual parameters, T_0 is the temperature at $p = 0$. The liquid spinodal CK has been constructed using experimental T,p,v data [2] in the region of the stable and metastable states of liquid argon (as envelope of the isochors of the liquid). The extension KD of the spinodal is less reliable.

The value $-p_*$ has the meaning of the characteristic internal pressure for a condensed phase in the low temperature limit

$$p_* =\left(\frac{\Delta u}{\Delta v}\right)_{T=0} ,$$ (2)

where $\Delta u = u_L - u_S$ and $\Delta v = v_L - v_S$ are changes of internal energy and volume in melting. The value $-p_*$ is close to the theoretical solidity of a liquid under homogeneous stretching.

The use of the low temperature asymptotic of the melting line is useful for generalize data on the melting of substances with the position of thermodynamic similarity [3]. The value of the limiting pressure can be used as a characteristic scale of pressure in descriptions of melting. If one introduces a shifted pressure scale

$$p' = p + p_*$$ (3)

A.R. Imre et al. (eds.), Liquids Under Negative Pressure, 69–80.
© 2002 *Kluwer Academic Publishers. Printed in the Netherlands.*

70

(individual for every substance) and reduced dimensionless temperature and pressure

$$\tilde{T} = T/T_0 \quad \text{and} \quad \tilde{p} = p'/p_* \tag{4}$$

then the Simon equation may be represented in the following form:

$$\tilde{p} = \tilde{T}^c \tag{5}$$

which reflects the idea of thermodynamic similarity for melting lines of different substances.

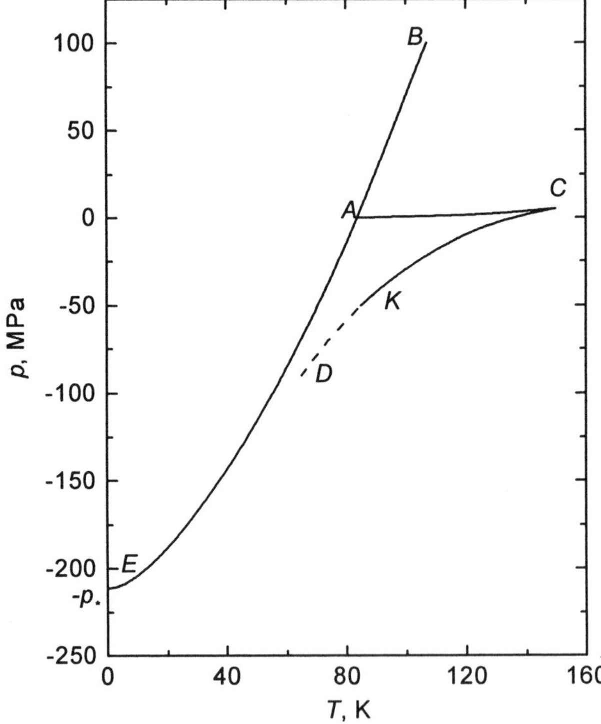

Figure 1. Melting line *AB* of argon with a metastable extension *AE* into the region of negativepressures; *CKD* is the spinodal of a stretched liquid; *AC* is the line of liquid-vapour phase equilibrium.

The aim of this paper is to analyse the behavior of some thermodynamic quantities of liquids on the solid-liquid (SL) phase equilibrium line, extended beyond the triple point. Use is made of experimental T, p, v data for liquids in the stable state and their extrapolation along chosen isolines into the region of metastability. We have restricted ourselves to consideration of normally melting substances, for which $dp/dT > 0$, $\Delta v > 0$.

2. The Boundary of Liquid Stability

The equilibrium of two phases in the region of negative pressure presupposes stability of each of them with respect to local perturbations of density or entropy. The condition of mechanical stability

$$-\left(\frac{\partial p}{\partial v}\right)_T > 0 \tag{6}$$

is to be fulfilled for each of the phases on the SL line including the metastable section of the line. The condition

$$\left(\frac{\partial p}{\partial v}\right)_T = 0, \tag{7}$$

corresponds to the boundary of stability (the spinodal).

An important qualitative conclusion follows from figure 1, namely, that the melting line and the spinodal come closer together as the tensile stress increases. The crystal phase also decreases its stability. That this is so may be seen from the pressure dependence of elasticity $-(\partial p/\partial v)_T$ on the melting line shown in figure 2.

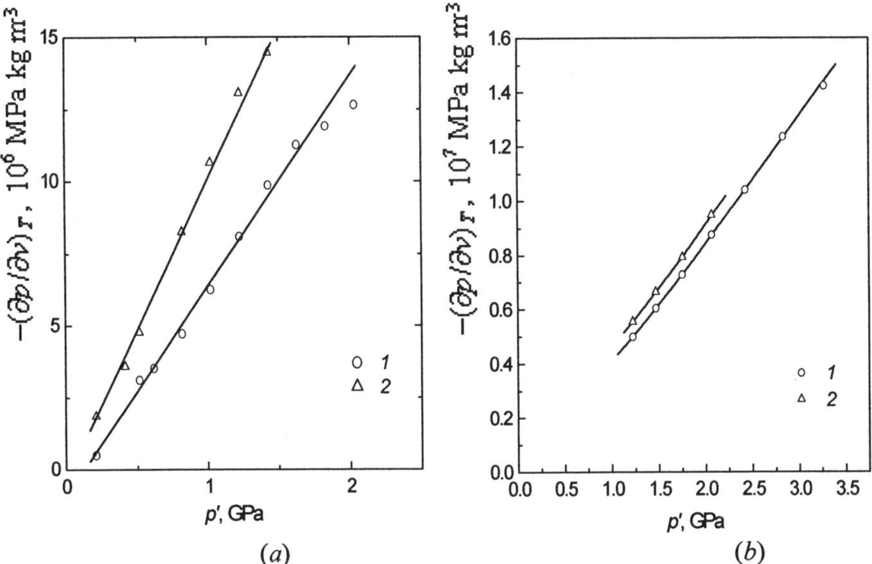

Figure 2. Behavior of the elasticity of liquid (1) and solid (2) argon (a) and sodium (b) on the melting line.

72

In figure 2 use is made of a shifted pressure scale $p' = p + p_*$. Data from references [4] and [5] were used in constructing the liquid and the crystalline branches of elasticity for argon and sodium. It can be seen that the stability boundaries of both phases $(\partial p / \partial v)_T = 0$ are reached in the vicinity of the initial point $p' = 0$, $T = 0$ of the melting line. The general character of the trend mentioned above is confirmed by comparison for different substances of the values of the limiting pressure $-p_* = p(0)$ on the melting line and the limiting pressure $p_{sp}(0)$ on the liquid spinodal. To retain uniformity in the approach to the evaluation of $p_{sp}(0)$ for substances of different nature, we turned to the V-d-W equation, according to which $p_{sp}(0) = -27 p_c$. This reveals the one-order and correlated character of the quantities p_* and $27 p_c$ in the series of such substances as inert and diatomic gases, organic liquids and metals. The quantities p_* and $27 p_c$ themselves in this series change by two decimal orders in going from neon, argon to copper and nickel. Figure 3 shows the correlation of these values for different substances.

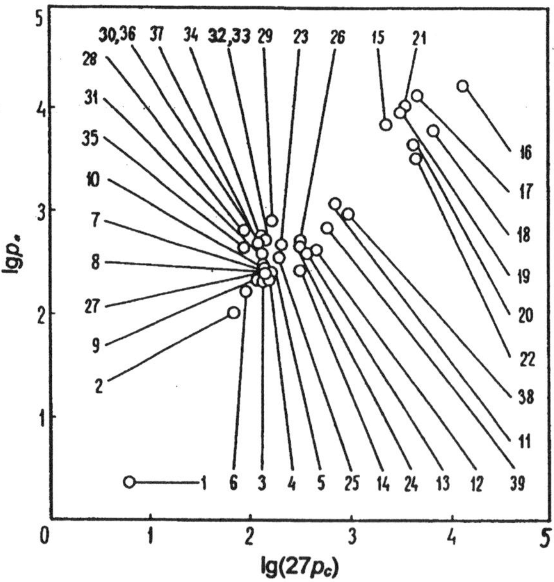

Figure 3. Correlation between the quantities p_* and $27 p_c$ for some substances (pressure in MPa): *1* – helium-4, *2* – neon, *3* – argon, *4* – krypton, *5* – xenon, *6* – nitrogen, *7* – oxygen, *8* – fluorine, *9* – methane, *10* – carbon tetrachloride, *11* – sodium, *12* – potassium, *13* – rubidium, *14* – cesium, *15* – aluminum, *16* – nickel, *17* – copper, *18* – zinc, *19* – silver, *20* – cadmium, *21* – gold, *22* – lead, *23* – chlorine, *24* – iodine, *25* – carbon dioxide, *26* – ammonia, *27* – ethane, *28* – pentane, *29* – chloroform, *30* – bromoform, *31* – benzene, *32* – aniline, *33* – nitrobenzene, *34* – monofluorobenzene, *35* – hexafluorobenzene, *36* – monochlorobenzene, *37* – monobrombenzene, *38* – sodium chloride, *39* – potassium chloride.

3. Behavior of the Internal Pressure

The internal pressure p_i of the isotropic phase is determined by the derivative of the internal energy u with respect to the volume:

$$p_i = (\partial u / \partial v)_T .\qquad(8)$$

In a thermodynamic equilibrium system the internal and the external pressure p are related by the following equation;

$$p_i = T\left(\frac{\partial p}{\partial T}\right)_v - p .\qquad(9)$$

The behavior of the internal pressure under changes of the system state reflects the change of the relationship between the forces of attraction ($p_i > 0$) and repulsion ($p_i < 0$) with position averaging of all particles. The internal pressures p_i in different states can be calculated from equation (9) if the appropriate thermal equation of state of the substance is known.

Using the Clapeyron-Klausius equation for the melting line

$$\frac{dp}{dT} = \frac{\Delta s}{\Delta v} ,\qquad(10)$$

where $\Delta s = s_L - s_S$, $\Delta v = v_L - v_S$ are entropy and volume changes in melting and the relation

$$T\Delta s = \Delta h = \Delta u + p\Delta v ,\qquad(11)$$

where h is enthalpy, we can introduce another quantity $\hat{p}$, which has the dimension of pressure and characterizes the phase transition:

$$\hat{p}_{SL} \equiv \left(\frac{\Delta u}{\Delta v}\right)_{SL} = T\frac{dp}{dT_{SL}} - p .\qquad(12)$$

The notation for the liquid-vapor (VP) phase transition is similar.

Taking into consideration the low-temperature region of metastable states of coexisting phases, we note that $\hat{p}_{SL}$, $p_{i,SL} \to p_*$ at $T \to 0$, whereas on the liquid-vapor equilibrium line $\hat{p}_{LV}$, $p_{i,LV} \to 0$ in the same limit, which follows from relations (1), (9), (12). At an arbitrary point of the melting line we have

$$\hat{p}_{SL} = cp_* + (c-1)p .\qquad(13)$$

74

The lines $p_{i,SL}(T)$, $p_{i,LV}(T)$ pertaining to the liquid intersect at the triple point.

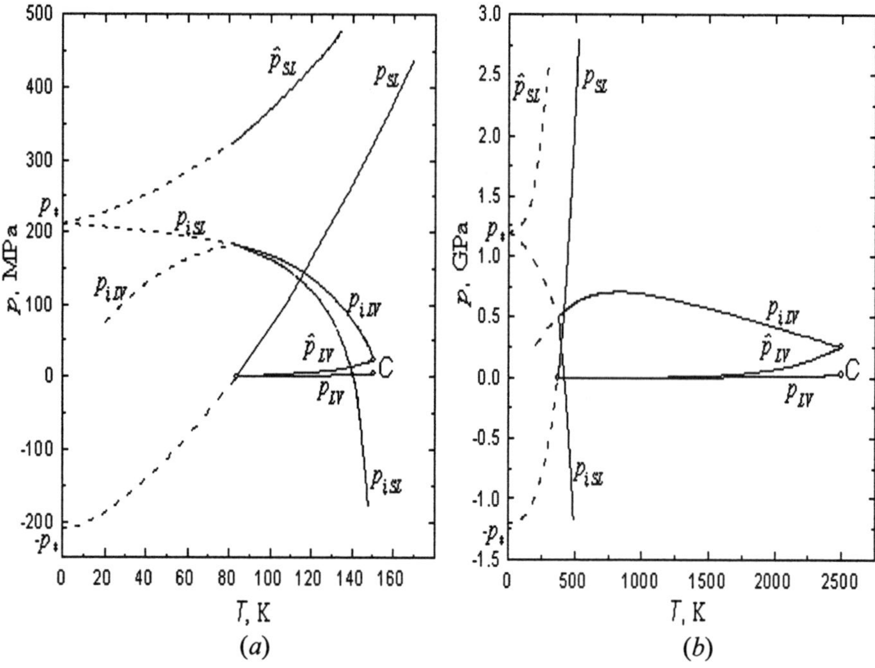

Figure 4. Behavior of the internal pressure p_i in liquid and quantity $\hat{p}$ by equation (12) on the lines of solid-liquid $p_{SL}(T)$ and liquid-vapor $p_{LV}(T)$ phase equilibrium for argon (*a*) and sodium (*b*), C is the critical point. The dashed sections of the curves show the extension beyond the triple point into the region of metastable states.

Figure 4 shows for argon and sodium the behavior of the quantities $p_{SL}(T)$, $p_{LV}(T)$, and also p_i and $\hat{p}$ for the liquid phase on the lines of the liquid-solid and liquid-vapor equilibrium. To construct the $p_i(T)$ and $\hat{p}(T)$ curves, use was made of the T, p, v data from reference [6] for argon and [5, 7] for sodium. Melting lines have been extended into the region $p < 0$ by equation (1).

From figure 4 it can be seen that the values of $\hat{p}_{SL}$ and $p_{i,SL}$ diverge rapidly with increasing temperature and pressure, due to the absence of an end point for the solid-liquid equilibrium of the critical-point type. The existence of a critical point for liquid-vapor equilibrium leads above the triple point to the approaching of the $p_{i,LV}$ and $\hat{p}_{LV}$ lines with increasing temperature and their convergence at the critical point. For SL equilibrium the values of $\hat{p}$ and p_i coincide only at $T \rightarrow 0$. On the whole line of LV equilibrium the internal pressure is positive, $p_i > \hat{p}_{LV} > 0$, whereas on the melting line the internal pressure with increasing temperature passes through zero and becomes negative. This supports the well-known view that a liquid-solid phase transition is not connected with the predominance of attractive forces in the molecular system, as is the case in the phenomenon of gas condensation.

4. Surface Tension

There is no simple and reliable method for measuring the surface tension σ along the melting curve at different temperatures. Theoretical surface tension estimates for the crystal-melt interface are very approximate. There are the estimations of the surface tension from experiments on spontaneous crystallization of supercooled liquid drops with the use of the homogeneous nucleation theory [8]. For stationary conditions the theory gives the following dependence of the rate of nucleation on the work of critical crystalline nucleus formation W_*:

$$J = N_1 B \exp(-W_* / kT), \tag{14}$$

where N_1 is the number of molecules in unit volume of the liquid, B is the kinetic factor, k is the Boltzmann constant.

Homogeneous nucleation theory was built in a thermodynamic approximation. A supercooled liquid and a crystalline nucleus are treated as volume phases, and interface effects are introduced through surface tension. For a spherical unstable equilibrium nucleus of radius r we have

$$W_* = \frac{4}{3}\pi r^2 \sigma = \frac{16}{3}\pi \frac{\sigma^3 v_S^2}{(\Delta\mu)^2} = \frac{16}{3}\pi \frac{\sigma^3 v_S^2}{(\Delta s)^2 (\Delta T)^2}. \tag{15}$$

Equations (15) are obtained with an accuracy to the first terms of the expansion of the $\Delta\mu(T,p)$ chemical potentials in powers of T and p [8]. A comparison of experiments on spontaneous crystallization with homogeneous nucleation theory makes it possible to estimate the surface tension σ, which is the only significant adjustable parameter in equation (15).

Data processing reveals two circumstances [8]: (1) the $J(T)$ dependence is satisfactorily described by homogeneous nucleation theory with a constant σ value surface tension, and (2) the dome-shaped $J(T)$ dependence is linearized in the $\log J$, $[T(\Delta T)^2]^{-1}$ coordinates. This implies the constancy of the complex quantity

$$Z = \frac{\sigma^3 v_S^2}{k(\Delta s)^2}, \tag{16}$$

which is the slope of the dependence of $\log J$ on $[T(\Delta T)^2]^{-1}$.

The Z complex can be reduced to the dimensionless form by dividing by T_0^3, where T_0 is the temperature on the melting curves of different substances at the corresponding points $p=0$:

$$Nc = \frac{\sigma^3 v_S^2}{kT_0^3 (\Delta s)^2}.$$ (17)

It was found that the complex $(Nc)^{1/3}$ for different substances differs insignificantly from each other. The table contains the values that determine Nc for several simple substances at the corresponding melting curve points at zero (atmospheric) pressure. The T_0, v_S and Δh values were taken from [9] for lithium and sodium and from [10] for the other substances. For σ the data from [8] were used for all substances except sodium and lithium [11] and argon. For argon the σ value was obtained from the condition $(Nc)^{1/3} = 0.5$, which corresponds to the mean dimensionless complex value for normally melting substances.

The use of complex (17) for determining σ at the crystal-liquid interface implies the use of a phenomenological thermodynamic similarity concept. Surface tension is assumed to be an explicit function only of temperature. Pressure corresponds to the temperature along the melting curve or its metastable continuation into the $p < 0$ region.

Admitting the $Nc = const$ condition not only close to $p = 0$ but also over large portions of substance melting curves we calculated the $\sigma(T)$ dependence for some substances on data for the entropy or enthalpy jump, Δs or $\Delta h = T\Delta s$, and the specific volume of the crystalline phase at different melting curve points.

TABLE 1. Values characterizing crystal-liquid equilibrium for different substances at atmospheric pressure; Nc is dimensionless complex (17)

Substance	T_0, K	v_S, 10^{-3} m³/kg	Δh, kJ/kg	σ, mJ/m²	$Nc^{1/3}$
Lithium	453,8	1,902	428,6	30	0,44
Sodium	370,8	1,019	115,2	20	0,50
Copper	1356	0,119	203,1	200	0,54
Silver	1235	0,102	104,7	143	0,55
Indium	429,8	0,139	28,4	31	0,49
Tin	505,0	0,139	60,7	60	0,54
Mercury	234,3	0,070	11,5	23	0,51
Lead	600,0	0,091	22,6	40	0,50
Argon	83,8	0,608	29,8	7	0,5
Tetrachloromethane	250,6	0,570	16,4	6,7	0,47
Benzene	278,6	0,990	128,2	21,7	0,54

The calculated surface tension dependences are shown in figures 5 for mercury and argon an in figure 6 for tin, lead, and sodium. The Δs and v_S values were taken from the literature for $p > 0$ and obtained from approximations correlated with the Simon equation for $p < 0$.

The surface tension values in the temperature region of experimental studies of the kinetics of nucleation are marked by squares. Circles correspond to the values calculated from the condition $Nc = const$ and the experimental Δh and v_S values. Continuous curves were obtained by extrapolating data for enthalpy jump and specific crystal volume. The temperature T_0 corresponds to the leftmost point (circle). The surface tension values marked by squares correspond to negative pressures because spontaneous crystallization was observed at a finite liquid supercooling and the projection of the figurative point at these T values fall onto the melting curve section with negative pressure.

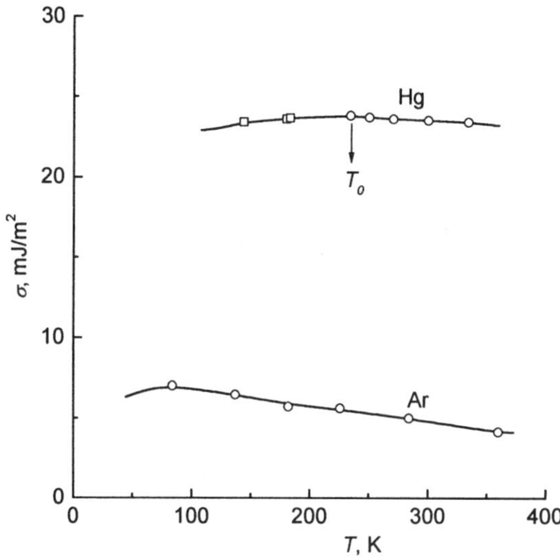

Figure 5. Temperature dependences of the surface tension of argon and mercury along the melting line.

The method that we use for estimating $\sigma(T)$ along the melting line gives surface tension values that decrease as temperature increases at positive pressures. At the same time the dependences contain maximum values determined by the behavior of the $(\Delta s / v_S)^{2/3}$ ratio in (17). Importantly, a decrease in the melting entropy jump at negative pressure (low temperature) follows from the Nernst theorem. The melting entropy jump decreases when the temperature tends to zero.

At high temperatures $(T > T_0)$ the surface tension decreases because of an increase in the amplitude of atomic oscillations, which results in a swelling of the interface and decreases σ. The decrease of the surface tension in lowering of temperature corresponds to approaching to the spinodal state of liquid and crystal on the melting line under their stretching. Retention of stability of the coexisting phases at zero temperature supposes retaining a non zero value of the surface tension.

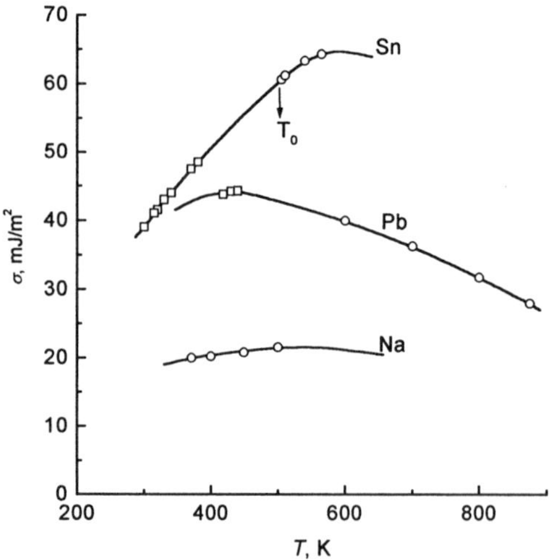

Figure 6. Temperature dependences of the surface tension of tin, lead and sodium along the melting line.

5. Viscosity of liquid phase on the melting line under negative pressures

When considering the extension of the crystal-liquid equilibrium line into the negative pressure region (low temperature) we should look at the behavior of the viscosity of a liquid phase on the melting line. Will there be a glass transition of the liquid at low temperatures? Glass-transition would create considerable kinetic difficulties for the structural relaxation of the liquid, and for the establishment of liquid-crystal equilibrium at low temperatures.

To analyse the behavior of the liquid viscosity along the melting line we chose substances for which there are experimental data near the melting line over a fairly wide range of temperatures and pressures. Using the Fulcher-Tamman approximation,

$$\eta = A \cdot \exp\left(\frac{B}{T - T_1}\right), \tag{18}$$

where the temperature T_1 corresponds to zero fluidity, A and B are constants, the viscosity of some liquids on the melting line has been calculated in a wide temperature range including the metastable extension of the melting line in the negative pressure region.

Figure 7 shows the temperature dependences of the viscosity for benzene on the melting line and on two isobars. As can be seen from the figure, the liquid viscosity on the melting line increases with increasing temperature. This leads to vitrification ($\eta = 10^{12}$ Pa·s) of the liquid phase at high temperatures. The glass-transition point for liquid benzene on the melting line corresponds to a pressure $p_g = 6.5$ GPa and a temperature $T_g = 83$ K.

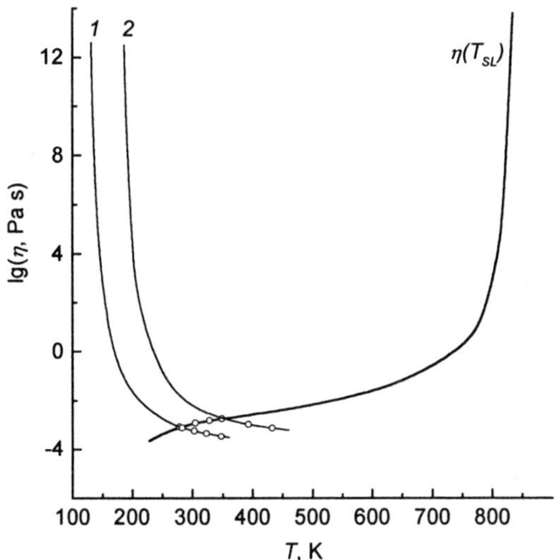

Figure 7. Behavior of the viscosity of benzene on the melting line $\eta(T_{SL})$ and on isobars $p = 0.1$ and 300 MPa (*1, 2* – curves). The points correspond to experimental data [12].

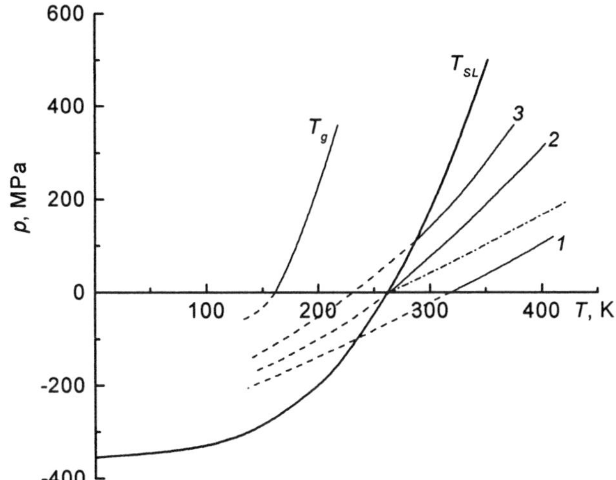

Figure 8. The melting line of dodecane T_{SL} and the constant viscosity curves *1 – 3* (*1* – 1 mPa·s, *2* – 3 mPa·s, *3* – 5 mPa·s) and the glass-transition line T_g ($\eta = 10^{12}$ Pa·s) with extension in the region of negative pressure. Dash-dot line is an isochore of the liquid, $v = 1.3$ cm^3/g .

For metastable extension of the melting line in the field of negative pressures (low temperatures) the maintaining of high fluidity of the liquid phase is characteristic.

Figure 8 shows the melting line for dodecane and the curves of constant viscosity, constructed with experimental data [13]. The conditions of the monotone behavior of the curves and their non intersection result in high fluidity of the liquid phase on the metastable section of the melting line in the region of negative pressures, that is the glass-transition line will never intersect the melting line.

Acknowledgement

The work was supported by the Russian Foundation for Basic Research (Grant No.00-02-16227)

References

1.   Simon F. E. and Glatzel G. (1929) Bumerkungen zur Schmezdruckkurve, *Z. anorg. Allgem. Chem.* **178**, 309 – 316.
2.   Baidakov V. G., Skripov V. P. and Kaverin A. M. (1974) Experimental investigation of liquid argon in the metastable state, *Zh. Eksp. Teor. Fiz.* **67**, 676 – 682. (Engl. Transl. (1975) *Sov. Phys.-JETP* **40**, 335 – 341).
3. Skripov V. P. and Faizullin M. Z. (1986) Melting lines of simple substances: thermodynamic similarity and behavior of thermal properties, *High Temp.-High Pressures* **18**, 1 – 12.
4.   Lahr P. H. and Eversole W. G. (1962) Compression isothermal of argon, krypton, and xenon through the freezing zone, *J. Chem. Eng. Data* **7**, 42 – 47.
5.   Makarenko I. N., Nikolaenko A. M., Ivanov V. A. and Stishov S. M. (1975) The equation of state of the alkaline metals: sodium, *Zh. Eksp. Teor. Fiz.* **69**, 1723 – 1733 (Engl. Transl. (1975) *Sov. Phys.-JETP* **42**, 875 – 885)
6.   Tegler Ch., Span R. and Wagner W. (1999) A new equation of state for argon covering the fluid region for temperatures from the melting line to 700 K at pressures up to 1000 MPa, *J. Phys. Chem. Ref. Data* **28**, 779 – 850.
7.   Pokrasin M. A., Roschupkin V. V., Fokin L. R. and Handamirova N. E. (1983) The interpolation equations and saturation pressure tables of sodium, potassium, rubidium, and cesium in the range of temperature from triple point to critical one. The saturation pressure equation of lithium up to 2500 K, *Thermophysical Properties of Substances and Materials (in Russian)*, Izdatelstvo Standartov, Moscow.
 8. Skripov V. P. and Koverda V. P. (1984) *Spontaneous Crystallization of Supercooled Liquids (in Russian)* Nauka, Moscow.
 9. Tonkov E. Yu. (1983) *Phase diagrams of compounds at high pressure (in Russian)* Nauka, Moscow.
10. Tonkov E. Yu. (1988) *Phase conversions of compounds under high pressure (in Russian)* Metallurgiya, Moscow.
11. Dokhov M. P. (1999) On modern state of theoretical investigations of crystal-melt interphase energy (in Russian) *Metally* **4**, 28 – 35.
12. Parkhurst H. J. and Jonas J. (1975) Dense liquids. II. The effect of density and temperature on viscosity of tetramethylsilane and benzene, *J. Chem. Phys.* **63**, 2705 – 2709.
13. Hogenboom D. L., Webb W. and Dixon J. A. (1967) Viscosity of several liquid hydrocarbons as a function of temperature, pressure, and free volume, *J. Chem. Phys.* **46**, 2586 – 2598.

LIQUID-LIQUID PHASE EQUILIBRIA IN BINARY MIXTURES UNDER NEGATIVE PRESSURE

ATTILA R. IMRE
KFKI Atomic Energy Research Institute, Materials Department
1525 Budapest, POB. 49, Hungary; E-mail: imre@sunserv.kfki.hu

Abstract. Several binary or multicomponent solutions exhibit liquid-liquid phase transition; i.e. changing the temperature and/or pressure and/or concentration etc., the initially homogeneous liquid can split into two or more liquid phases. In this paper we would like to give examples where the extension of liquid-liquid solubility branches below p=0 gives us new information about the studied binary and quasi-binary mixtures.

1. Introduction

Although p=0 is not a special point of liquids, i.e. their properties do not change abruptly at that point, there are some phenomena which can be seen only (or can be seen better) under negative pressure. In this paper we would like to show a few examples, where - although crossing p=0 does not change any physical properties of a liquid abruptly – interesting phenomena can be seen in the region of negative pressures.

Our main interest is the liquid-liquid equilibrium in binary (or quasi-binary) liquids, when – under some circumstances – the homogeneous mixture can split into two phases and vice versa. A short overview about liquid-liquid equilibrium can be find in the following chapter. In most of the following examples long-chain molecules (polymers, proteins) will be involved, but some of these phenomena also can be seen in small molecule solutions.

First, we would like to present some results concerning the merging of virtually separated (i.e. separated on positive pressures) liquid-liquid equilibrium curves [1]. Second, we will explain, why polymer mixtures (polymer+polymer) cannot be compatibilized by increasing the pressure while the solubility of polymers in solutions (polymer+solvent) can be increased by pressure [2]. Overcoming this difficulty, solubility islands hiding mainly below p=0 can be used for compatibilization of polymer blends [3,4]. Fourth, an interesting phenomenon, the hiding solubility branch – or negative saturation curve - will be presented; this phenomenon had been predicted by Timmermans and Lewin [5] in 1953 but found only recently by Rebelo and his co-workers and will be presented in the following paper [6].

In biological systems we can find several strongly pressure-dependent phenomena; one of them is the protein denaturation. Proteins can be denaturized by heating, by cooling, by increasing the pressure – and also by stretching them, i.e. under negative pressure [7]. Some bacteria can be killed in similar ways [7]. "Phase diagrams" -

A.R. Imre et al. (eds.), Liquids Under Negative Pressure, 81–94.

concerning protein denaturalization or bacteria inactivation – where p<0 parts can be interesting will be shown.

Finally an alternative explanation – involving negative pressure generated in capillaries [8] - will be given for the strange phase equilibrium curve of liquid ^{3}He+^{4}He mixture in aerogels [9].

2. Liquid-liquid equilibrium

Concerning binary mixtures with limited solubility, two major groups exist: weakly interacting systems and strongly interacting systems. Their miscibility behaviour is very different. Schematic representation of liquid-liquid equilibria (at p=constant) in these two types of systems can be seen on Figure 1. In a weakly interacting system we have two two-phase regions separated by a one-phase region, while in strongly interacting systems a closed two-phase island can be seen in a one-phase background; at constant concentration and pressure we would see two one-phase region separated by a two phase one in this later case. In some case, only part of these loci can be seen; other parts might be masked by freezing, by thermal degradation etc.

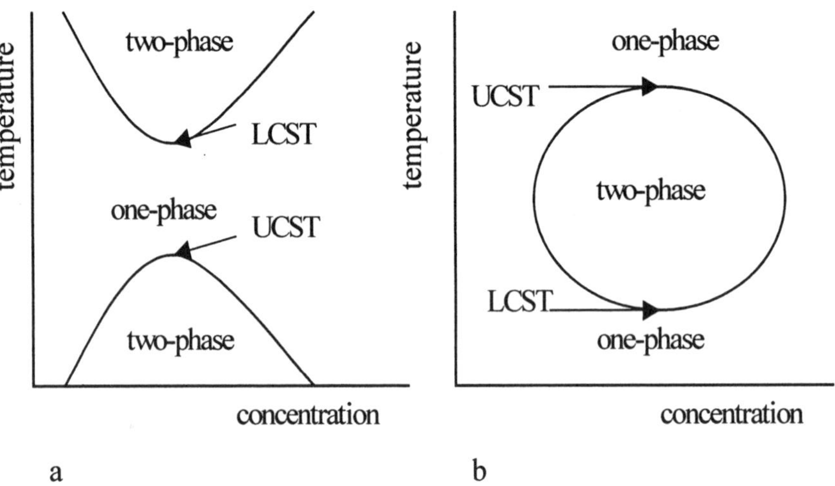

Figure 1. Liquid-liquid phase equilibria in binary liquid mixtures. a: weakly interacting system, b: strongly interacting system.

For example polystyrene solutions (polystyrene in cyclohexane or in benzene etc.) are weakly interacting solutions, while a lot of aqueous solutions (water+polyethylene glycol, water+3-methyl pyridine, water+poly-n-isopropylacrylamide etc.) are strongly interacting systems.

3. Pressure dependence of miscibility in weakly interacting binary mixtures

3.1. LIQUID-LIQUID PHASE EQUILIBRIA OF BINARY POLYMER SOLUTIONS

The pressure dependence of the liquid-liquid loci of binary linear homopolymer (like polystyrene, polyethylene etc.) solutions seems to be a little bit disturbing. Most solution can be described by two separated branches representing UCST and LCST. The lower branch (UCST) usually has negative slope, while the upper one (LCST) has always positive slope (Figure 2, solid parts of lines A and B). We have to mention here, that in several cases, one of the branches can be hidden by freezing or by thermal degradation. Also in a few cases, UCST might be curved or might have positive slope, but our statement (UCST with negative and LCST with positive slope) is true in most

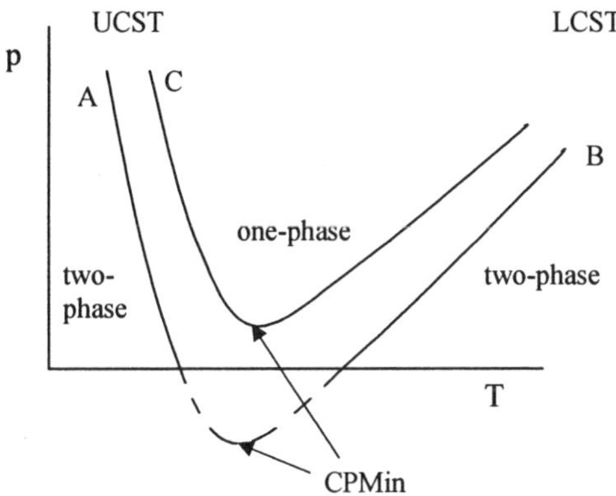

Figure 2. Experimentally obtained pressure dependencies of liquid-liquid locus in (temperature, pressure) space. For further explanation, see text.

cases. Classical theory [10] can explain only the existence of the UCST, therefore these two branches are usually explained on different theoretical background. In the sixties a few polymer solutions were found (like polystyrene/acetone or polystyrene/ diethylether) where this kind of UCST and LCST were found with low molar mass polymers, but increasing the molar mass, a new kind of dependence was found with joined UCST and LCST (Figure 2, curve C). The point where the two branches are joined is a so-called double critical point (or hypercritical point) – to distinguish it from the other double critical points, we can call it Critical Pressure Minimum (CPMin).

In the mid-seventies Wolf suggested [11], that A+B type phase diagrams could be extended into the region of negative pressures, and therefore two types, the one with virtually separated UCST and LCST (A+B type) and the one with joined UCST and LCST (C type) are just one, only in the first case the bottom of the locus is hidden in the metastable liquid region. The experimental proof of his hypothesis was given almost two decades later by Imre and Van Hook [1, 12], and by Rebelo and his co-workers [13]. An experimental phase diagram of polystyrene/propionitrile solution can be seen on Figure 3. The phase diagram was determined visually in glass Berthelot tubes.

As we mentioned above, there are some solutions where the slope of the UCST is positive. In some solutions (like polystyrene/cyclohexane or a non-polymeric n-alkane/cyclohexane) UCST flips into positive-sloped to negative-sloped by changing

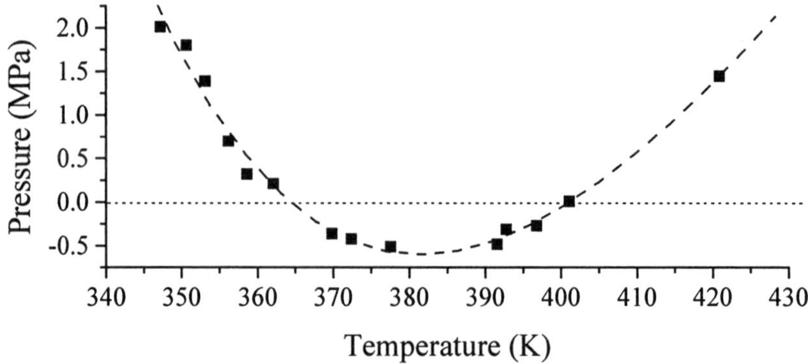

Figure 3.: Liquid-liquid equilibrium in polystyrene/propionitrile solution (after Ref. [12]).

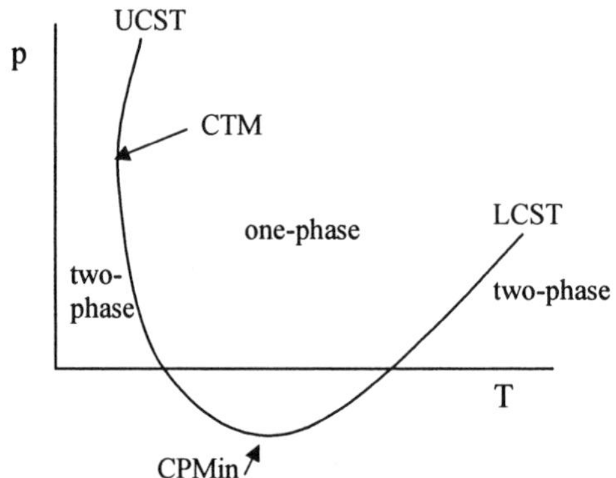

Figure 4. General liquid-liquid locus of binary linear homopolymer solutions (after Refs. [17, 18]).

the chain length of the chain molecule component. An important clue for the origin of this flip was given in the early nineties independently by three groups [14, 15, 16]. They studied the polystyrene/methylcyclohexane system, which was considered as a system with negative-sloped UCST. They studied the system up to 200 MPa and found curved UCST with an extremum on it. They concluded that UCSTs can have a so-called double critical point, where the slope of the UCST changes. They also found that the pressure where that double critical point is located, depends on the molecular weight. In their study all solutions exhibited double critical point on positive pressure. Following their study, Imre and Van Hook introduced a so-called master-curve for the pressure dependence of the liquid-liquid locus (Fig. 4) [17, 18]. Their curve exhibited two double-critical points; one located on the UCST (called Critical Temperature Minimum, TMC) and one at the joining of the UCST and LCST (called Critical Pressure Minimum, CPMin). They suggested that for most systems, CPMin is hiding below p=0, while CTM can be hidden by freezing or also by the liquid-vapour metastability. Having a CTM below p=0, only positive-sloped UCSTs can be seen in "regular" (i.e.

p>0) experiments; having CTM on moderate positive pressure, curved UCSTs can be seen, while having it above the experimental limit or below the freezing curve one can see only negative-sloped UCSTs. The first experimental example where both double critical points were seen (on p>0) had been given by Imre and Van Hook [17], while the first example where CTM was above p=0 but CPMin was seen below p=0 was presented by Rebelo et al. [19].

3.2. LIQUID-LIQUID PHASE EQUILIBRIA OF BINARY POLYMER BLENDS

For polymer blends (i.e. mixtures of two polymers) usually only one branch (a UCST or an LCST) can be seen. There is a "crucial" difference concerning the behaviour of the UCST of a "typical" homopolymer solution and "typical" homopolymer blend, namely that for solutions, the slope of the UCST is usually negative (i.e miscibility can be increased by pressure), while for almost all weakly interacting homopolymer blends the slope of the UCST is positive (i.e. miscibility cannot be increased by pressure). This is very bad: several polymers used in billions of tons per year (like polystyrene, polyethylene, etc.) are not compatible on atmospheric pressure and they cannot be compatibilized even by raising pressure.

This difference is strange because there is no sharp distinction between blends and solutions. Blends are the mixture of two long-chain molecules while solutions are mixtures of a long-chain and a small molecules, but "long" and "small" are relative. Mixing a polymer with alkanes we can consider the systems as solutions (where the alkane is the solvent) or – having really long-chain alkanes – we can consider them as blend (where the long alkane is the second polymer, namely polyethylene). Therefore – being no sharp distinction between blends and solutions – we cannot expect really big difference in the pressure dependence of their UCSTs.

In the nineties a handful of polymer blends were found, where the slope of the UCST was negative (see references in Ref. [4]). Later some of these blends turned out to have slightly curved UCST with an extremum, a CTM. Based on these facts, Imre et al. proposed that their "master curve" used previously to describe the pressure dependence of the liquid-liquid locus can be applied for blends too [2]. In blends,

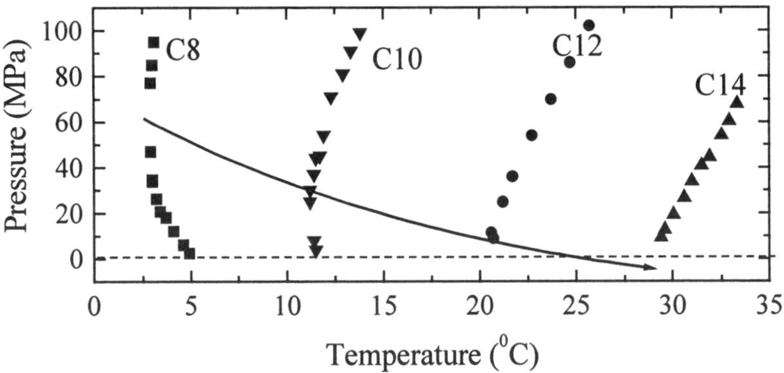

Figure 5. The movement of CTM with chain length in oligostyrene/n-alkane system. C8, C10, C12 and C14 mark octane, decane, dodecane and tetradecane, respectively (from Ref. [2]).

CPMin (almost) always located below p=0, additionally the LCST or the UCST (at least one of them) are almost always masked by the freezing or thermal degradation. The difference in the slope of the UCST can be easily explained by a chain-length dependent CTM; having two components with very different chain length (like a polymer solution), CTM is located above p=0, but decreasing the difference between the chain length of the two components (like in the mixture of two long-chain polymers) CTM shifts below zero. An example of this behaviour can be seen in the mixture of polystyrene and n-alkanes [2]; increasing the chain length of the alkane (and keeping the chain length of the polystyrene) we can see how the CTM moves below p=0 (Fig. 4).

3.3. MISCIBILITY ISLANDS IN QUASI-BINARY POLYMER BLENDS

The phase diagram shown on Fig. 1/a is an idealized one. Concerning polymers, there is no binary solution or blend; every real polymer sample has polymer chains with various length, instead of having molecules with one well-defined length. This so-called polydispersity causes some distortion on the shape of the liquid locus in (concentration, temperature) space (Fig. 1/a). In polydisperse system it is possible to get phase diagrams with two maxima (for low temperature branch) and with two minima (for high temperature branch). Having a UCST with CTM, the phase diagram in (concentration, pressure) space (T=constant) will be similar to the one in (concentration, temperature) space: single humped in ideal case but double humped in polydisperse case (Figure 6). Approaching the CTM by decreasing the temperature (Figure 6/a), the upper and lower branch can merge, forming first a miscibility peninsula and then a miscibility island (Figure 6/b) [3, 4]. Similar island in a polymer solutions in the vicinity of CPMin was found by Rebelo and Van Hook [20].

The existence of this miscibility island might give us a new way to compatibilize virtually immiscible polymers. From experimental point of view, immiscible refers that the two polymers cannot form homogeneous mixture, except in the extremely low or

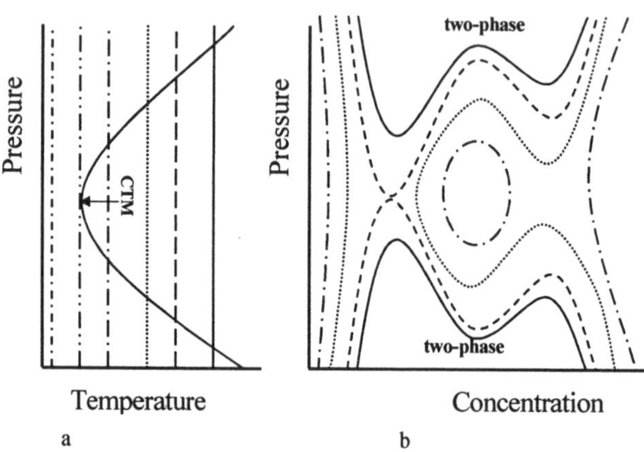

Figure 6.: Liquid-liquid locus of a binary polymer blend or solution with CTM and the formation of a miscibility island.

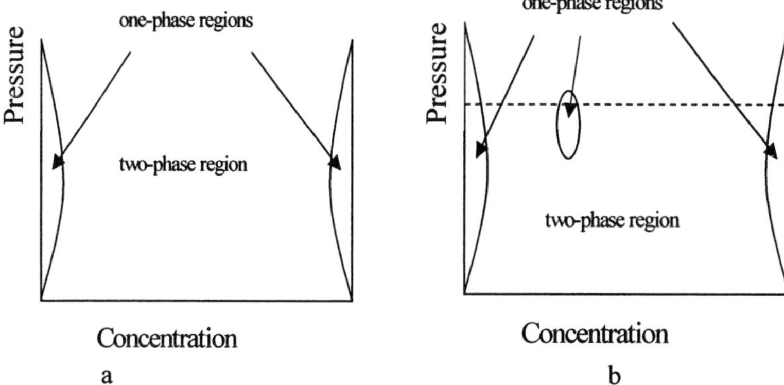

Figure 7.: Liquid-liquid equilibria in immiscible (a) and virtually immiscible (b) polymer mixtures. Dashed line represents p=0.

high concentration range. On Figure 7/a a schematic phase diagram of an immiscible polymer blend can be seen. On Figure 7/b a similar phase diagram can be seen with a small miscibility island. Although for polymer blends CTM is usually below p=0 - it means only that the middle of that small miscibility island is also below p=0 – but the top of the island can be in positive pressures, i.e. it might be usable to mix two virtually incompatible polymers.

4. Pressure dependence of miscibility in strongly interacting binary mixtures

Strongly interacting binary mixtures (for example aqueous solutions of polyethylene glycol etc.) are homogeneous at lower temperatures and they split into two liquid phases by increasing the temperature. In some cases a second phase transition can be seen by increasing the temperature further, but in a lot of systems this second transition is hidden by boiling or by degradation. The pressure dependence of these liquid-liquid loci can be seen in Figure 8.

Good – and well-known – examples for these kinds of solutions are the aqueous solutions of the methylpyridines. There are three different methylpyridenes: 2-methylpyridine (2MP), 3-methylpyridine (3MP) and 4-methylpyridine (4MP) (see references in Refs. [21, 22, 23]).

It has been known for almost a century that mixing these methylpyridines with water (W) or heavy water (HW), very different phase diagrams can be obtained. 2MP with water (2MP/W) gives us only one liquid-liquid locus located on high pressure (Figure 8., line D), while with heavy water (2MP/HW), one can obtain a high pressure branch located below the previous one (C) and a very small low pressure branch (C*). "Small" refers that the part of C* above p=0 is very small. Knowing that liquids can endure negative pressure, one can see that in the first case (2MP/W) the lower branch (D*) was simply hiding below p=0. In case of 3MP, the 3MP/W mixture shows only a high pressure branch (like D), but the 3MP/HW shows two separate branches – but now they

88

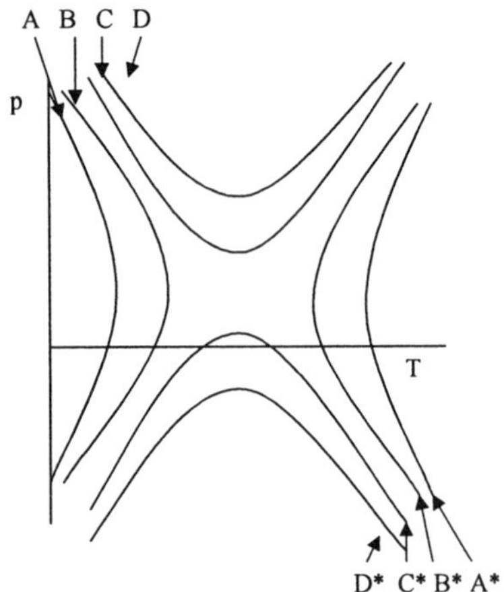

Figure. 8.: Schematic representation of the liquid-liquid loci of strongly interacting binary mixtures.

are B and B* type. With 4MP, 4MP/W gives D while 4MP/HW gives C type. It is possible to add salt or use W/HW mixtures as solute; in that case, by changing the salt amount or changing the W/HW ratio, it is possible to go smoothly from type D+D* (and only D is seen on normal pressures) to A+A*. During this transition we can see an immiscibility island in (concentration, temperature) diagram emerging below p=0 (as C* reaches p=0) then join with the upper branch (at the cross-section of the dashed lines) then split and turn to B+B* then A+A* [21, 22, 23].

The first attempt to reach D* (which is completely below p=0) in 3MP/W system had been done in the fifties by Timmermans and Lewin [5] and was unsuccessfull. Later Schneider proposed several times that D* branch had to be there (see references in ref. [23]), but no one was able to reach it. Finally it was reached successfully by Rebelo and his co-workers and will be presented in the following paper [6].

Having a polymer solution which exhibits closed-loop phase diagram (Fig.1/b) on some molecular weight, the whole process can be seen better, by changing the chain lenght of the polymer. Based on the results of Takahashi et al. [24], Rebelo et al. [19, 25] and Imre and Wolf [26], the aqueous solution of poly-N-isopropylacrylamide (PNIPAM) would show exactly the previously described A+A* to B+B* to C+C* to D+D* transition, without adding any third component, only by changing the chain length of the polymer.

5. The effect of negative pressure in biological systems

5.1. PROTEIN DENATURATION UNDER NEGATIVE PRESSURE

It has been know for a long time that proteins can be denaturized by heating, by cooling or by pressurizing [7]. In (temperature, pressure) space an approximately elliptical phase diagram can be constructed (Fig. 9) about the denaturation of proteins in aqueous

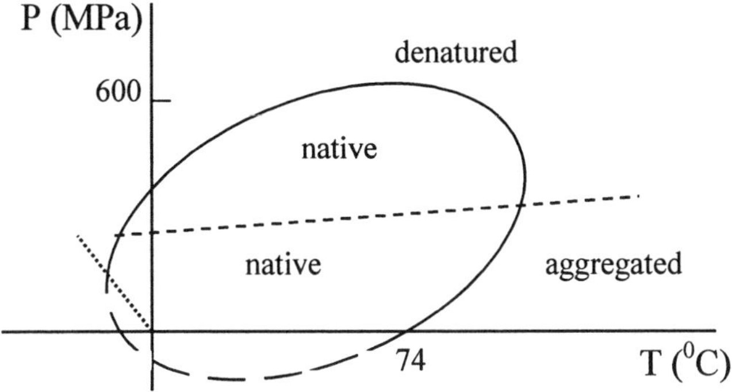

Figure 9.: Schematic phase diagram suggested for myoglobin (from Ref. [7]). Dotted line represents the freezing line.

solutions. In experimentally obtained phase diagrams the lowest part usually hides below p=0, and therefore is thought to be missing [7, 27].

It can be clearly seen that the already observed part (solid part of the elliptic curve) can be extended below p=0. In that case the bottom can be located around -200 MPa, which is well below the stability line of water. Choosing different protein and/or changing the pH or adding salt the phase diagram can be shifted and in some cases the negative pressure part of the diagram can be moved to the experimentally obtainable negative pressure range.

The possibility of this "stretched" denaturation is not only theoretically interesting. Although normally biological systems (like our body) are not under negative pressures (or at least not under big negative pressures), there is one situation where the aqueous protein solutions in our body can face big negative pressure: the medical (diagnostical and therapeutical) ultrasound. Diagnostic ultrasound can produce negative pressures down to -3 MPa [28], while therapeutic ultrasound (like ultrasound used again kidney-stones) can produce positive pressure waves as big as 43 MPa and negative ones as deep as -14 MPa (this asimetry between the positive and negative values is caused by the cavitation) [29]. Although ultrasound can produce negative pressure only for very short periods, but these periods can be repeated again and again, therefore – although there is no experimental evidence for it – we cannot exclude the possibility that medical ultrasound might cause protein denaturation.

We were not able to find any example in the literature for protein denaturation by negative pressure.

5.2. MICROORGANISMS UNDER NEGATIVE PRESSURE

Heat can destroy microorganisms, see for example the Pasteurization of milk. But sometimes the heat needed for the destruction of the microorganisms also destroys the proteins (see previous chapter), therefore recently more and more foods are "Pasteurized" by pressurization, when microorganisms are destroyed by high pressure. A so-called stability diagram of the E. coli can be seen in Fig. 10 [7, 30]: the line represents an isokinetic line, where the number of bacteria decreases with two orders of magnitude in 5 minutes. This diagram also seems to be elliptical, therefore it is natural

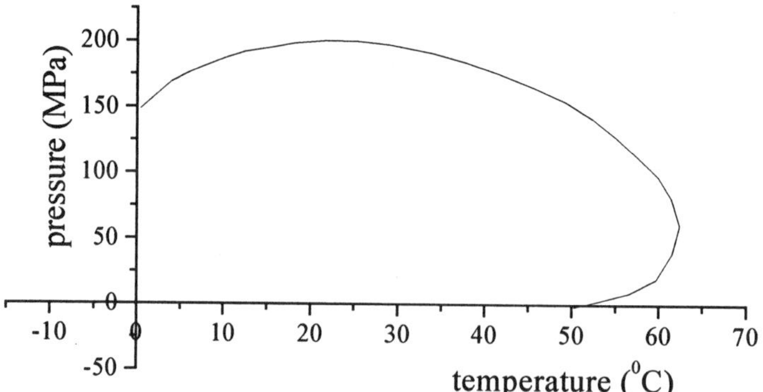

Figure 10.: Schematic stability diagram of *E. coli* (from Ref. [7]).

to extend it below p=0. In this case the bottom of the elliptic diagram is located on more accessible pressure range, around -20 MPa.

We would not recommend to cure a person with E. coli infection (which causes diarrhea) by putting him/her into a Berthelot tube and generate negative pressure in his/her body – that cure would have some side effects, like drowning …, but it might be possible to sterilize food by applying negative pressure (positive pressure would do the same, but in this case an order of magnitude higher positive pressure is necessary to kill the bacteria), for example by using high intensity ultrasound.

We were not able to find any example in the literature for bacteria destruction by negative pressure.

6. The phase diagram of ^{3}He+^{4}He mixture in porous materials

The behaviour of liquids and phase diagrams of liquids in porous materials differ remarkably from their bulk behaviours [31]. In most cases this difference is explained by the interaction of the wall with the liquid; while in bulk liquid only a very small amount of the atoms or molecules are connected with the wall, in capillaries almost all molecules interact with the wall.

Due to the meniscus, liquids confined into very narrow capillaries can be stretched, i.e. they can be under negative pressure [8]. Although this fact is known, there are only a very few attempts where the difference between bulk and capillary liquids was explained by the negative pressure [32, 33, 34].

An interesting phase diagram of ^{3}He+^{4}He mixture was published several years ago by Kim et al. [9], it is shown schematically on Fig. 11. Similar phase diagram was published later in a different porous material by Pobell and his co-workers [35], demonstrating that the effect is independent from the material of the confining solid phase, it is generated by the geometry (confinement). It can be seen on Figure 11 that the bulk system (solid lines) has a tricritical point (at 0.872 K and at 0.669 volume fraction of ^{3}He), where the line separating the superfluid/coexistence regions join to the line separating the normal fluid from the other two regions. In the coexistence region

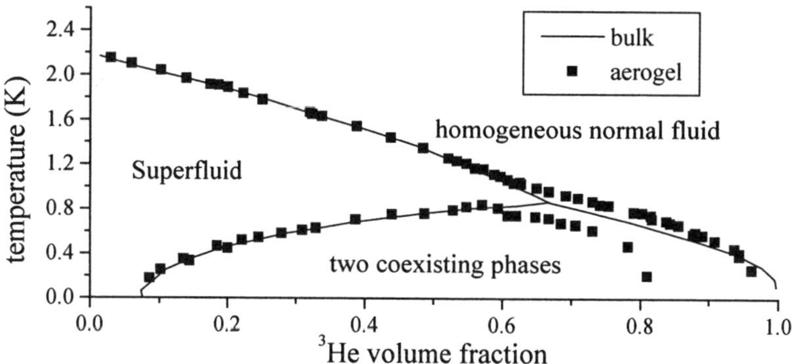

Figure 11.: Phase diagram of the ^{3}He+^{4}He mixture in bulk and in aerogel (from Ref. [9]).

one has two coexisting liquid phases; a ^{3}He-rich and a ^{4}He–rich. When the mixture is in aerogel, the phase diagram will be very different (see squares on Fig. 11); one of the most remarkable differences is the disappearance of the tricritical point (open-up). This diagram has been explained by several different ways; by the instability of the tricritical point [9] and by the "quenched randomness" [36] by the confinement [37].

It is well known, that the stability limit of the two isotopes of the helium is less than -1 MPa (-3 bar for He3 and -9 bar for He4) [38], which means that generating even small negative pressure the system will be close to its instability, i.e. it will be very sensitive for pressure.

To generate negative pressure in a liquid, we need curved meniscus, i.e. superfluid helium – even in nanopores – would not experience any stretching. But it is possible to generate negative pressure in the normal fluid and in the coexisting phase region.

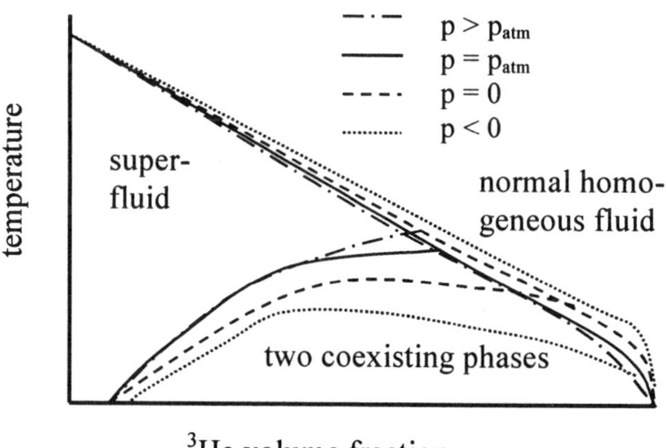

Figure 12.: A possible explanation (based on the effect of the negative pressure) for the phase diagram of helium mixture in confined systems.

Negative pressure in ^{4}He in capillaries was generated first by Lilly and Hallock (without giving the actual value) [39].

It is well-known for regular solutions that (moving on the saturation curve) approaching the liquid-vapour critical point of the more-volatile substance one can generate liquid-liquid phase transition. This phase transition is caused by the increasing molar volume difference of the compounds. Approaching any points of the spinodal curve would cause the same effect. Generating even a small negative pressure in He3-He4 mixture, the system would be very close to the spinodal of He3 (but still far from the spinodal of He4), therefore the increasing molar volume difference between He3 and He4 should initiate phase separation, i.e. while on a given temperature the bulk mixture would be homogeneous, the mixture in a capillary would split into two phases. This effect could be seen as a shift of the border between the homogeneous normal fluid and two coexisting phases regions.

We would like to propose a scenario, where the tricritical point moves with pressure (Fig. 12). In that case that "open-up" would be partially explainable by the shift of the tricritical point down to zero temperature (although the complete disappearance of the tricritical point would not be explainable). To check the validity of this scenario, one would need bulk phase diagrams on higher pressure in the 0-1 K temperature range with ^{3}He rich mixture. We were not able to find this kind of data in the literature, only some data measured on lower temperature (0-0.25 K) by Yorozu et al. [40].

7. Conclusions

In this paper several examples have been given to demonstrate that the extension of liquid-liquid equilibrium into the range of negative pressures might show us several interesting phenomena.

- The merging of virtually separated liquid-liquid equilibrium curves in polymer solution has been presented.
- The phenomenon, why polymer miscibility can be increased by applying positive pressure, while compatibility decreases in polymer blends by doing the same can be explained by the existence of a double critical point, where pressure induced miscibility turns to pressure induced immiscibility. For most solutions this point is above p=0, but in blends it can shift below zero.
- In polydisperse polymer blends and solutions a miscibility island can be seen around the above mentioned double critical point. Although for blends the center of this island should be located in the region of negative pressures, some parts can reach the positive pressure region, giving a new opportunity to enhance compatibility in polymer blends.
- For some aqueous solutions, an island of immiscibility can be found below p=0.
- Proteins can be denaturized by applying heat, cold or pressure. An alternative way - denaturation by stretching was shown in this paper.
- Some microorganisms can be destroyed by pressurization. Extrapolating their stability diagram, it can be seen that some of them can be destroyed by applying relatively small negative pressure.
- Finally an alternative explanation was given to the phase equilibrium of ^{3}He+^{4}He mixture in porous materials.

Although we know that p=0 is not a special point in liquids, there are several phenomena - some of them have been presented in this paper - which can be seen - or can be seen better - under negative pressure.

Acknowledgement

This work was partially supported by the Hungarian Research Fund (OTKA) under contract number F 034333. The author was supported also by the Bolyai Research Fellowship.

References

1. Imre, A., Van Hook, W.A. (1998) Liquid-liquid equilibria in polymer solutions at negative pressure, *Chem. Soc. Rev.,* **27**, 117-123
2. Imre, A.R, Melnichenko, G., Van Hook, W.A. and Wolf, B.A. (2001) On the effect of pressure on the phase transition of polymer blends and polymer solutions: Oligostyrene-n-alkane systems, *Phys. Chem. Chem. Phys.,* **3**, 1063-1066
3. Imre, A.R., Van Hook, W.A. and Wolf, B.A. (2002) Liquid-liquid phase equilibria in polymer solutions and polymer mixtures, *Macromolecular Symposia,* in press
4. Imre, A.R., Kraska, T. and Yelash, L.V. (2002) The effect of pressure on the liquid-liquid phase equilibrium of two polydisperse polyalkylsiloxane blends, *Phys.Chem.Chem.Phys,* **4**, 992-1001
5. Timmermans, J. and Lewin, J. (1953) A forgotten theory: the "negative saturation curve", *Discuss. Faraday Soc.,* **15**, 195-201
6. Rebelo, L.P.N., Visak, Z..P. ans Szydlowski, J., this book, next paper
7. Smeller, L. (2002) Pressure-temperature phase diagrams of biomolecules, *Biochim. Biophys. Acta,* **1595**, 11-29
8. Machin, W.D. (1998) A simple method for the generation of negative pressure in liquids, *Can. J. Chem.,* **76**, 1578-1580
9. Kim, S.B., Ma, J. and Chan, M.H.W. (1993) Phase diagram of ^{3}He-^{4}He mixture in aerogel, *Phys. Rev. Lett.,* **71**, 2268-2271
10. Flory, P.J. (1953) *Principles of Polymer Chemistry,* Cornell University Press, Ithaca.
11. Wolf, B.A. and Blaum, G. (1976) Pressure Influences on the True Cosolvency - Measured and Calculated Solubility Limits of Polystyrene in Mixtures of Acetone and Diethylether, *Makromol. Chem.,* **177**, 1073-1088
12.. Imre, A., Van Hook, W.A. (1994) Polymer-Solvent Demixing Under Tension. Isotope and Pressure Effects on Liquid-Liquid Transitions. VII. Propionitrile-Polystyrene Solutions at Negative Pressure, *J. Polym. Sci. B.,* **32**, 2283-2287
13. Rebelo, L.P.N., Visak, Z.P. and Szydlowski, J. (2002) Metastable critical lines in (acetone+polystyrene) solutions and the continuity of solvent-quality states, *Phys. Chem. Chem. Phys.,* **4**, 1046-1052
14. Hosokawa, H., Nakata, M. and Dobashi, T. (1993) Coexistence curve of polystyrene in methylcyclohexane. VII. Coexistence surface and critical double point of binary system in T–p–ϕ space, *J. Chem. Phys.,* **98**, 10078-10084
15. Wells, P.A., de Loos, Th. W. and Kleintjens, L.A. (1993) Pressure pulsed induced critical scattering: spinodal and binodal curves for the system polystyrene + methylcylcohexane, *Fluid Phase Eq.,* **83**, 383-390
16. Vanhee, S., Kiepen, F., Brinkmann, D., Borchard, W., Koningsweld, R. and Berghmans, H. (1994) The System Methylcyclohexane/Polystyrene. Experimental Critical Curves, Cloud-Point and Spinodal Isopleths, and their Description with a Semi-phenomenological Treatment, *Macromol. Chem. Phys.,* **195**, 759-780
17. Imre, A.R., Melnichenko, G. and Van Hook, W.A. (1999) Liquid-liquid equilibria in polystyrene solutions: the general pressure dependence, *Phys. Chem. Chem. Phys.,* **1**, 4287-4292

18. Imre, A.R., Melnichenko, G. and Van Hook, W.A. (1999) A Polymer-Solvent System with Two Homogeneous Double Critical Points: Polystyrene (PS)/(n-heptane+methylcyclohexane), J. Polym. Sci. B., 37, 2747-2753

19. Rebelo, L.P.N., Visak, Z.P., de Sousa, H.C., Szydlowski, J., de Azevedo, R.G., Ramos, A.M., Najdanovic-Visak, V., da Ponte, M.N. and Klein, J. (2002) Double critical phenomena in (water plus polyacrylamides) solutions, Macromolecules, 35, 1887-1895

20. Rebelo, L.P.N. and Van Hook, W.A.(1993) An unusual phase diagram: the polystyrene/acetone system in its hypercritical region; near tricritical behavior in a pseudo-binary solution, J. Polym. Sci. B. , 31, 895-897

21. Schneider, G. M. (1972) Phase behavior and critical phenomena in fluid mixtures under pressure, Ber. Bunsen Ges., 76, 325-331

22. Narayanan, T. and Kumar, A. (1994) Reentrant phase transitions in multicomponent liquid mixtures, Phys. Rep., 249, 135-218

23. Schneider, G.M. (2002) Aqueous solutions at pressures up to 2 GPa: gas-gas equilibria, closed loops, high-pressure immiscibility, salt effects and related phenomena, Phys. Chem. Chem. Phys., 4, 845-852

24. Otake, K., Karaki, R., Ebina, T., Yokoyama, C. and Takahashi, S. (1993) Pressure Effects on the Aggregation of Poly(N-isopropylacrylamide) and Poly(N-isopropylacrylamide-co-acrylic acid) in Aqueous Solutions, Macromolecules, 26, 2194-2197

25. de Azevedo, R. G., Rebelo, L.P.N., Ramos, A.M., Szydlowski, J., de Sousa, H.C. and Klein, J. (2001) Phase behavior of (polyacrylamides+water) solutions: concentration, pressure and isotope effects, Fluid Phase Eq., 185, 189-198

26. Imre, A.R. and Wolf, B.A., unpublished result

27. Heremans, K. and Smeller, L. (1998) Protein structure and dynamics at high pressure, Biochim. Biophys. Acta, 1368, 353-370

28. Whittingham, T.A. (2001) Estimated fetal cerebral ultrasound exposures from clinical examinations, Ultrasound Med. Biol., 27, 877-882

29. Bailey, M.R., Blackstock, D.T., Cleveland, R.O. and Crum, L.A. (1998) Comparison of electrohydraulic lithotripters with rigid and pressure-release ellipsoidal reflectors. I. Acoustic fields, J. Acoust. Soc. Am., 104, 2517-2524

30. Ludwig, H., Scigalla, W. and Sojka, B. (1996) Pressure and temperature inactication of microorganisms, in. J.L. Markley, D.B. Northrop and C.A. Royer (eds.), High Pressure Effects in Molecular Biophysics, Oxford University Press, New York, 346-363

31. Gelb, L.D., Gubbins, K.E., Radhakrishnan, R. and Sliwinska-Bartkowiak, M. (1999) Phase separation in confined systems, Rep. Prog. Phys., 62, 1573-1659

32. Machin. W.D. and Stuckless, J.T. (1985) Capillary-condensed Water in Silica Gel, J. Chem. Soc. Faraday Trans., 81, 597-600

33. Zhang, J., Liu, G. and Jonas, J. (1992) Effects of Confinement on the Glass Transition Temperature of Molecular Liquids, J. Phys. Chem., 96, 3478-3480

34. Jackson, C.L. and McKenna, G.B. (1996) Vitrification and Crystallization of Organic Liquids Confined to Nanoscale Pores, Chem. Mater., 8, 2128-2137

35. Hohenberger, Th., König, R. and Pobell, F. (1998) Phase Separation of Liquid ^{3}He-^{4}He Mixtures in Porous Vycor Glass, J. Low Temp. Phys., 110, 579-584

36. Falicov, A. and Berker, A.N. (1995) Correlated Random-Chemical-Potential Model for the Phase Transitions of Helium Mixtures in Porous Media, Phys. Rev. Lett., 74, 426-429

37. Pricaupenko, L. and Treiner, J. (1995) Phase separation of Liquid ^{3}He-^{4}He Mixtures: Effect of Confinement, Phys. Rev. Lett., 74, 430-433

38. Maris, H. and Balibar, S. (2000), Negative Pressures and Cavitation in Liquid Helium, Physics Today, 53, 29-34

39. Lilly, M.P. and Hallock, R.B. (1998) Compressibility of ^{4}He Liquid Confined in Nuclepore, J. Low Temp. Phys., 110, 555-560

40. Yorozu, S., Hiroi, M., Fukuyama, H., Akimoto, H., Ishimoto, H. and Ogawa, S. (1992) Phase separation curve of ^{3}He-^{4}He mixtures under pressure, Phys. Rev. B, 45, 12942-12948

MEASURING THE PROPERTIES OF LIQUIDS AND LIQUID MIXTURES AT ABSOLUTE NEGATIVE PRESSURES

L.P.N. REBELO, Z.P. VISAK, J. SZYDLOWSKI, H.I.M. VEIGA, R. GOMES DE AZEVEDO, P.F. PIRES AND M. NUNES DA PONTE
Instituto de Tecnologia Química e Biológica, ITQB2, Universidade Nova de Lisboa, Apartado 127, 2780-901 Oeiras, Portugal
E-mail: luis.rebelo@itqb.unl.pt

Abstract. An overview of experimental work recently performed in the Lisbon laboratories is provided. Several physical properties of liquids as well as those of the coexistence between two liquid phases and between liquid and solid have been determined at absolute negative pressures. Examples for this include thermal pressure coefficients, loci of temperature of maximum density, negatively sloped melting lines, speed of propagation of low-intensity ultrasound waves, and (p, T, x) conditions of occurrence of liquid-liquid phase separation. The major achievements that emerged within the scope of this research program were: (i) the first experimental evidence for the existence of a TMD in supercooled and stretched gallium, and (ii) the first experimental detection of phase separation phenomena in molecular liquid mixtures under tension.

1. Introduction

Absolute negative pressure regimes in the liquid state constitute a subclass of liquids'superheating or of simultaneous superheating and supercooling [1], which, in turn, are special cases of the global phenomenon of metastability [2]. Figure 1(a) reproduces the textbook case that schematically illustrates the one-component phase diagram under totally stable conditions. We are interested in exploring those events where the liquid invades the thermodynamic territory belonging to the gas (superheating) and/or the solid phase (supercooling). In other words, liquid states now occupy the major part of the whole area of the p-T projection of the phase diagram (Figure 1(b)). From the view's point of thermodynamics only regimes corresponding to higher temperatures (or lower pressures) than those of the mechanical spinodal line are inaccessible to the liquid. As the liquid approaches this border its isothermal compressibility tends to diverge to infinity. If it could cross the border, the liquid would present a negative compressibility, and, thus, would be exploring unreachable totally unstable liquid states. This is the territory only accessible for a more expanded state - the gas phase.

We are particularly interested in reaching regimes where the liquid is stretched up to tensions corresponding to p < 0, and, thus have inserted in Figure 1(b) the line-border p = 0. As is well known and will be many times experimentally illustrated in this work, crossing the p = 0 line has no special striking consequences on the liquids behavior for

95

A.R. Imre et al. (eds.), Liquids Under Negative Pressure, 95–108.
© 2002 *Kluwer Academic Publishers. Printed in the Netherlands.*

its properties are merely the smooth continuation of those at positive pressures. The only novel feature introduced by extending the phase diagram to conditions where p < 0, is the appearance of a "forbidden land" located at temperatures higher than those of the mechanical spinodal locus. Neither the liquid nor the gas, irrespective of any degree of metastability, can occupy this "land". For the former it would have meant a negative value for the compressibility, while for the latter one would have reached unphysical extreme expanded states (of negative volume). In other words, one cannot stretch a gas phase.

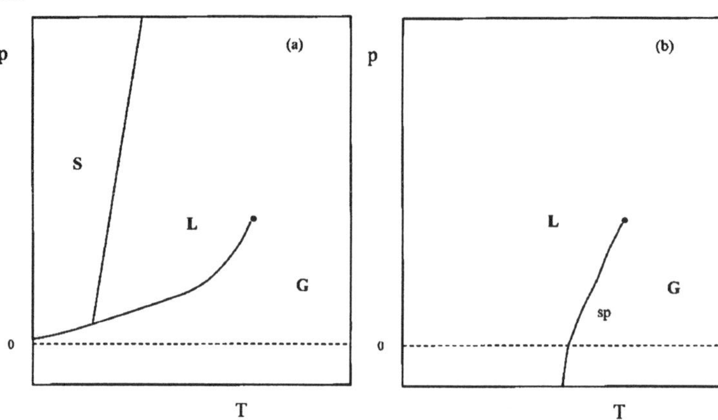

Figure 1. Totally stable (a) and metastable (b) – from the liquid's perspective - schematic phase diagrams of a pure substance.

The measurements herein described correspond to the achievement of static and isotropic tension and can be divided into two types: (i) those performed in microsamples (typically less than 0.005 cm^3) and (ii) those using macroscopically-sized volumes between 0.2 and 0.6 cm^3. In the former case, the liquids' container constitutes itself a Bourdon-type gage [3], and, thus, negative pressures can be "directly" monitored following the strategy proposed by Henderson and Speedy [4],

$$p = a + b\theta + c\theta^2 \qquad (1)$$

where θ is the angle of rotation of the glass helix, which is calibrated against pressure. For macroscopically-sized samples, negative pressures have to be estimated [5-7] from the experimental value of the thermal pressure coefficient, γ_V, of the given isochore at positive pressures or from a reliable equation of state.

$$p = p_{fill} + \int_{T_{fill}}^{T} \gamma_V \, dT \qquad (2)$$

Alternatively, one could have used a set-up of strain-gauges mounted on the exterior of the liquid's container [8] but this can only be used with limited types of materials. In all cases (therefore, irrespective of the sample's volume and type of property to be measured) one uses the Berthelot's recipe to achieve isotropic tensions [1,3,6]. The exception is the determination of negatively sloped melting lines where a more complicated p-T cycle [3] has to be executed.

In the following sections we will present and discuss results obtained both for pure substances (loci of temperature of maximum density (TMD) for water, heavy water and gallium, melting line of gallium, and speed of propagation of ultrasound for organic substances), and binary mixtures (liquid-liquid phase transitions for polystyrene + acetone and water + 3-methylpyridine).

2. Experimental, Results, and Discussion

2.1. ONE-COMPONENT SYSTEMS

2.1.1. Measurements Using the Berthelot-Bourdon Type of Cell

We designed and built a Berthelot-Bourdon type of cell, which was inspired in one previously described by Henderson and Speedy [4]. There was an improvement in the

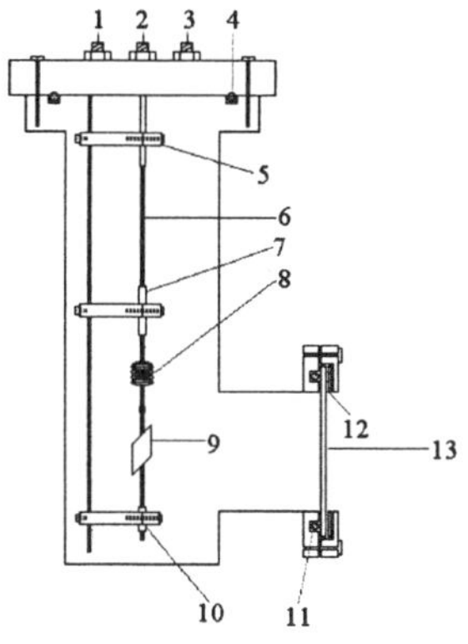

Figure 2. Schematic diagram and photograph of the cell. 8: glass capillary helix; 9: mirror; 13: silica window.

temperature control and stability [3]. The liquid is contained in a Pyrex glass capillary in helix form. The mechanical properties of the helix are sensitive to pressure, and induce rotation (of a θ-angle) in any plane containing the central axis of the helix. That plane is typically a small mirror attached to the bottom of the helix. A He-Ne laser beam, reflected off of the mirror, is imaged on a large curved screen about 3 meters from the cell. The helix is housed in a stainless steel cylinder containing a silica window (see Fig. 2). It is well known that the major key for success in tension a liquid lies in using as small samples of it as possible. Figure 3 is a photograph that compares the typical I.D. (~ 0.1 mm) of our laboratory glass capillaries with those found in nature. Nature is by far more efficient and resourceful. The example herein given is of the xylem of a mangrove tree (I.D. ~ 0.005 mm) where sap in transported (at negative pressure) to the

98

leaves. The negative pressure is needed to compensate for the osmotic pressure provoked by the highly salted water where the mangrove tree's roots are immersed.

Figure 3. Diameters of xylem of mangrove tree (0.005 mm) and laboratory glass capillary (0.1 mm) compared.

Initial tests to our apparatus [3] were performed using organic solvents and water.

The methodology used produces quasi-isochores which, for the organic solvents, are almost linear lines in the p-T projection but they are highly curved (presenting minima) for water or aqueous solutions due to the presence of the locus of temperature of maximum density (TMD). In most cases, that minimum of the quasi-isochore coincides with the TMD within experimental precision. Different isochores are obtained by manipulating the total amount of liquid contained in the constant-volume glass capillary. Figure 4 shows a typical quasi-isochore obtained [3] with a liquid presenting TMD (D_2O) and compares results obtained by other authors with ours for normal and heavy water.

Our main purpose though was to study the element gallium. Gallium is an "abnormal" substance. It has plenty of stable (Ga(I) or Ga(α), Ga(II), and Ga(III)) and metastable (Ga(β), Ga(γ), and Ga(δ)) solid forms [9-11] and one of the broadest atmospheric pressure liquid ranges (Tmp = 302.9 K, Tbp = 2 676 K). Its liquid form is very easy to supercool [2] ($T_H/Tmp \sim 0.5$), presenting a homogeneous nucleation limit of about 150 K. Similarly to water, this limit may be precluding the possibility of achieving a singularity [11] which can occur in this temperature neighboring region. Its unusual behavior extends to the presence of a negatively sloped melting line for the Ga(α)-liquid equilibrium. In contrast, the equilibrium between either metastble Ga(β) or metastable Ga (γ) and supercooled liquid evolves with a normal (positive) slope.

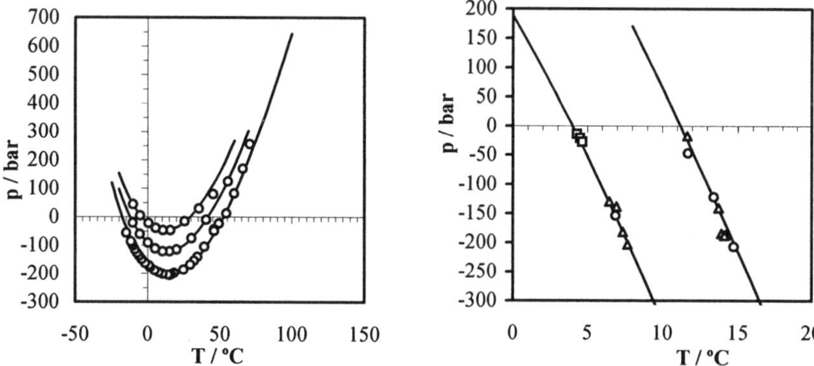

Figure 4. Quasi-isochores for D_2O and TMDs of H_2O (left) and D_2O (right) compared

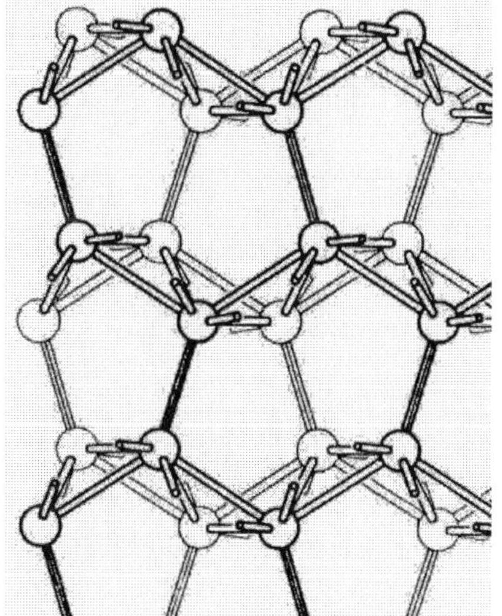

Figure 5. The Ga(α) structure. Adapted from ref. [10].

Ga(α) has a rather opened structure [9,10] (base centered orthorhombic (eco(8))) where each atom has only one nearest neighbor ("Ga$_2$" molecules). For practical purposes of tensioning, liquid gallium presents the nice peculiarity of wetting well glass (forming a mirror). The second most important condition for a successful stretching experiment is thus fulfilled – good adhesion between the liquid and the walls of its container.

We determined [3] the continuation of the negatively sloped melting line of Ga(α) up to a tension corresponding to about -100 bar, and found a slope of -515 ± 25 bar. K^{-1}. More interesting was the observation [12] of a complete set of eight isochores showing minima in the p-T projection, and, thus, change in the sign of the isobaric expansion coefficient, α_p. To the best of the authors' knowledge this constitutes the first experimental evidence for the existence of a TMD in gallium. As far as our experimental capabilities and uncertainties are concerned the TMD could only be detected in the double metastable region of simultaneously supercooled and stretched states ($-5 < t/°C < -17$; $-125 < p/bar < -325$) and evolves with a negative slope of approximately -15 barK^{-1} (in water, in the positive and negative pressure neighborhood region of p = 0, the slope of the TMD is approximately -50 barK^{-1}).

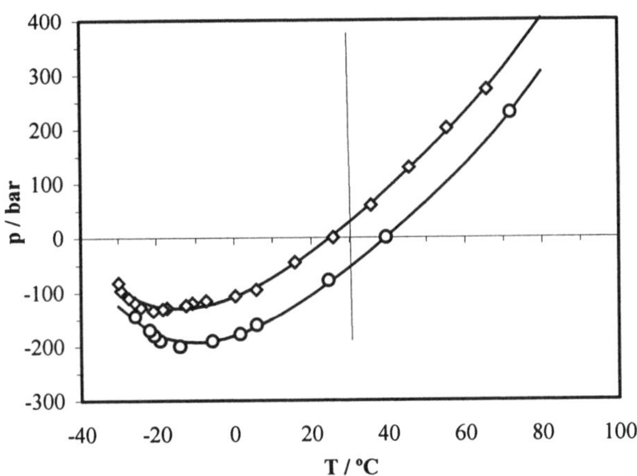

Figure 6. Quasi isochores for liquid gallium probing two points of its TMD. The "vertical" line locates the melting locus.

Whether the TMD of gallium also exists at positive pressures (although in the supercooled regime) is still an open question. Figure 6 shows two TMD points, one in the middle range of the experimental conditions, the other obtained at the lowest possible temperature. Note that in the latter case the temperature at which the glass capillary is completely filled with liquid (interception between the isochore and the vapor pressure line – nominal p = 0 line) is already slightly lower than the melting temperature. This was the only successful run amongst several where we were able to fill the capillary in the supercooled regime avoiding solidification. Nevertheless, if one trusts that a smooth linear extrapolation of our results is valid, then the TMD should cross the p=0 line and start to appear at positive pressures for a supercoling of about 55 K. In order to know whether the TMD crosses the melting line of Ga(α) to the non-supercooled regime, a qualitative extrapolation of our results tells us that if that occurs is at some Kbars of tension. We were unable to reach tensions higher than about -325 bar, which correspond for the TMD to temperatures of c.a. $-5°C$ (therefore still 35 K of supercooling).

Figure 7 shows schematically the phase diagram of gallium between 1 and -1 Kbar, enhancing our determinations of the continuation of the melting line of Ga(α) and of its TMD in the double metastable regime.

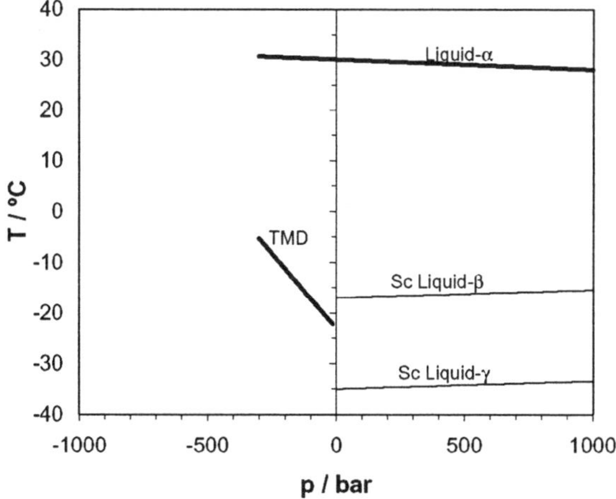

Figure 7. The phase diagram of gallium between moderate pressures and tensions

TABLE 1. Anomalous liquids. Experimentally known or predicted (by simulations) "abnormal" density behavior. TmD – temperature of *minimum* density.

Substance	Neg. Sloped T_m	TMD	L-L Equil. (possibly)	Open Structure
H_2O (and D_2O)	√	√	√	tetrahedral
SiO_2 (silica)		√	√	tetrahedral
GeO_2			√	tetrahedral
BeF_2		√ (and TmD)	√	tetrahedral
Ga (gallium)	√	√		b.c. orthor.
C	√		√	tetrahedral
Si (silicon)	√	√	√	tetrahedral
Ge	√		√	tetrahedral
P	√		√	P4(w), b.c.o.
Sb	√			
Bi	√		√	
S	√	(TmD)	√	
Se	√		√	

It should be noted that water also presents a good portion of its TMD in the supercooled regime. Taking into account that the approximate slopes of its TMD and melting line are, respectively, - 50 and – 135 barK^{-1} (and that there is a 4 K temperature shift between them at atmospheric pressure), above 317 bar (and below – 2.3 °C) the water TMD lies in the supercooled region. TMDs and negatively sloped melting lines are expected to be closely related [2,13]. In turn, these density anomalies are often associated with open structures [14]. It has been proposed [13,15] that the corresponding effective isotropic potential able to rationalize density anomalies is one presenting two wells (see Figure 8). Note in Figure 8 the analogy (double "bumps") between the doubly welled potential and those pair potentials typically used for metals such as gallium. Gallium surely fits well within the "family" of anomalous liquids with abnormal density behavior (see Table 1).

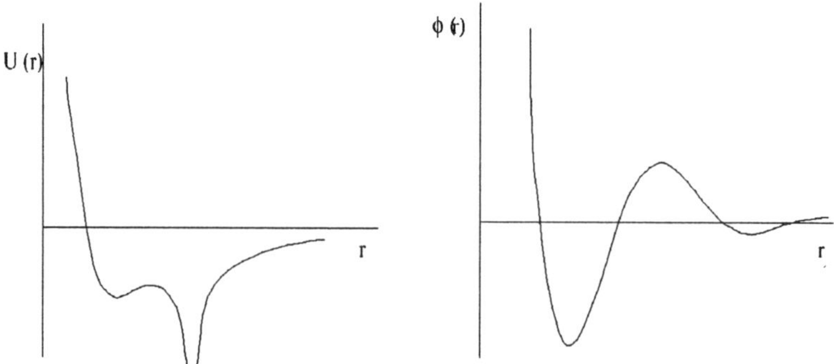

Figure 8. Isotropic double welled potential [15] and metalic pair potential of gallium [9] compared.

2.1.2. *Measurements Using a Miniaturized Cell for the determination of Speed of Ultrasound*

Measurements of thermodynamic properties of liquids under tension are usually limited to a few physical quantities. This limitation lies on two major restrictions. (i) the very small size of the samples and (ii) the need for avoiding external perturbations to a system that is inherently in a metastable status. The challenge is thus to envisage a method that simultaneously permits us to use minimal quantities and does not perturb significantly the system.

The speed, *u*, of propagation of ultrasound waves in a liquid media (combined with molar density data, ρ) is a very powerful tool for the determination of a wide variety of thermodynamic quantities [16]. In particular, one can access values of compressibility, β_S and β_T, expansivity, α_p, and heat capacities, C_p and C_v:

$$\beta_S = 1/\left(\rho M u^2\right); \quad C_p = T\rho^{-1}\alpha_p^2/\left(\beta_T - \beta_S\right); \quad C_V = C_p\beta_S/\beta_T \qquad (3)$$

We have just designed, built and tested a micro thick-walled stainless steel (SS) cell for the measurement of the speed of very low-intensity ultrasound waves in liquids at extreme (p,T) conditions. The method is a non-intrusive one, for the transducers (piezoelectrics) are located on the external walls of the SS block (see Fig. 9).

The wave travels the wall, then the liquid, and finally reaches the second transducer. Measuring the "time of travel" (time of flight, t) permits indirect access to the value of the speed of sound (u) in the liquid. This can be obtained by summing up the times of flight in the SS (known) and in the liquid (unknown) providing the geometry of the cell has been accurately determined, or, more accurately, by calibration against fluids of known speed of sound. We have elected the latter option. The apparatus typically works in the (270 < T/K < 350; - 100 < p/bar < 2000) range. Figure 10 shows a high accuracy calibration run using six fluids. Once a calibration matrix t = f(u,T,p) is obtained for a given cell, values of u can be obtained for conditions outside the calibration range by smooth extrapolation.

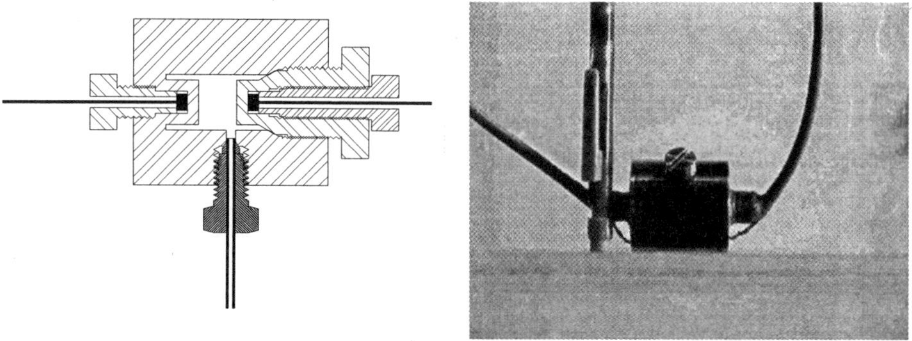

Figure 9. Miniaturized cell for the speed of ultrasound measurements using a non-intrusive method. Approximate internal volume of 0.2 cm^3.

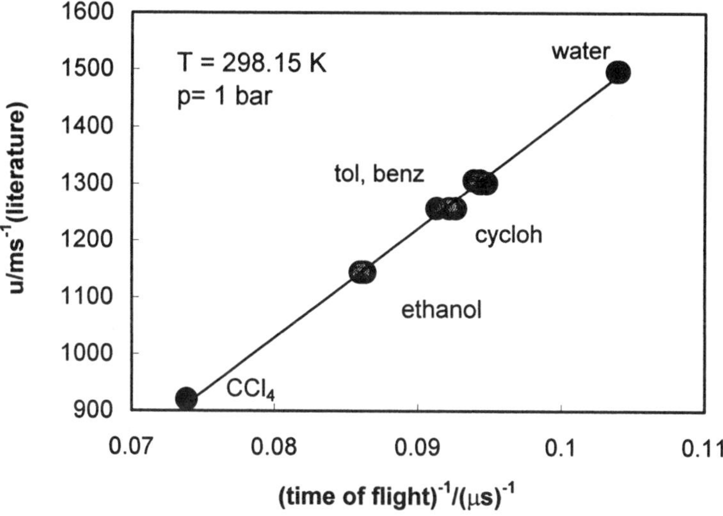

Figure 10. Calibration of the speed of sound versus the time of flight.

Preliminary runs [17] have permitted us to obtain the speed of sound of ethanol and CCl$_4$ up to tensions of about − 20 bar (see Fig. 11). So far, the negative pressures achieved are modest. But these preliminary results clearly demonstrate that this method can be very useful for the determination of a wide variety of thermodynamic properties in the negative pressure regime.

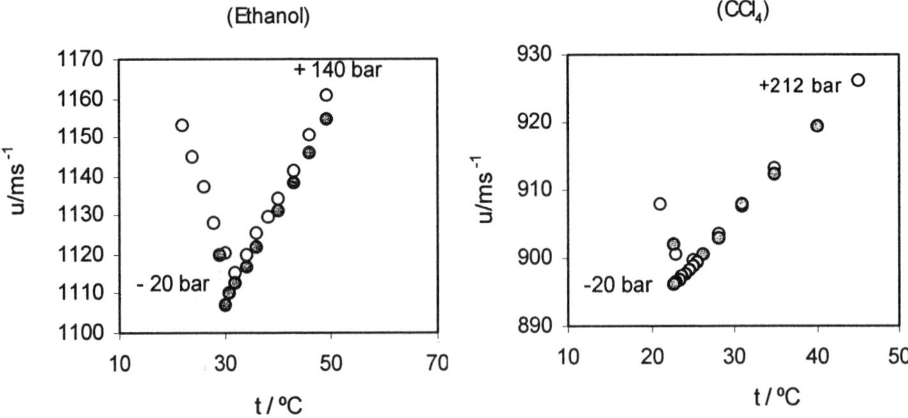

Figure 11. Results of Berthelot p-T cycles for the speed of sound. (●): cooling branch, (o): heating branch.

2.2. BINARY MIXTURES

Phase-separating solutions illustrate negative pressure better than pure substances do, for in the latter case one cannot visually detect any phenomenon. In the case of solutions, negative pressure effects can provoke phase transition, which induces turbidity or cloudiness in the liquid.

Among binary liquid mixtures, polymer solutions and solutions of methyl-pyridine and water have deserved particular attention during the recent years. Both type of solutions show a wide variety of phase behavior, in respect to the influence of pressure, temperature, concentration, isotope effects and (in the polymer solution case) change in the polymer molecular weight (M_w). Several recent articles [5-7,18,19] have investigated liquid-liquid equilibrium (LLE) in polymer solutions at negative pressures. On the other hand, for the aqueous methyl-pyridine solutions, LLE was studied only at positive pressures [20,21]. To the best of the authors' knowledge negative pressures in non-polymer solutions have not been reported so far.

In this paper we present the attempts to achieve absolute negative pressures in two binary mixtures exhibiting partial miscibility: Polystyrene (PS) +acetone and 3-methyl pyridine (3-MP)+ water. These experiments were performed in Pyrex-glass capillaries. In contrast to tensioning pure liquids, a macroscopically sized volume (~0.3-0.6cm^3) of liquid sample was used in order to enable the observation of the phase transition phenomenon. The detailed description of the experimental set-up and procedure used to obtain tensions was reported earlier [6,7]. At this point, it should be noted that negative pressures were not directly measured (see Eq. 2), but estimated using the thermal pressure coefficient of the liquid (a comprehensive explanation was given [7] in a previous paper).

In the case of PS+acetone mixtures, we report liquid-liquid equilibrium (LLE) p-T data at two different molecular weights of PS - 25,000 and 16,600. The first one was used in a 1.8 wt.% solution (off-critical concentration) whereas for the other one a near-critical concentration of 18 wt.% was chosen. The combinations of molecular weights and concentrations permit us to fine-tune experimentally reachable negative pressures. In the case of the critical concentration, two double criticalities of different kinds were observed: a pressure-double critical point (p-DCP) in the stable region of high-pressures, and a temperature-double critical point (T-DCP) in the metastable regime of negative pressures. The p-DCP is associated with a change in the sign of the excess volume as pressure is varied, while at the T-DCP neighboring region a change in the sign of the excess enthalpy should occur with temperature change. As far as our knowledge is concerned, this constitutes the first experimental observation of a polymer solution presenting simultaneously two types of double critical points with one penetrating the metastable region of negative pressures.

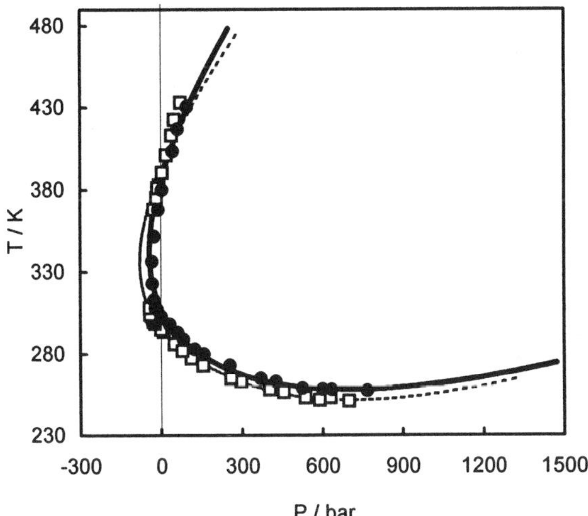

Figure 12. Experimental (p,T) data points of L-L phase transitions in (acetone + polystyrene) solutions. Isopleths for 1.8 wt.% PS-25,000 (□) and 18 wt.% PS-16,600 (●). Curves (...., —) represent a scaling fit [23].

Figure 12 gives a graphical presentation of the L-L phase transition data in p-T coordinates, for isopleths at two different molecular weights. Two extrema are observed for both concentrations, off critical and critical. In this figure, LLE data at positive pressures are reported as well. It has to be pointed out that the data at positive pressures were not achieved using the same experimental method used to achieve tension. The experimental procedure for obtaining data at positive pressure was explained in a recent paper [22].

Figure 13 gives a close-up presentation of our data up to some 100 bar, with emphasis on the negative pressure region. Data points in the negative pressure region show a good match with those at positive pressures (obtained by a much more precise

106

technique), which is reassuring. A detailed explanation of the underlying reason for the interruption of the p-T data points of the diluted solution can be found in ref. [7].

For the 3-MP+water solutions, T-P phase transition curves were determined in the negative pressure region for a near critical (~30 wt.% of 3-MP) and off-critical composition of 3-methyl pyridine (~27wt.% of 3-MP). High negative pressures were obtained (over –200 bar) when a solution of H_2O and D_2O was used as a solvent. In some experiments high tensions were held for a certain period of time despite the fact that the system already entered the two-phase region. Moreover, during one run we succeeded in putting under tension a solution that was even heterogeneous since the beginning of the Berthelot cycle. Two phases were completely separated: top 3-MP rich phase and bottom H/D water rich one. Once the system was under tension, it was possible to slowly follow the phase transition in the 3-MP rich phase, up to the moment where the system relaxed and returned to positive pressure. This is the first time that negative pressure studies in non-polymer solutions are reported.

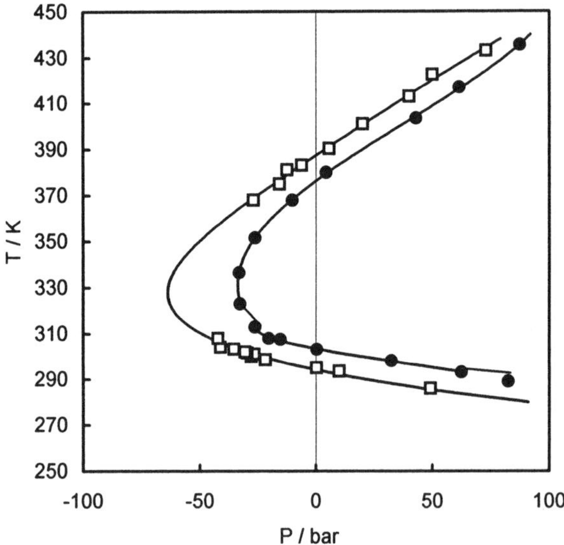

Figure 13. A close-up of Figure 12 in the neighboring region of absolute negative pressures. The full lines merely constitute visual guides to the data. (□), 1.8 wt.% PS-25,000 ; (●), 18 wt.% PS-16,600.

Figure 14 presents LL phase transition data for off critical and critical 3-MP concentrations. The influence of solvent deuterium content on the LL phase behavior was also investigated. A strong isotope effect on the double critical point was detected. For pure water as a solvent, the high-pressure [21] and the low-pressure (negative) critical loci are far apart, leaving a large solubility window opened. The increase of deuterium content to approximately 21 wt.% of D_2O in a (H_2O+D_2O) mixed solvent makes the two loci to join at a quadruple critical point, and, thus, completely closing the solubility window.

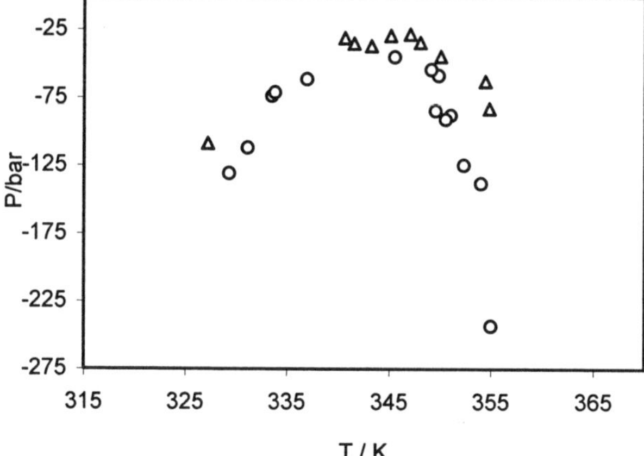

Figure 14. p-T phase transition data in the metastable region of negative pressures:(Δ)- 30 wt% of 3-MP in a solvent (12.9 wt.% D_2O in (D_2O+H_2O)); (o)- 27 wt % of 3-MP in a solvent (14.9 wt% D_2O in (D_2O+H_2O))

In spite of the very complicated and unusual phase behavior of this system, we have just succeeded in rationalizing it using a simplified g^E-model [24]. The results [17] clearly show that the interpretation of the main features of the phase diagram can be performed under the assumption that the molar excess enthalpy of the solutions is isotope independent (and changing sign as temperature is varied), while entropic effects are sensitive on isotopic substitution.

3. References

1. Imre, A., Martinás, K., Rebelo, L.P.N. (1998) Thermodynamics of negative pressures in liquids, *J. Non-Equilibrium Thermodyn.* **23**, 351-375; Rebelo, L.P.N., Van Hook, A., Ye, Y.J. (1993) Liquids at negative pressures (transl.), *Química* **51**, 33-42.
2. Debenedetti, P.G. (1996) *Metastable Liquids*, Princeton Univ. Press, Princeton, N.J.
3. Veiga, H.I.M., Rebelo, L.P.N., Nunes da Ponte, M., Szydlowski, J. (2001) Water and gallium at absolute negative pressures. Loci of maximum density and of melting, *Int. J. Thermophys.* **22**, 1159-1174.
4. Henderson, S.J., Speedy, R.J. (1987) Temperature of maximum density in water at negative pressure, *J. Phys. Chem.* **91**, 3062-3068; Henderson, S.J., Speedy, R.J. (1987) Melting temperature of ice at positive and negative pressures, *J. Phys. Chem.* **91**, 3069-3072.
5. Imre, A., Van Hook, W.A. (1998) Liquid-liquid equilibria in polymer solutions at negative pressure, *Chem. Soc. Rev.* **27**, 117-123.
6. Visak, Z.P., Rebelo, L.P.N., Szydlowski, J. (2002) Achieving absolute negative pressures in liquids:precipitation phenomena in solution, *J. Chem. Educ.* **79**, 869-873
7. Rebelo, L.P.N., Visak, Z.P., Szydlowski, J. (2002) Metastable critical lines in (acetone + polystyrene) solutions and the continuity of solvent quality states, *Phys. Chem. Chem. Phys.* **4**, 1046-1052.
8. Ohde, Y., Komori, K., Nakamura, T.,Tanzawa, Y., Nishino, Y., Hiro, K. (2001) Effects of gas transports in metals on negative pressures in water in Mo/Cu Berthelot tubes, *J. Phys. D: Appl. Phys.* **34**, 1717-1726.
9. Young, D.A. (1991) *Phase Diagrams of the Elements*, Univ. California Press, Berkeley, California.

10. Donohue, J. (1974) *The Structures of the Elements*, John Wiley & Sons, N.Y.

11. Bosio, L., Windsor, C.G. (1975) Observation of a metastability limit in liquid gallium, *Phys. Rev. Lett.* **35**, 1652-1655; Bosio, L., Cortes, R., Defrain, A. (1973) Chaleur massique du gallium liquide en surfusion et des phases cristallines β et γ, *J. Chimie Phys.* **70**, 357-359.

12. Manuscripts in preparation.

13. Buldyrev, S.V., Franzese, G., Giovambattista, N., Malescio, G., Sadr-Lahijany, M.R., Scala, A., Skibinsky, A., Stanley, H.E. (2002) Models for a liquid-liquid phase transition, *Physica A* **00**, 00-00 (in press).

14. Poole, P.H., Grande, T., Angell, C.A., McMilan, P.F. (1997) Polymorphic phase transitions in liquids and glasses, *Science* **275**, 322-323.; Angell, C.A., Bressel, R.D., Hemmati, M., Sare, E.J., Tucker, J.C. (2000) Water and its anomalies in perspective: tetrahedral liquids with and without liquid-liquid phase transitions, *Phys. Chem. Chem. Phys.* **2**, 1559-1566; Rebelo, L.P.N., Debenedetti, P.G., Sastry, S. (1998) Singularity-free interpretation of the thermodynamics of supercooled water. Thermal and volumetric behavior, *J. Chem. Phys.* **109**, 626-633.

15. Stanley, H.E., Buldyrev, S.V., Canpolat, M., Mishima, O., Sadr-Lahijany, M.R., Scala, A., Starr, F.W. (2000) The puzzling behavior of water at very low temperature, *Phys. Chem. Chem. Phys.* **2**, 1551-1558; Franzese, G, Malescio, G., Skibinski, A., Buldyrev, S.V., Stanley, H.E. (2001) Generic mechanism for generating a liquid-liquid phase transition, *Nature* **409**, 692-695; Mishima, O., Stanley, H.E. (1998) The relationship between liquid, supercooled and glassy water, *Nature* 396, 329-335.

16. Streett, W.B., Ringermacher, H.I., Burch, J.L. (1972) Sound velocity, adiabatic compressibility, and specific heat of liquid and dense fluid krypton, *J. Chem. Phys.* **57**, 3829-3838.

17. Work in progress.

18. Imre, A., Van Hook, W.A. (1994) Polymer-solvent demixing under tension. Isotope and pressure effects on liquid-liquid transitions, VII. Propionitrile-polystyrene solutions at negative pressure, *J. Polym. Sci: B: Polym. Phys.* **32**, 2283-2287.

19. Imre, A., Van Hook, W.A. (1997) Continuity of solvent quality in polymer solutions. Poor-solvent to θ-solvent continuity in some polystyrene solutions, *J. Polym. Sci: B: Polym. Phys.* **35**, 1251-1259.

20. Narayanan, T., Kumar, A. (1994) Reentrant phase transitions in multicomponent liquid mixtures, *Phys. Rep.* **249**, 135-218

21. Schneider, G. (1963) Druckeinfluß auf die entmischung flüssiger systeme, *Zeitschrift Phys. Chem. Neue Folge* **37**, 333-352; *ibid*: **39**, 187-197.

22. Rebelo, L.P.N., Visak, Z.P., de Sousa, H.C., Szydlowski, J., de Azevedo, R.G., Ramos, A.M., Najdanovoc-Visak, V., Nunes da Ponte, M., Klein, J. (2002) Double critical phenomena in (water + polyacrylamides) solutions, *Macromolecules* **35**, 1887-1895.

23. Imre, A.R., Melnichenko, G., Van Hook, W.A. (1999) Liquid-liquid equilibria in polystyrene solutions: the general pressure dependence, Phys. Chem. Chem. Phys. 1, 4287-4292.

24. Rebelo, L.P.N. (1999) A simple gE-model for generating all basic types of binary liquid-liquid equilibria and their pressure dependence. Thermodynamic constraints at critical loci, Phys. Chem. Chem. Phys. 1, 4277-4286.

LIMITING TENSILE STRENGTH OF CAPILLARY-CONDENSED LIQUIDS

WILLIAM D. MACHIN
Department of Chemistry, Memorial University of Newfoundland
St. John's, Newfoundland, Canada
A1B 3X7

Abstract. Condensation of vapours into mesoporous adsorbents occurs spontaneously at pressures less than their saturation vapour pressure. The thermophysical properties of the resulting capillary liquids, density, thermal expansivity, isothermal compressibility and heat capacity are similar to those of the corresponding bulk liquid. However, the liquid-vapour interface may be highly curved resulting in a large tension or negative pressure within the capillary liquid. Since the liquid is thermodynamically stable, homogenous nucleation of the vapour phase is suppressed but increasing the tension may eventually lead to its rupture by spinodal decomposition. Conditions under which vapours condense into small pores and the requisite adsorbent properties will be discussed. Experimental results for several vapours and adsorbents will be presented.

1. Introduction

Capillary condensation of vapours into mesoporous adsorbents occurs spontaneously at pressures less than the bulk saturation vapour pressure. At low pressures the adsorbed fluid is present as a surface film, but when condensation begins the pores fill with liquid, reaching a plateau at higher pressures when all pores are completely filled with capillary liquid. Condensation-evaporation cycles are usually irreversible and exhibit hysteresis with evaporation occurring at lower pressures than condensation [1], as shown in Figure 1. Typically, such hysteresis loops have an upper closure point, above which the isotherm is reversible, and a lower closure point below which it is also reversible. The former corresponds to the point at which all pores have been filled with condensate and the pressure at which this occurs is determined by the size of the largest pores. While the position of the lower closure point of the hysteresis loop may be determined by the size of the smallest pores, this point is frequently determined, not by pore size, but by the limiting tensile strength (spinodal limit) of the capillary liquid [1]. In this paper we consider the conditions required to induce spinodal failure in capillary-condensed liquids and present several examples of spinodal decomposition in capillary liquids.

2. Results and Discussion

Capillary condensation in mesopores, *i.e.* pores with diameters greater than *ca.* 2 nm [2], is usually described with reference to the Kelvin equation [1],

109

A.R. Imre et al. (eds.), Liquids Under Negative Pressure, 109–115.
© 2002 *Kluwer Academic Publishers. Printed in the Netherlands.*

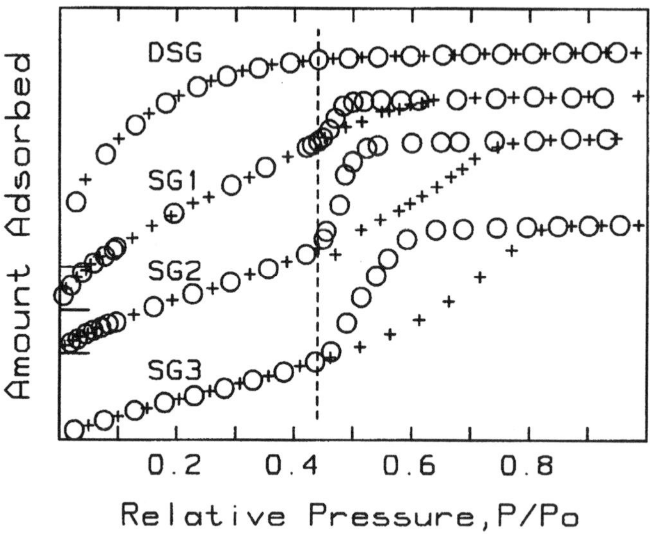

Figure 1.: Adsorption isotherms: Butane on four silicas at 273.15 K; +, adsorption; O, desorption. The capillary liquid region for each isotherm occurs at high relative pressures where the isotherm is nearly horizontal. Pore volume V_P, is calculated from the amount adsorbed at $P/P^o = 1$. Note that the lower closure point of each hysteresis loop occurs near $P/P^o = 0.44$. Surface areas and pore volumes are listed in Table 1. Isotherms are offset for clarity. See also [3].

$$r_k = \frac{-2\gamma\cos\theta}{RT\Delta\rho\ln(f_v/f^0)} = r_p - t \qquad (1)$$

where r_p is the pore radius, t is the thickness of the adsorbed film on the pore wall and r_k, the Kelvin radius, is the radius of curvature of the hemispherical liquid-vapour interface (see Figure 2). Surface tension, γ, and density ρ, of the capillary condensate are assumed to be equal to bulk values, and

$$\Delta\rho = \rho(liquid) - \rho(vapour) \qquad (2)$$

Relative fugacity, f_v/f^0, is the ratio of the vapour fugacity f_v at pressure P_v to the fugacity f^0 of the saturated vapour at P^0. For wetting liquids the contact angle, θ for the liquid on the pore wall is zero. Commonly, experimental conditions are such that $\rho(vapour) << \rho(liquid)$ and $f_v = P_v$ so that eq.1. reduces to

$$r_k = \frac{-2\gamma V_L}{RT\ln(P_v/P^0)} = r_p - t \qquad (3)$$

where V_L is the liquid molar volume. When the relative pressure, P_v/P^0, is such that $r_k>r_p-t$, the pore will fill spontaneously with capillary liquid and when $r_k< r_p-t$ it will evaporate.

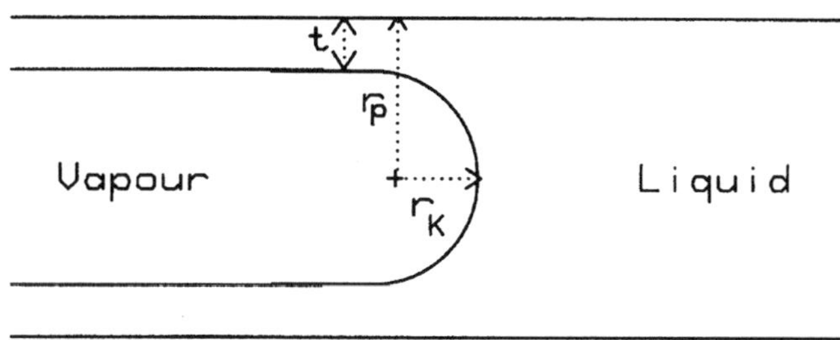

Figure 2.: Model cylindrical pore; r_p is the pore radius, t is the adsorbed film thickness, and r_k is the radius of curvature of the hemispherical liquid-vapour interface.

There is, however, an alternative mechanism for capillary evaporation. At all $P_v/P^0<1$ capillary liquid is under tension (negative pressure) since a pressure difference, Δp , must exist across the curved liquid-vapour interface, *i.e.*

$$\Delta p = P_v - P_L = \frac{2\gamma}{r_k} - \left(\frac{RT}{V_L}\right)\ln\left(\frac{P_v}{P^0}\right) \tag{4}$$

where P_L is the pressure within the capillary liquid. At conditions well removed from the critical point, $P_v< |P_L|$ hence

$$P_L = \left(\frac{RT}{V_L}\right)\ln\left(\frac{P_v}{P^0}\right) \tag{5}$$

Therefore, when P_L is greater than the spinodal decomposition pressure, P_s, and, $r_k\leq r_p-t$ the pore will empty by evaporation according to eq. 1 (Kelvin evaporation), but if $P_L\leq P_s$ and $r_k > r_p-t$ then the pore empties by spinodal decomposition. One consequence of this latter condition is that for each liquid there is a relative pressure below which capillary liquid cannot exist. The position of the lower closure point of the hysteresis loop should therefore be dependent only on the properties of the liquid, and should be relatively constant for different adsorbents, if appropriately sized pores are present in each. This phenomenon is illustrated in Figure 1 for butane in four silica gels having different pore size distributions. The three silicas that exhibit hysteresis all have a lower closure point located near $P_v = 0.44\ P^0$. The properties of these silicas are summarized in Table 1 and their preparation has been presented elsewhere [3]. Included in this table are the average, or geometric pore radii r_g defined as $r_g=2V_p/S$ where V_p is the pore volume and S the surface area of each adsorbent. The relative pore size, r_g/d_e, where d_e is the molecular diameter of butane [4], 0.586 nm, is also listed. Note that DSG, with the smallest pore size, does not exhibit hysteresis. Also shown are estimates of the number of molecules, z, in an average pore. Substitution of the appropriate quantities into eq. 5 yields $P_L = -19.3$ MPa as an estimate of the spinodal limit of capillary-condensed butane at 273 K.

TABLE 1. Adsorption Isotherms, Butane on Silica at 273.15 K.

Silica	Surface Area S/m^2g^{-1}	Pore Volume V_p/cm^3g^{-1}	Pore Radius r_g/nm	r_g/d_e	z
DSG	530	0.299	1.13	1.93	200
SG1	325	0.303	1.86	3.17	1500
SG2	280	0.288	2.06	3.52	2000
SG3	295	0.285	1.93	3.29	2000

Another estimate of spinodal limit may be obtained from an expression for homogeneous nucleation of the vapour phase at negative pressure given by Fisher [5],

$$\Delta p = \left[\frac{16\pi\gamma^3}{3kT \ln\left(kTz/h\right)} \right]^{1/2} = P_v - P_F \qquad (6)$$

where k is the Boltzmann constant, and h the Planck constant. Since capillary liquids are thermodynamically stable with respect to bulk liquid, and the number of molecules within a given pore is relatively small, then homogenous nucleation should be supressed until the spinodal limit is reached. Δp increases very slowly as z decreases, so that P_F should be almost independent of pore size and be close to the spinodal limit, P_S. For butane at 273 K, eq. 6 gives $P_F = -20.7$ MPa for $z = 200$ and $P_F = -20.1$ MPa for $z = 2000$. Both values are in excellent agreement with the spinodal limit estimated from the lower closure point of the hysteresis loop (eq. 5), $P_L = -19.3$ MPa.

Berezin [6] has derived a simple expression for the spinodal limit,

$$\Delta p = \frac{0.833\gamma}{d_B} = P_v - P_B \qquad (7)$$

where d_B is the thickness of a single layer of liquid. Assuming this to be the effective diameter of a molecule, then $d_B=(6V_L/\pi L)^{1/3}$ where L is Avogadro's number, or for butane, $d_B= 0.689$ nm and $P_B = -18.0$ MPa. This result is also in good agreement with the limit estimated from the lower closure point of the hysteresis loops. Berezin's equation also yields some insight into the conditions required to observe spinodal decomposition. From the Laplace equation [1]

$$\Delta p = \frac{2\gamma}{r} \qquad (8)$$

and eq. 7 it is readily shown that at the spinodal limit $r_k = 2.4\ d_B$, which suggests that pores having diameters equal to five molecular diameters are required. If only larger pores are present then all pores will empty by Kelvin evaporation, and if only smaller pores are present then hysteresis may be absent. This may explain why DSG does not show hysteresis with butane (see Figure 1) but does with smaller molecules such as xenon and water [7,8].

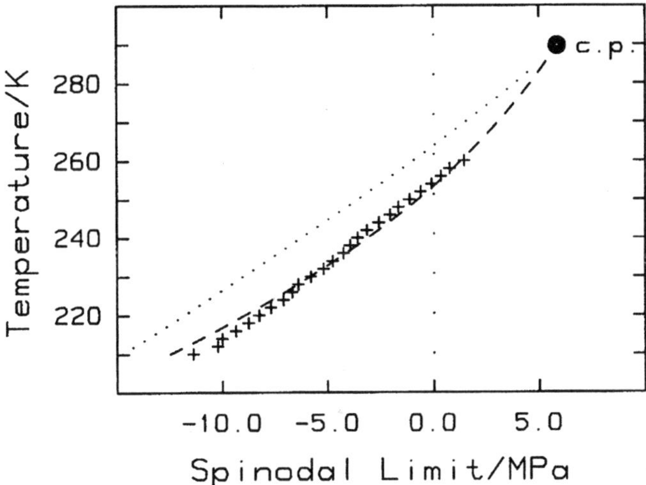

Figure 3.: Spinodal limit; xenon adsorbed on SG3, 210 K to 260 K; +, experimental values calculated from the lower closure point of each hysteresis loop (eq 1 and 4); - - -, limit calculated using eq. 6; ...., limit calculated using eq. 7. See [9] for the adsorption isotherms.

When suitable adsorption isotherms are available over a wide range of temperatures, the spinodal line as a function of temperature may be determined. This is shown in Figure 3 for xenon in SG3 from 210 K to 260 K. The general shape of these isotherms is the same as that for butane on SG3 (see Figure 1) but the hysteresis loop shrinks and the lower closure point moves to higher relative fugacities as temperature increases [9]. Also shown in Figure 3 are spinodal lines calculated from eq. 6 and eq. 7. Note that the former equation yields results in excellent agreement with experimental values over the entire temperature range whereas the equation proposed by Berezin tends to predict lower pressures.

Of all liquids, water is probably the most interesting showing exceptional properties under normal conditions and even more exceptional properties when stretched or supercooled [10]. An equation of state for liquid water at negative pressures has been proposed by Speedy [11], as

$$1 - \frac{P_L}{P_M} = Q\left[\left(\frac{V_M}{V_L}\right) - 1\right]^2 \tag{9}$$

where Q is a temperature dependent constant and V_M is the liquid molar volume at the limiting negative pressure, P_M. We have found that using the values for Q and V_M given

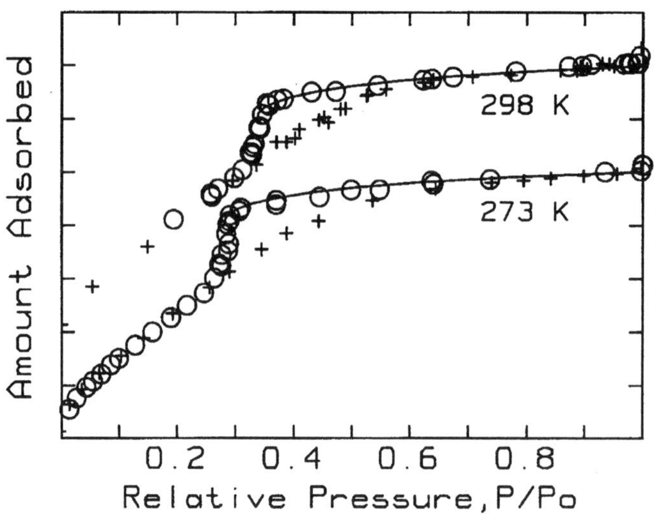

Figure 4.: Water adsorbed on DSG at 273.65 K and 298.15 K; +, adsorption; O, desorption. Isotherms are offset for clarity. The solid line through the capillary liquid plateau is calculated using eq. 9 with the parameters listed in Table 2. See text for details.

by Speedy, but with modified values for P_M, this equation provides an excellent description of the entire capillary liquid region for water adsorbed in DSG at 273.65 K and 298.15 K (see Figure 4 and [8]). These values for P_M and those calculated from the position of the lower closure point of the hysteresis loop (P_L), together with the parameters reported by Speedy are summarized in Table 2. We have also found that eq. 9 provides an excellent description of other capillary liquids [12].

A basic assumption when using this method to study the properties of capillary liquid is that its density is the same as that of bulk liquid. Calculation of the adsorbent pore volume is based on this assumption, which is supported by the observation that different liquids tend to give the same pore volume for a given adsorbent [1, 7-9, 12, 13].

TABLE 2. Water Adsorbed on DSG. Equation of State Parameters

Temperature	Lower Closure	P_L (eq.5)	P_F (eq.6)	P_B (eq.7)	P_M (eq.9)	V_M (eq.9)
K	P/P_0	MPa	MPa	MPa	MPa	$cm^3 mol^{-1}$
298.15	0.32	−157	−204	−232	−150	22.32
273.65	0.26	−171	−229	−244	−160	22.06

Furthermore, since the pore volume for a given liquid is also constant over a wide temperature range [7, 9, 13, 14] the isobaric thermal expansivity of the capillary liquid must be the same as that of bulk liquid. As well, we have shown that the isothermal compressibility and heat capacity of several capillary liquids are essentially the same as the corresponding bulk liquid properties [7, 12-14]. One major difference between bulk and capillary liquid is that the later is always associated with a large adsorbent interface (surface area, S) and a corresponding excluded volume V_X. This volume is that part of the pore volume unavailable to the centers of the adsorbed molecules by virtue of their finite size, *i.e.* $V_X=S(d_e/2)$, and this may be a significant fraction of the total pore volume. We have ignored this factor in the present work but it should be noted that its inclusion may affect some estimates for the spinodal limit [13, 14].

Finally, we note that condensation-evaporation studies on capillary liquids offer several advantages: (i) the liquid is thermodynamically stable with respect to bulk liquid, (ii) pressure within the liquid is readily changed by adjusting the pressure of the external vapour phase, and (iii) adsorbent and liquid can be thoroughly degassed prior to use. Adsorbents must have a significant fraction of their pore volume in pores that are approximately five times the diameter of the adsorbed molecules, and for practical reasons, the adsorbate must be sufficiently volatile to permit ready handling under vacuum.

Acknowledgement

This work was supported by a grant from the Natural Sciences and Engineering Research Council of Canada.

References

1. Gregg, S.J. and Sing, K.S.W. (1982) *Adsorption, Surface Area and Porosity*, 2nd ed. AcademicPress, London.
2. Sing, K.S.W., Everett, D.H., Haul, R.A.W., Moscou, L., Pierotti, R.A., Rouquerol, J. and Siemieniewska, T. (1985) Reporting Physisorption Data for Gas/Solid Systems, *Pure Appl. Chem.*, **57**, 603-619.
3. Machin, W.D. and Golding, P.D. (1989) Surface Area and Pore Structure of Four Silica Gels, *Langmuir*, **5**, 608-612.
4. Ben-Amotz, D. and Herschbach, D.R. (1990) Estimation of Effective Diameters for Molecular Fluids, *J. Phys. Chem.*, **94**, 1038-1047.
5. Fisher, J.C. (1948) The Fracture of Liquids, *J. Appl. Phys.*, **19**, 1062-1067.
6. Berezin, G.I. (1998) The Spinodal and Surface Tension of Liquids, *Zh. Fiz. Khim.*, **72**, 758-759, in translation, *Russian J. Phys. Chem.*, **72**, 663-664.
7. Machin, W.D. and Golding, P.D. (1990) Adsorption of Xenon on Silica Gel, *J. Chem. Soc., Faraday Trans.*, **86**, 171-174.
8. Machin, W.D. and Stuckless, J.T. (1985) Capillary-condensed Water in Silica Gel, *J. Chem. Soc., Faraday Trans. 1*, **81**, 597-600.
9. Machin, W.D. (1994) Temperature Dependence of Hysteresis and the Pore Size Distributions of Two Mesoporous Adsorbents, *Langmuir*, **10**, 1235-1240.
10. Stanley, H.E., Buldyrev, S.V., Canpolat, M., Mishima, O., Sadr-Lahijany, M.R., Scala, A. and Starr, F.W. (2000) The Puzzling Behavior of Water at Very Low Temperature, *Phys. Chem. Chem. Phys.*, **2**, 1551-1558.
11. Speedy, R.J. (1982) Stability-Limit Conjecture, An Interpretation of the Properties of Water, *J. Phys. Chem.*, **86**, 982-991.
12. Machin, W.D. and Stuckless, J.T. (1985) Pressure-Volume Behavior of Two Capillary Liquids, *J. Colloid Interface Sci.*, **108**, 46-49.
13. Machin, W.D. and Golding, P.D. (1987) Properties of Capillary-condensed Benzene, *J. Chem. Soc., Faraday Trans. 1*, **83**, 1203-1212.
14. Machin, W.D. and Golding, P.D. (1987) Adsorption of n-Butane on Silica Gel, *Langmuir*, **3**, 346-349.

ON THE TRICRITICAL POINT OF THE ISOTROPIC – NEMATIC TRANSITION IN A ROD-LIKE MESOGEN HIDDEN IN THE NEGATIVE PRESSURE REGION

SYLWESTER J. RZOSKA AND ALEKSANDRA DROZD-RZOSKA
Silesian University, Institute of Physics, ul. Uniwersytecka 4, 40-007 Katowice, Poland

Abstract. A presentation of experimental results suggesting fluidlike, spinodal and tricritical nature of the isotropic – nematic transition are given. Results presented based on broad-band dielectric spectroscopy and strong electric field induced changes of dielectric permittivity. Results above as well the obtained systematic changes of the temperature discontinuity of the isotropic – nematic transition suggest the possible existence of a tricritical point which may be experimentally reached due the application of negative pressures. The presented picture for the I-N transition correlates with the recent general model for liquid-liquid transition proposed by Tanaka [Phys. Rev. E68, 6968 (2000)].

1. Introduction

Although the validity of the negative pressure idea is well recognized for liquids and critical, binary mixtures or blends [1-5], the first experimental negative pressure result for a liquid crystalline compound was only recently reported. Manjuladevi et al. [6] placed droplets of two commercial mesogens, p-cyanophenyl trans-4-butyl cyclohexane carboxylate and a mixture of n-octylcyanobiphenyl (8CB) and octyloxycyanobiphenyl (8OCB) in a matrix of a glassformer (sucrose). Next they reported the occurrence of negative pressures from -0.22 MPa to -1.22 MPa, manifested in the shift of the nematic – isotropic (I-N) clearing temperature. The values of negative pressures were estimated from dependencies of dT_{IN}/dP known for positive pressures. The application of negative pressure for a set of liquid systems such as water, polymer blends, critical mixtures lead to an essentially new insight into their properties, particularly concerning equations of state and loci of critical or spinodal lines. Hence, the question arises whether such results may be expected for liquid crystalline compounds. Surprisingly, the situation for the I-N transition, presumably the simplest phase transition in liquid crystalline compounds, is puzzling. In the important monographs on the physics of liquid crystals the situation for the I-N transition seems to be clear. It is presented as one of the best examples of the application of a simple mean field description, most often in terms of the phenomenological Landau – de Gennes (LdG) model. For the isotropic phase it assumes the following description of the free energy density [7-10]:

$$g(P,T) - g_o(P,T) = \frac{3}{4} AS^2 - \frac{1}{4} BS^3 + \frac{9}{16} CS^4 \tag{1}$$

117

A.R. Imre et al. (eds.), Liquids Under Negative Pressure, 117–125.
© 2002 *Kluwer Academic Publishers. Printed in the Netherlands.*

118

where $g_0(P,T)$ is the isotropic part of the free energy density, S denotes the scalar parameter to which the orientational order parameter is reduced due to the rod like symmetry of molecules. Coefficients B and C are constant. The coefficients $A=a_T(T-T^*)$ for $P=const.$ and $A=a_P(P-P^*)$ for $T=const.$ (T^*, P^*) are coordinates of a hypothetical point of continuous phase transitions and the supercooled/superpressed isotropic phase.

The point (T^*, P^*) cannot be reached experimentally since a discontinuous transition at $T_{IN}>T^*$ or $P_{IN}<P^*$ occurs. The value of $\Delta T^*(P=0.1MPa)=T_{IN}-T^*$ is a basic parameter characterizing the I-N transition. The relatively weak discontinuity of the I-N transition results in the appearance of the nematic phase precursors in the isotropic liquid. Their appearance is easily registered by inhomogeneity-sensitive methods as the Kerr effect (KE), Cotton –Mouton effect (CME), light scattering (I_L) or recently the nonlinear dielectric effect. Successful parameterization of temperature behavior of $I_L(T)$, $KE(T)$, $CME(T)$ is taken as the main argument for the mean field nature of the I-N transition [7-9]. However, the full set of available experimental results is considered the agreement between predictions of the mean-field description and experiments is surprisingly poor. This disagreement is particularly visible when a set of novel high-resolution experiments conducted in the last decade is taken into account (Table I in ref. [11] and refs. [12-14])). In the years 1995/96 a novel proposal for the description of the I-N transition was postulated. Its was analyzed using the equation of state resembling the ones applied for the gas-liquid critical point or the critical consolute point in binary mixtures. This gave $\Delta T^*=1$-3K, $\gamma=1$ and $\alpha=1/2$ for the susceptibility and the specific heat critical exponents and for the dimensionality $d=3$ in very good agreement with the experiment [15, 16]. The recent comparison of the dynamic properties near the critical consolute point and for the I-N transition also pointed to the clear application of the fluidlike hypothesis for these properties [13, 14]. Hence one may put forward a hypothesis of a possible common description for such different systems as binary mixtures near the critical consolute point or one-component fluids near the gas - liquid critical point and mesogenic compounds near the I-N transition. It should be noted that there is recent experimental evidence for the tricritical nature of the I-N transition [12, 17]. The fluidlike description made both the "critical" and the "tricritical" hypothesis possible [18].

Despite some progress in the description of the I-N transition the question still remains why the discontinuity ΔT^* of the I-N transition is so small. This question was not put forward probably due to the phenomenological character of the existing theories.

In this paper results are presented for the spinodal, tricritical nature of the I-N transition suggesting that the application of negative pressure may cause that $\Delta T^* \approx 0$, i.e. it reaches the limit of the continuous tricritical I-N transition. The results presented were based on studies of dielectric permittivity and the nonlinear dielectric effect (NDE) in 5CB, known as the most tested "classical" liquid crystalline compound. The mentioned $NDE=\Delta\varepsilon^E/E^2$ describes changes of the dielectric permittivity ($\Delta\varepsilon^E$) induced by the application of a strong DC electric field E. The applied experimental techniques are described in details in refs. [10, 13, 19]. All novel results presented in this paper were obtained for 4-cyano4-n-alkylbiphenyls (nCB: C_nH_{2n+1}-Ph-Ph-CN, Ph denotes the phenyl ring) synthesized by Krzysztof Czuprynski and Roman Dabrowski Dept. of Chemistry at the Military Technical University (WAT), Warsaw, Poland. They were purified to decrease the residual ionic impurities to the lowest possible level.

2. Results and Discussion

Fig. 1 shows results of measurements of static dielectric permittivity, a magnitude directly related to the order parameter tensor $Q_{ij}=\varepsilon_{ij}=(S/2)[3n_in_j-\delta_{ij}]$ which is the symmetric traceless tensor. For rod-like molecules, as for 5CB, only two components of directors n_i, n_j should be taken into account and hence $\varepsilon_{||}=\varepsilon_{kk}$ and $\varepsilon_\perp=\varepsilon_{kk}=\varepsilon_{ij}$ [7]. For 5CB the "molecular" anisotropy is reached for $T<<T_{IN}$ when $S\rightarrow1$ and then to $\Delta\varepsilon=\varepsilon_{||}-\varepsilon_\perp=9$-10 [9].

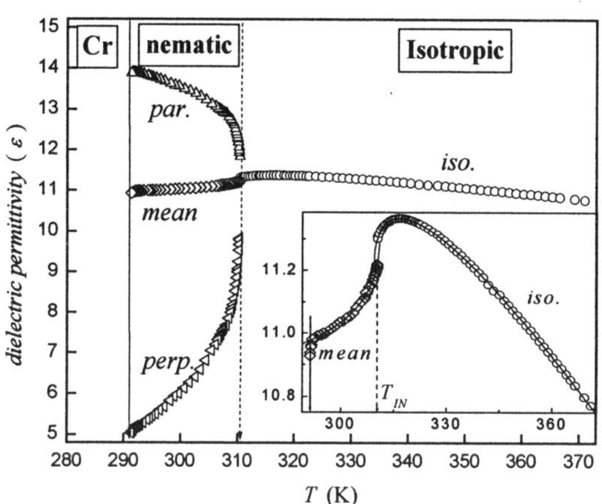

Figure 1. Results of measurements of static dielectric permittivity (f= 10 kHz) in the isotropic and nematic phase, respectively. In the nematic phase sample was oriented in the way parallel and perpendicular to the long axis of the molecule using a strong magnetic field (equal to about 1 *Tesla*). The inset shows, in the enlarged scale, the behavior of the mean and of the isotropic dielectric permittivities. Solids lines are parameterized by relations (2a) and (2b), respectively. The dashed arrow shows the nematic clearing temperature (I-N transition) and the sold arrow points to the crystallization of the sample.

The inset in Fig. 1 shows in an enlarged scale of two dependencies from Fig. 1. They are the temperature dependencies of the static dielectric permittivity in the isotropic phase and of the mean permittivity in the nematic phase. It is clearly visible that solid lines are well parameterized by (the inset in Fig.1):

$$\varepsilon_{iso.}(T)=\varepsilon^* + a^+\left|T-T^*\right| + A^+\left|T-T^*\right|^{1-\alpha} \tag{2a}$$

$$\varepsilon_{mean}(T)=\varepsilon^{**} + a^-\left|T^{**}-T\right| + A^-\left|T^{**}-T\right|^{1-\alpha'} \tag{2b}$$

where $\varepsilon^* = 11.19\pm0.02$, $a^+ = -0.022\pm0.001$, $A^+ = 0.12\pm0.01$ $T^* = 309.25K\pm0.2$, $\alpha = 0.49\pm0.02$ for the isotropic (liquid) phase and $\varepsilon^{**} = 11.28\pm0.06$, $a^- = 0.0054\pm0.001$, $A^- = -0.11\pm0.03$, $T^{**} = 310.9K\pm0.2$, $\alpha' = 0.51\pm0.06$ for the mean permittivity in the nematic phase. The mean permittivity is defined by $\varepsilon_{mean} = \varepsilon_\perp/3 + 2\varepsilon_{||}/3$, to take into account the rod-like symmetry of molecules.

The form of these dependencies is isomorphic to relations known for critical, binary mixtures in the homogeneous phase and for the diameter of the coexistence curve. The validity of these dependencies was recently shown for yet another nematogen, 7OCB (n-heptyloxycyanobiphenyl) [12].

On approaching T_{IN} changes of $\Delta\varepsilon(T)$ follows the scalar order parameter $S(T)$ and hence from relation (1):

$$\Delta\varepsilon(T) = \Delta\varepsilon^{**} + B\left|T^{**} - T\right|^{\beta} \tag{3}$$

where $\Delta\varepsilon^{**}$ and T^{**} are extrapolated values from data in the nematic phase which for the mean-field approximation may be considered as the limit of the superheated nematic phase.

The discontinuity of the I-N transition causes that the application of relation (1) may be difficult particularly if there not enough data available in the immediate vicinity of T_{IN} [12]. Fig. 2 shows, following ref. [12], that this problem may be overcome by considering the derivative of experimental data $\Delta\varepsilon(T)$, which reduce the fit only to T^{**} parameter.

$$\frac{d\Delta\varepsilon(T)}{dT} \propto \left|T^{**} - T\right|^{\beta-1} \tag{4}$$

The exponent $\beta=0.25\pm0.04$ obtained is definitely beyond the simple mean-field models ($\beta=0.5$) and agrees fairly well with the expectations for the tricritical model: $\beta=1/4$ [18].

Despite results presented above a question may arise whether indeed anomalies shown in relations (3) or (4) have the mentioned relationship to the behavior observed in critical mixtures.

The positive answer suggests the validity of the equation for amplitudes of specific heat above and below the critical point, also denoted by A^+ and A^-, and the critical exponent α [20]:

$$\frac{A^+}{A^-} \approx 1 - 4\alpha \tag{5}$$

The obtained values of exponent α are characteristic for the tricritical model (TCP) as well as for the set of other models [9]. However, the obtained value $\left|A^+/A^-\right| \approx 1$ was calculated theoretically just for TCP model [21]. The obtained ratio of discontinuities:

$$\frac{\Delta T^*}{\Delta T^{**}} = 0.45 \pm 0.1 \tag{6}$$

can be also obtained theoretically from the fluidlike theoretical analysis within the TCP model [18].

It is noteworthy that dependence (5) is also valid when substituting values of parameters A^+, A^-, ΔT^* and ΔT^{**} obtained for 7OCB from ref. [12].

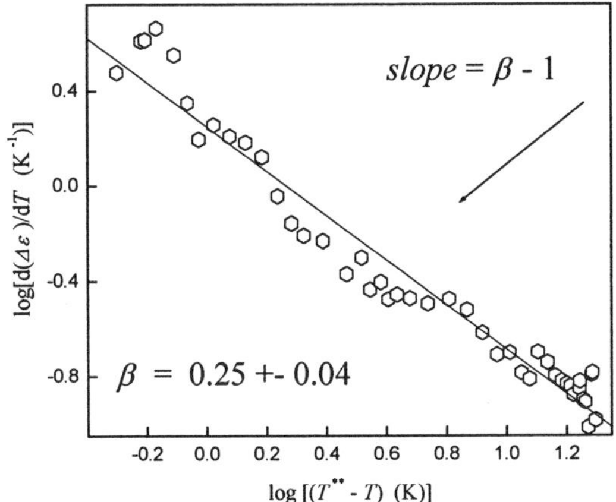

Figure 2, The temperature derivative of the order parameter experimental dependence based on data for the nematic phase presented in Fig. 1.

Figure 3 shows the evolution of ΔT^* for the whole series of *n*-cyanobiphenyls (*n*CB). In this series only the Cr-N-I phase sequence occurs up to 7CB. For 8CB, 9CB and 11CB the sequence is Cr-SmA-N-I, (SmA denotes the Smectic A phase), the width of the N phase is equal about 12 K in 8CB, 1.5 K in 9CB and 0.5 K in 11CB. For 10CB, 12CB and 14CB the sequence is Cr-SmA-I. For 4CB, 3CB and shorter molecules, the crystallization directly from the isotropic phase takes place (Cr-I sequence). Fig. 3 contains data from ref. [10] with novel results for 14CB, 4CB. Regarding 14CB, it is a new, hardly tested up to now member of *n*CB series with the isotropic – Smectic A (I-SmA) transition [details will be published soon].

Results for 4CB was obtained from static dielectric permittivity and the static nonlinear dielectric effect (*NDE*) measurements. They are presented in Fig. 4. It is clearly visible that the pretransitional behavior, associated with the appearance of prenematic fluctuations, dominates the properties of the isotropic phase even if crystallization occurs instead of the I-N transition.

Fig. 4 clearly shows different molecular origins of linear and nonlinear dielectric permittvities ($\varepsilon_{iso.}(T)$ and $NDE(T)$) dependencies. Changes of $\varepsilon_{iso.}(T)$ are associated with the cancellation of permanent dipole moments ordered in an antiparallel way in the prenematic fluctuations. The number of molecules associated with them strongly increase on cooling and hence $\varepsilon(T)$ drops significantly. The nonlinear dielectric effect only registers the appearance of inhomogeneities, i.e. prenematic regions in the fluidlike surrounding. It is described by 10, 13]:

$$NDE(T) = C \langle \Delta M^2 \rangle_V \chi = \frac{A_{NDE}}{\left(T - T^*\right)^\gamma} = \frac{2}{3a} \varepsilon_0 \frac{(\Delta\varepsilon)^2}{T - T^*} \tag{7}$$

It is noteworthy that for *NDE* the analysis is reduced only to the linear regression of direct experimental data.

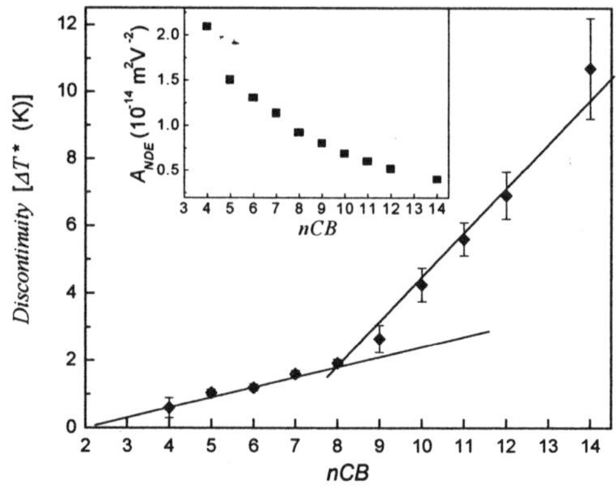

Figure 3 , The evolution of discontinuity of the I-N transition (*ΔT**) in the whole available nCB homologous series. Solid lines are guides for eyes. The inset shows the analogous evolution of the *NDE* amplitude, proportional to the amplitude of susceptibility.

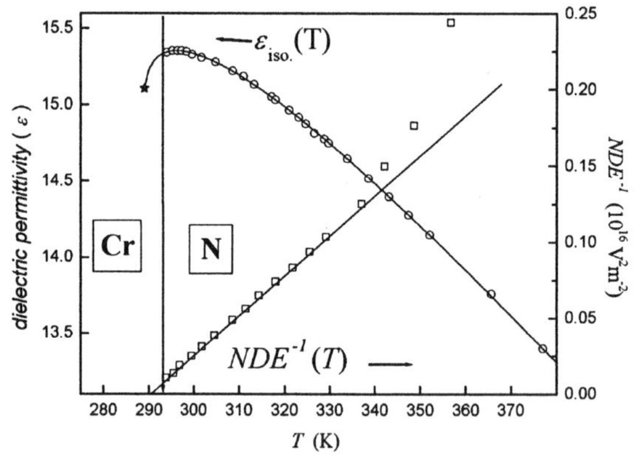

Figure 4 , Results of static dielectric permittivity (f = 20 kHz) and static nonlinear dielectric effect (f = 25 kHz) measurements in the isotropic phase of 4CB, up to the crystallization limit. Solid line are parameterized by relations (2a) and (8). For dielectric permittivity $\varepsilon^* = 15.01 \pm 0.08$, $a^+ = -0.04 \pm 0.01$, $A^+ = 0.20 \pm 0.03$, $T^* = 289.2K \pm 0.2$, $\alpha = 0.49 \pm 0.1$. For *NDE* the slope of the straight line determines the amplitude from relation (8): $A_{NDE} = 210(10^{-16} m^{-2}V^{-2})$, the exponent $\gamma = 1$. Extrapolation into the crystallization shows the position of the spinodal temperature.

The systematic decrease of $\Delta T^*(T)$ on decreasing value of n (or the length of the molecule) may be associated with the fact that for shorter nCB molecules the volume available for rotation of rod-like molecules in the nematic phase increases. Consequently the distance between symmetries of the isotropic and nematic phase decreases. Unfortunately, for 4CB the crystallization on cooling in the isotropic occurs despite the existence of strong precursors of the hypothetical nematic phase.

However, one may expect that reaching the hypothetical limit $\Delta T = 0$ may be possible by applying negative pressures. This may be particularly effective in 4CB due to a very strong dependence of $dT_{Cr}/dP \approx 1\ KMPa^{-1}$ whereas dependencies for $T^*(P)$ and $T_{IN}(P)$ in nematic nCB (known for 5CB, 6CB, 7CB) are much weaker [10, 11, 16]:

$$T_{IN}(P) = T_{IN}(P = 0.1MPa) + 0.37P \tag{8}$$

$$T^*(P) = T^*(P = 0.1MPa) + 0.355P \tag{9}$$

Extending this relation to 4CB one may expect $\Delta T^* \approx 0$ at about $-(40 - 60)$ MPa.

Taking into account the fluidlike, spinodal, tricritical character of the I-N transition $\Delta T = 0$ point will be probably associated with the end of the spinodal line. Below this point continuous changes, without a phase transition, from the nematic phase may be possible, similarly as for the gas-liquid critical point. Results for temperature studies in disorder-induced surrounding, where also low-density state of 5CB was induced, are worth mentioning here. Under such condition, a continuous "stretched" phase transformation from the nematic to the isotropic phase was observed [7, 22].

The hypothesis for the fluidlike character of the I-N with the tricritical point hidden in the nematic pressure region coincides well with the recent general model for liquid-liquid transitions proposed by Tanaka [23]. He assumed that in any liquid, apart from a density related local order parameter, additional cooperative medium ranged ordering, i. e. unique locally favored structures in a normal liquid surrounding, exist. This assumption made a preliminary proposal for a common description of such different phenomena as liquid-liquid transition in one-component liquids, critical phenomena or the so-called Fisher cluster in supercooled liquids possible. The latter is related to some excess properties for certain physical magnitudes observed well above the glass transition temperature, probably due to the appearance of glasslike inhomogeneities in the fluidlike surrounding. Tanaka related this phenomenon to the hypothetical spinodal line below T_g, associated with a critical point hidden in the negative pressure region. In the opinion of the authors the I-N transition may be included into Tanaka's model. Fig. 5 shows the behavior of dielectric relaxation times for peaks of dielectric loss curves in isotropic 5CB. It is clearly visible that it is definitely non-Arrhenius to be expected within the simple mean-field picture. The results obtained can be well portrayed by the Vogel-Fulcher-Tamman (VFT) [24] relation:

$$\tau = \tau_o^{VFT} \exp\left(\frac{DT_0}{T - T_0}\right) \tag{10}$$

where $\tau_o^{VFT} = 0.172ns$, $T_o = 218K \pm 5$ and $D = 2.26 \pm 0.2$. Basing on this dependence the glass temperature may expected at about $T_g \approx 240K$.

Such a description coincides with the broadening of the dielectric loss curves on cooling in the isotropic nCB. The VFT dependence and the non-Debye character of loss curves are commonly taken as canonical features of supercooling, glassforming liquids [24]. The "glassy" dynamics of isotropic 5CB is also supported by the possibility of describing the relaxation times within the mode-coupling theory (MCT):

$$\tau = \tau_0^{MCT}(T - T_X)^{-g} \tag{11}$$

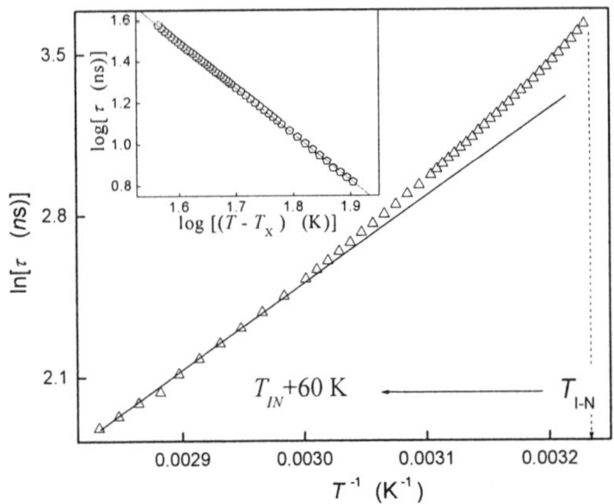

Figure 5 Temperature dependencies of dielectric relaxation times in the isotropic phase of 5CB, taken from recuprocals of peaks frequencies of dielectric loss curves given in ref. [13,14]. Solid curve is described by the VFT dependence (1). The straight line presents the Arrhenius dependence.The inset shows the MCT "critical-like" (relation (11)): the exponent $g = 2.22 \pm 0.05$, $\ln \tau_0^{MCT} = 5.06$ and for the crossover MCT "critical-like" temperature $T_X = 273K \pm 1$.

It is worth recalling that Tanaka [23] predicted that in supercooled liquids pressure should increase the distance from the hypothetical spinodal curve (the appearance of Fisher inhomegeneities should drop with rising pressure). Moreover, the evolution of relaxation times associated with these inhomogeneities should be described by the power exponent $y = 1$ and they should be associated with the non-conserved order parameter $z = 2$. All these properties clearly occur in the isotropic phase of nCB.

In the opinion of the authors the above may suggest that the I-N transition can be included in Tanaka's picture as a specific glassy liquid for which the glassy state cannot be reached due to an earlier phase transition. For a nematic liquid crystalline compound the hypothetical spinodal curve is located well above the hypothetical glass transition.

Concluding we present arguments for the fluidlike, spinodal and tricritical nature of the I-N transition based on novel and most recent reference results. They may suggest that owing to the application of negative pressures reaching a continuous I-N transition for rod-like calamitic liquid crystals, never observed up to now in temperature or (positive) pressures tests, may be possible. Moreover, it has been shown such hypothesis correlates with the recent general model for liquid-liquid transition proposed by Tanaka.

3. Acknowledgement

The presented research was supported by KBN Poland, grant P03B 022 020.

4. References

1. Netz, P. A., Starr, F. W., Stanley and H. E., Barbosa, M. C. (2001), Static and dynamic properties of stretched water, J. Chem. Phys. 115, 344-348.
2. Mercury, L., and Tardy, Y., Negative pressure of stretched liquid water (2001) Geochemistry of soil capillaries, *Geochim. et Cosmica Acta,* **65**, 3391-3408.
3. Imre, A., and Van Hook, W. A. (1998) Liquid-liquid equilibria inpolymer solutions at negative pressure, *Chem. Soc. Rev.* **27**, 117-123.
4. Imre, A., Martinas, K. and Rebelo, L. P. N. (1998) Thermodynamics of negative pressures in liquids, *J. Non-Equilib. Thermodyn.* **23**, 351-375.
5. Imre, A. R., Melnichenko, G., Van Hook and W. A., Wolf, B. A. (2001) On the effect of pressure on the phase transition of polymer blends and polymer solutions: oligostyrene - n-alkane systems, *Phys. Chem. Chem. Phys.* **3**, 1063-1066.
6. Manjuladevi, V., Pratibha, R., and Madhusudana, N. V. (2002) Phase transitions in liquid crystals under negative pressures, *Phys. Rev. Lett.* **88**, 055701(1-4).
7. Vertogen, G.,and de Jeu, W. H. (1986) *Thermotropic Liquid Crystals, Fundamentals – Springer Series in Chemical Physics* (Springer Verlag, Berlin, 1986).
8. D. Demus, D., Goodby, J., Gray, G. W., Spiess, H. W. , and Vill, V., (editors) (1998*) Handbook of Liquid Crystals vol. 1, Fundamentals* (Wiley-VCH, Weinheim).
9. Chandrasekhar, S. (1994) *Liquid Crystals* (Cambridge Univ. Press., Cambridge).
10. Drozd-Rzoska, A., Rzoska, S. J., and Czuprynski, K. (2000) Phase Transitions from the isotropic to liquid crystalline mesophases studies by linear and nonlinear dielectric permittivity, *Phys. Rev.* **E61**, 5355-5360
11. Drozd-Rzoska, A., Rzoska, S. J., and Ziolo, J., (2000) The fluidlike and critical behavior of the isotropic – nematic transition appearing in linear and nonlinear dielectric studies, *Acta Phys. Polon.* **A98**, 431-437.
12. Rzoska S. J., Zioło, J., Sułkowski, W., Jadżyn, J. and Czechowski, G., (2001) Fluidlike behavior of dielectric permittivity in a wide range of temperature above and below the nematic-isotropic transition (2001) *Phys. Rev.* **E64**, 052701(1-4).
13. Drozd-Rzoska, A., and Rzoska, S. J. (2002) Complex dynamics of isotropic 4-cyano-4-*n*-pentylbiphenyl in linear and nonlinear dielectric relaxation studies, *Phys. Rev.* **E65**, 041701(1-9).
14. Rzoska, S. J., Paluch, M., Drozd-Rzoska, A., Janik, P., Zioło, J., and Czuprynski, K. (2002) Glassy and fluidlike behavior of the isotropic phase of mesogens in broad-band dielectric, *Europ. Phys. Journal E* in print: ref MsE01177.
15. Mukherjee, P. K. (1996) Critical region at the nematic - isotropic phase transition, *Mod. Phys. Lett.* **B10**, 771-775.
16. Drozd-Rzoska, A., Rzoska, S. J. and Ziolo, J. (1996) Phys. Rev. E Critical behaviour of dielectric permittivity in the isotropic phase of nematogens, *Phys. Rev.,* **E54** , 6452 - 6456.
17. Marinelli, M. and Mercuri, F., Effects of fluctuations in the orientational order parameter in the cyanobiphenyl (*n*CB) homologous series, *Phys. Rev.* **E61**, 1616 -1628 (2000).
18. Mukherjee, P. K. (1998) Evidence of the tricritical behavior at the nematic isotropic transition, *Int. J. Mod. Phys.* **B12**, 1585-1599.
19. Rzoska, S. J., Drozd-Rzoska A., Górny, M., Jadżyn, J., and Zioło, J. (2002) Nonlinear dielectric effect in superpressed chiral isopentylcyanobiphenyl (5*CB), *J. Non-Cryst. Solids,* Special issue for the 4th Int. Disc. Meeting on Complex Liquids, Hersonissos 2001.
20. Hohenberg, P. C., Aharony, A., Halperin, B. I.. and Siggia, E. D., Two-scale-factor universality and the renormalization group, *Phys. Rev.* **B13**, 2986 -2991(1976).
21. Stine, K., and Garland, C. W. (1989) Calorimetric study of nematic to smectic-*A* tricritical behavior, *Phys. Rev.* **A39**, 3148-3156 .
22. Bellini, T., Clark, N. A., Degiorgio, V., Mantegazza, F., and Natale, G. (1998) Light-scattering measurements of the nematic correlation length in a liquid crystal with a quenched disorder *Phys. Rev.* **E57**, 2996-3006 (1998).
23. Tanaka, H. (2000) General view of the liquid – liquid transition, *Phys. Rev.* **E62**, 6968-6976
24. Dhont, E. (2001) *The Glass Transition: Relaxation Dynamics in Liquids and Disordered Materials,* Springer, Berlin.

NEGATIVE PRESSURE DEVELOPMENT DURING CRYSTALLIZATION OF POLYMERS

A. GALESKI AND E. PIORKOWSKA
Centre of Molecular and Macromolecular Studies, Polish Academy of Sciences, 90 363 Lodz, Poland

1. Introduction

Polymers crystallize from melt usually in the form of polycrystalline aggregates growing outwards from primary nuclei. In the absence of temperature gradient spherulites grow radially until the impingement with neighbors. In most cases polymer crystallization is associated with a decrease of the specific volume by several percents and the resulting shrinkage is compensated by a flux of a polymer melt towards the crystallization front.

In the course of crystallization spherulites occlude pockets of melts. In such regions the volume change can not be compensated by the inflow of melt thus, leading to a negative pressure buildup. It follows then that the polymer crystallization in bulk or in thin films between stiff plates is associated with a local development of negative pressure. The negative pressure increases with the progression of crystallization until the melt in the locked pocket is fully crystallized or a cavitation of melt occurs. The cavitation releases the stresses, instead it results in the formation of holes between spherulites. Both the negative pressure and/or the cavities weaken a polymer, therefore the occluded pockets of melt are weak spots of spherulitic structure.

While the effect of positive pressure on polymer crystallization and melting was studied by many authors in the past, the knowledge about the negative pressure effect is rather scarce, since the application of negative pressure to polymer in the conditions allowing for studies of crystallization is difficult. The weak spots are natural laboratories allowing to study the cavitation and crystallization under negative pressure.

Apart from the cavitation, the negative pressure lowers the equilibrium melting temperature of a polymer, hence it decreases the supercooling and leads to changes in crystallization and supermolecular structure of a polymer. In the paper we describe the results of the studies of formation, crystallization and cavitation in pockets of melt occluded between spherulites during polymer crystallization.

2. Volume deficiency at polymer interfaces

The scheme of polymer melt occlusion between spherulites is shown in Fig.1a. During the crystallization with a free surface the volume deficiency in weak spots results in thinning of a polymer and the formation of holes [1] which is clearly seen in the micrograph in Fig.1b. When a polymer crystallizes in contact with another molten polymer, the negative pressure causes the influx of the melt between the crystallizing spherulites, schematically shown in Fig.2a. Thus, the final interface has a complicated shape with intrusion of the second polymer between the spherulites of first crystallized

A.R. Imre et al. (eds.), Liquids Under Negative Pressure, 127–136.
© 2002 *Kluwer Academic Publishers. Printed in the Netherlands.*

128

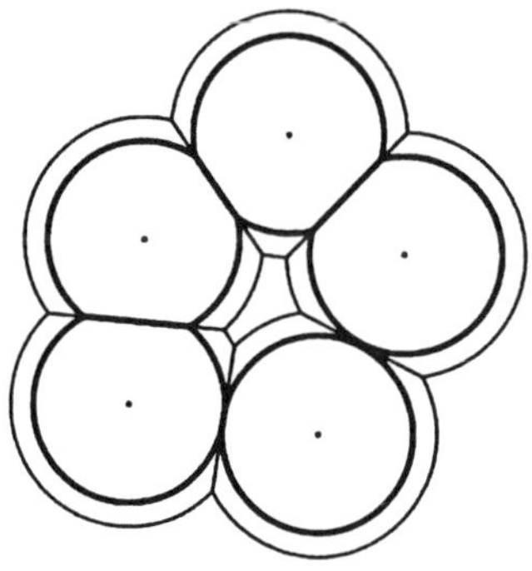

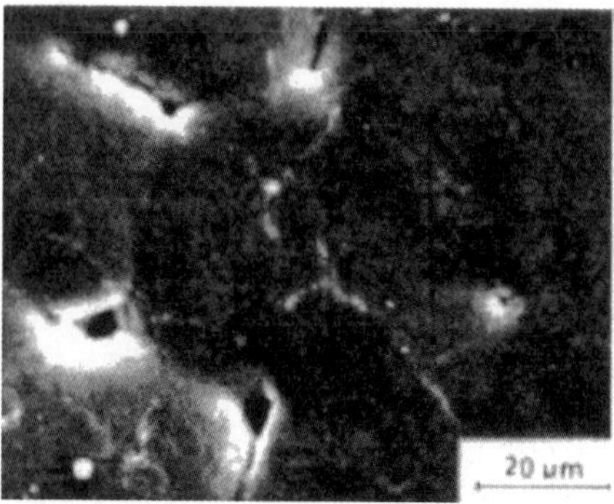

Figure 1. Weak spots in thin film: scheme of weak spot formation. (a), scanning electron micrograph of POM film crystallized isothermally with free surface (b) [1].

component extending on millimeter distance (Fig.2b). Such phenomenon was first found by us [2] at interfaces of isotactic polypropylene (iPP) with high density polyethylene (HDPE) and low density polyethylene (LDPE) and poly(ethylene oxide) (PEO). The same effects was described later by Wool [3] for interfaces of iPP with HDPE and linear low density polyethylene (LLDPE). Due to influxes the bonding between two polymers is much stronger than by the adhesion alone. It leads to erroneous results of adhesion force measurements if the influxes are not accounted for. The influxes of molten polymer between the spherulites of the crystallizing component explains also the often observed emulsification effect at the interface between crystallizable polymers.

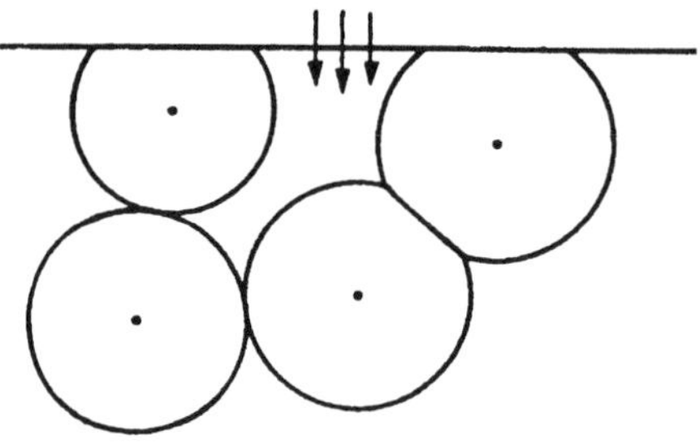

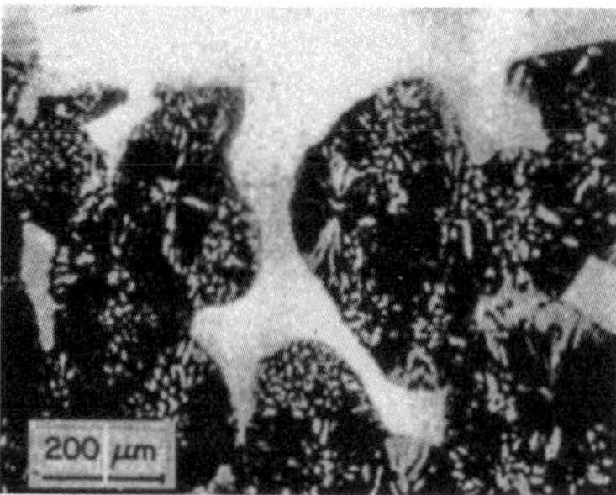

Figure 2. Influx of a polymer melt between spherulites of the other crystallizing polymer [2]: schematic representation (a), light micrograph (crossed polars) of iPP/LDPE sandwich section.

The formation of influxes takes place also in blends and can cause deformation of the inclusions of the second polymer dispersed in the crystallizing matrix. The effect is schematically illustrated in Fig. 3a and demonstrated for the blend of iPP and LDPE in micrograph in Fig. 3b [2].

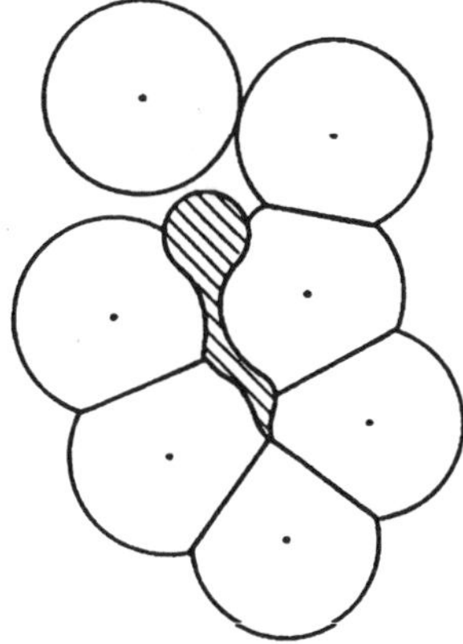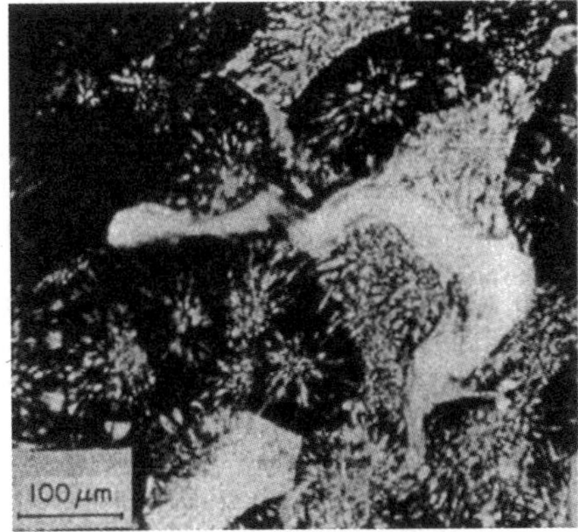

Figure 3. Deformation of inclusion in blends during crystallization of a polymer matrix [2]: schematic representation (a), light micrograph (crossed polars) of deformed LDPE inclusion in iPP matrix (b).

3. Crystallization in weak spots

The spherulite growth rate of a polymer having the glass transition temperature T_g and equilibrium melting temperature $T_m^°$, under atmospheric pressure, is described as a function of the temperature T by the Hoffman formula [e.g. 4]:

$$G = G_o \exp\{-U[R(T-T_g+30K)]^{-1}\} \exp\{-K_g[RT\Delta Tf]^{-1}\} \qquad (1)$$

where U is the activation energy of polymer chains transport, ΔT denotes the supercooling ($T_m^°- T$), K_g and G_o are the coefficients dependent on the crystallization regime; K_g is proportional to $\sigma\sigma_e T_m^°\Delta h_f^{-1}$, where σ and σ_e are the lateral and fold surface energies, Δh_f is the heat of fusion per unit volume of a crystal at $T_m^°$, f is a correction term originating from the dependence of heat of fusion on T equal to $2T(T+T_m^°)^{-1}$ and usually being of the order of unity.

The lamellae thickness, L, is described by the relation [e.g.4]:

$$L = \gamma \, [2\sigma_e T_m^°(\Delta h_f \quad \Delta T)^{-1}+\delta] \qquad (2)$$

where and γ denotes the degree of thickening and $\delta \ll 2\sigma_e T_m^°(\Delta h_f\Delta T)^{-1}$ at low and medium supercooling.

Although in general the pressure influences the surface energies , the heat of fusion, Δh_f and the glass transition temperature, T_g [e.g.5,6], the most significant effect on the spherulite growth rate G is due to the pressure dependence of $T_m^°$. The change in $T_m^°$ is reflected primarily in the change of isothermal supercooling. Thus, the negative pressure decreases the spherulite growth rate and increases L by lowering the isothermal supercooling.

The decrease of the spherulite growth rate, G, observed in weak spots by light microscopy during isothermal crystallization of iPP, poly(methylene oxide) (POM) and poly(ethylene adipate) (PAE) [7,8]. The spherulite growth rate G measured for the spherulites forming a weak spot decreases in the same way. In contrary, if a spherulite participates in a formation of several weak spots the growth rate of its fragments being

130

within different weak spots decreases differently with time. It must be pointed out that the source of a negative pressure in a weak spot is the volume deficiency which increases with the progression of crystallization. The progression of crystallization can then be defined as the ratio of momentary volume of melt in the weak spot, S, to the initial volume of that weak spot, S_i, at the moment of its occlusion by spherulites. The exemplary dependencies of the spherulite growth rate on the progression of crystallization in weak spots, S/S_i, are shown in Fig. 4. In the absence of cavitation the spherulite growth rate, G, diminishes continuously with the progression of crystallization until the end of crystallization in the weak spot. If cavitation occurs the negative pressure is released and the spherulite growth rate abruptly increases. This rapid elevation of the spherulite growth rate after cavitation proves that the drop of observed in weak spots originates primarily from the negative pressure. The decrease of the growth rate can serve as the measure of negative pressure. Knowing the dependence (1) and assuming that the drop of the spherulite growth rate is caused primarily by the decrease of isothermal supercooling one can calculate the depression of equilibrium melting temperature, $\Delta T_m^{\,o}$, from the momentary value of the spherulite growth rate.

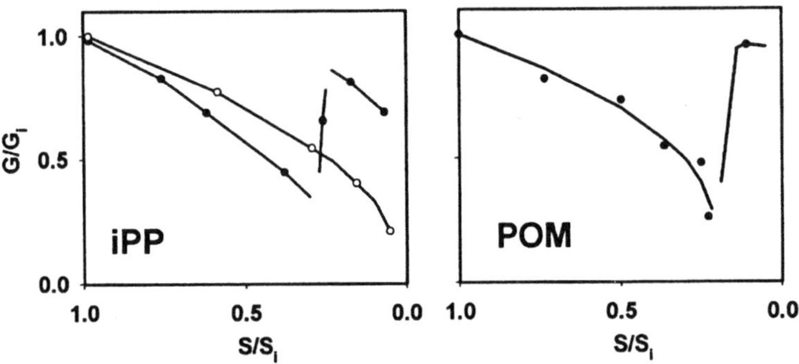

Figure 4. Ratio of momentary growth rate over initial growh rate plotted against progression of crystallization in weak spots in iPP [7] and POM films during isothermal crystallization.

Apart from the decrease of spherulite growth rate, the reduced supercooling increases the lamellae thickness. The melting temperature of polymer crystals, T_m, depends on their thickness, L, and it is expressed by the well known relation:

$$T_m = T_m^{\,o}[1 - 2\sigma_e(\Delta h_f L)^{-1}]\qquad(3)$$

When the stresses in the crystallized polymer are released, for example by removing the upper glass in the case of films or by sectioning a polymer bulk, the thicker crystals melt at a higher temperature. This is because if the melting is performed at atmospheric pressure T_m is governed by the lamellae thickness while the value of $T_m^{\,o}$ in eq.(3) is now taken under atmospheric pressure.

The increase of T_m in weak spots was studied in thin films of POM, PAE, PEO , iPP of two different brands and HDPE by the polarized light microscopy [7,9,10]. The increase in T_m around holes was also found in thin sections of 2 mm thick plates of isothermally crystallized   iPP [10]. The effect is clearly demonstrated in the micrographs of iPP thin film and of a section of iPP bulk sample shown in Fig .5. The samples were crystallized isothermally and then cooled to room temperature. The melting of sections and films after removal of an upper glass was studied. In order to avoid the influence of annealing on melting the samples were heated initially at a high rate and only last few degrees before reaching the desired temperature were passed at a reduced rate. The dependence of T_m on ΔT, measured experimentally, allowed to determine ΔT corresponding to the elevated T_m and to calculate the depression of equilibrium melting temperature , $\Delta T_m^{\,o}$.

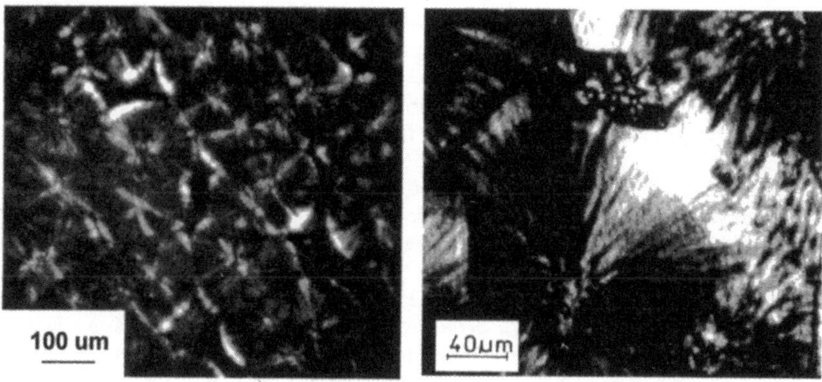

Figure 5. Light micrographs (crossed polarizers) of isothermally crystallized iPP during melting: thin film (left) [9] and thin section (right) [10].

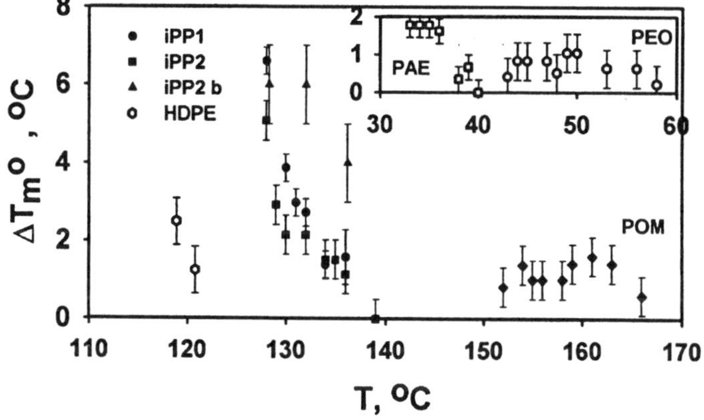

Figure 6. $T_m^{\,o}$ decrease measured by studies of melting behaviour in weak spots in films and in thin sections (iPP2b), data from [9,10].

Extrapolation of the dependence of $T_m^{\,o}$ on the positive pressure to negative pressure range allows for the determination of the negative pressure value in weak spots crystallized at various temperatures. Based on $\Delta T_m^{\,o}$ the negative pressure is calculated either from the depression of the spherulite growth rate or from the elevation of T_m. We have to point out that the measurements of T_m have a limited application since the dependence of L on ΔT weakens with the increasing ΔT. However, the measurements of involving a direct observation of crystallization are not possible in the case of crystallization in bulk. We have to point out that the $\Delta T_m^{\,o}$ values determined for weak spots crystallizing without cavitation are most probably somewhat lower than the real

132

final values, since the precise measurements at the very end of solidification are difficult. In each polymer studied varying values of $\Delta T_m^{\,o}$ were found in weak spots for each crystallization temperature which indicates diverse levels of negative pressure. The dependence of the maximum $\Delta T_m^{\,o}$ values on crystallization temperature, T, measured based on the elevation of T_m are collected in Fig.6. The highest values were found in iPP. In the case of iPP these values are in agreement with those determined from the depression of the spherulite growth rate [7,8] and correspond to negative pressure values of nearly -20 MPa. In sections of bulk samples the decrease of $\Delta T_m^{\,o}$ with crystallization temperature is weaker than in thin films but this might originate from considerably larger measurement error. Nevertheless, the highest value of $\Delta T_m^{\,o}$ is the same for both thin sections and thin films of the same brand of iPP. The largest negative pressure values determined from the melting behavior in other polymers were -12 and -6 MPa for PAE and POE, respectively, and -10 MPa in both HDPE and POM. In all polymers studied the decrease of the maximum negative pressure levels with the increase of the crystallization temperature was found, especially remarkable in iPP, PAE and PE. In both iPP and PE the significant negative pressure values were determined in weak spots during crystallization in the regime III. With the increase of the crystallization temperature a partial relaxation of the negative pressure in weak spots becomes more efficient. When the spherulite growth rate decreases, the negative pressure buildup is slower. In spite of higher final crystallinity the deformation of surrounding spherulites is easier. This might be also connected with the increasing number of tight folds on the crystal surface, thus a decreased number of cilia and tie molecules. The bonding of the amorphous phase to crystalline phase as well as bonding between crystalline lamellae becomes weaker which allows for easier displacements. Also the secondary crystallization phenomena are delayed at the elevated crystallization temperature.

4. Weak spots content and sizes

The spherulite centers are distributed randomly within a polymer, hence the weak spots are of various sizes and shapes. The computer simulation of isothermal spherulitic crystallization described in [1,11] demonstrated that the total content of weak spots is governed primarily by the dimensionality of crystallization and to some extent by the time distribution of primary nucleation.

The weak spots constitute 13% of the surface area of instantaneously nucleated polymer film. The sporadic nucleation at a constant rate during the entire crystallization reduces the weak spot content to 10%. In bulk samples the weak spot content equals 0.5% and 0.1% the material volume in the case of instantaneous and sporadic nucleation, respectively. However, in bulk the linear dimensions of largest weak spots are also large and approaching the size of a mean spherulite.

The sizes of weak spots are closely related to sizes of spherulites. In bulk only the largest weak spots reach the size of average spherulite. In thin film they are larger than the average spherulite. The exemplary size distributions of weak spots are shown in Fig.7. Since the intensity of primary nucleation controls the spherulite sizes it influences also the sizes of weak spots; they can be decreased by artificial nucleation.

We have to note that the actual size of hole due to cavitation should not exceed several percents of a weak spot initial size. In polymer bulk much larger holes are frequently observed as it is shown in Fig.8. That is because in the computer simulation a uniform nucleation of spherulites inside a sample was assumed as well as the unlimited melt supply to growing spherulites from outside. In real polymer portions the nucleation can occur first at the peripheries of a polymer portion and/or at material borders being as the result of temperature gradient and/or the contact with foreign substances. The material borders limit also the amount of melt available to growing spherulites. Hence, the computer simulation sets only the lower limit for the weak spot content in a crystallizing polymer.

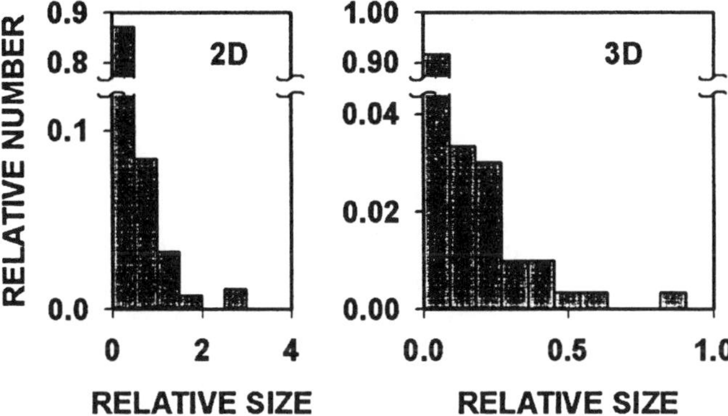

Figure 7. Size distributions of weak spots formed during 2D and 3D crystallization. Size scaled to the size of an average spherulite. Size is calculated as a radius of a circle or sphere having a surface area or a volume of a weak spot or spherulite.

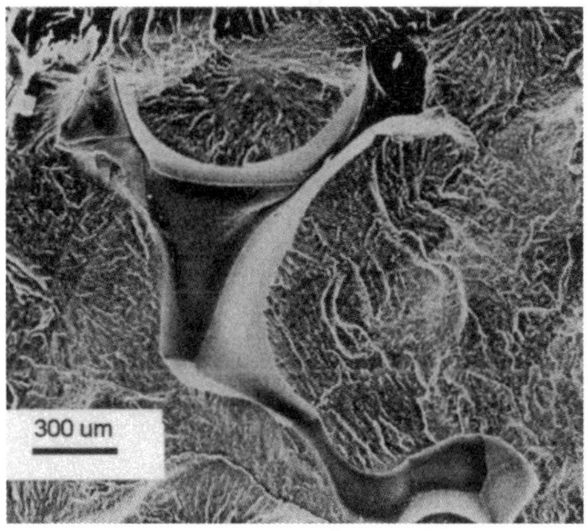

Figure 8. Scanning electron micrograph of fracture surface if iPP plate crystallized isothermally at 134 °C; large cavities between spherulites are visible.

5. Acoustic emission due to cavitation

The smallest number of spherulites required for occlusion of melt is 3 in thin films which can be considered as a two-dimensional sample. In a polymer bulk the smallest number of spherulites necessary for confinement of a portion of melt is 4. Three and four are also the numbers of spherulitic growth fronts which can meet at a point, on a plane and in space, respectively. However, many more spherulites can be involved in

134

the occlusion of melt. Even in those larger weak spots formed by several spherulites the respective number, 3 or 4, of spherulites meet at points at the end of crystallization. Therefore, the time distribution of formation of contact point of tetrads of spherulites should be close to the time distribution of cavitation events. The mathematical description of the formation of spherulitic structure [13,14] based on the probability theory allows to predict the formation of such points during polymer crystallization. The normalized time distribution of contact points formation between tetrads of spherulites for instantaneously nucleated bulk crystallization is expressed by the equation:

$$H(t_h)dt_h = 1.5 \ (\ln 2) t_h^{\ 8} \ 2^{\ -z} \ dt_h \tag{4}$$

where $z = t_h^{\ 3}$ and t_h is the ratio of time over the crystallization half-time.

The cavitation occurring in weak spots releases abruptly the negative pressure which is a source of acoustic waves. The acoustic emission was recorded during crystallization of iPP and POM using wide band sensor, 1kHZ-10 MHz, with low threshold level of 22dB [12,13]. Fig.9 demonstrates the comparison of the curve described by eq.(4) and the time distribution of acoustic events recorded during isothermal crystallization of iPP at 120°C. The good agreement was obtained between

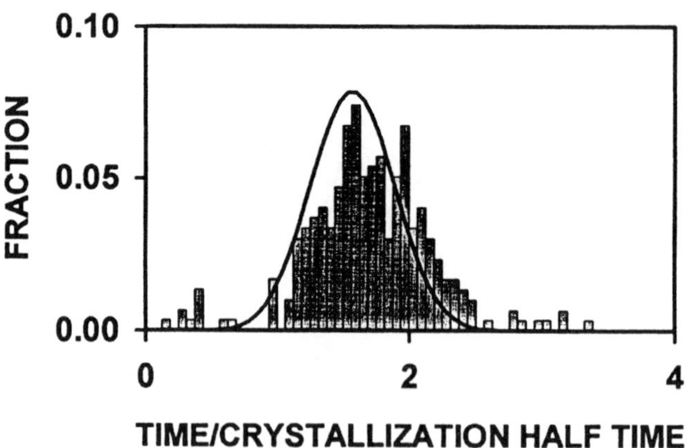

Figure 9. Comparison of relative number of acoustic events recorded during crystallization of iPP at 120°C and the function $H(t_h)dt_h$ calculated according to eq.(4); data from [12,13].

the theoretical predictions and the experimental data. The cavitation occurs before the melt is fully crystallized in weak spots but the decrease of the spherulite growth rate under negative pressure is responsible for the apparent time lag between acoustic events distribution of the time distribution of the formation of contact points of tetrads of spherulites as seen in Fig. 9.

The number of recorded events plotted against time during iPP crystallization at various temperatures is shown in Fig.10. The number of recorded events increases with the decreasing crystallization temperature. More intense spherulite nucleation at lower crystallization temperature results in the increased number of spherulites reflected in larger number of weak spots. However, further increase of the number of spherulites by addition of nucleating agents suppresses the cavitation, - the weak spots become too small to cause cavitation. Small weak spots allow for easier relaxation of negative pressure by the deformation of surrounding spherulites. This conclusion was supported by further studies of cavitation during isothermal crystallization of iPP [8].

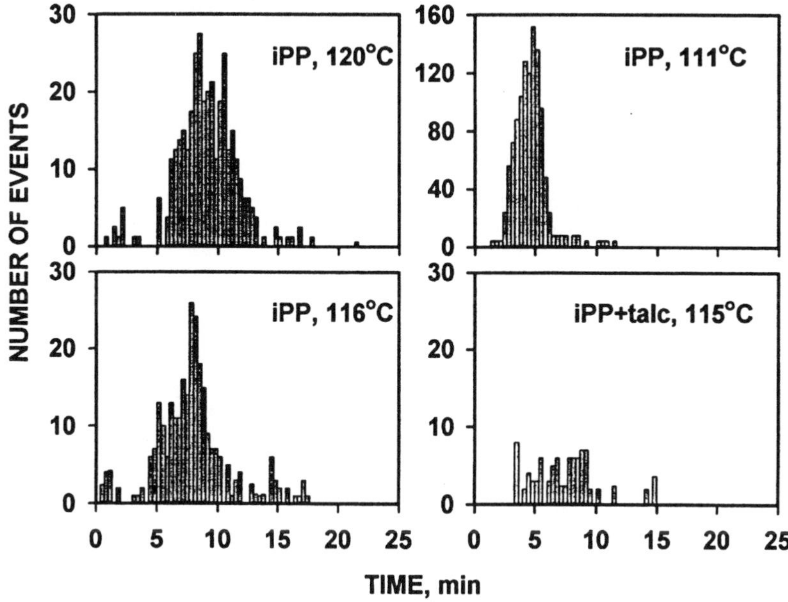

Figure 10. Number of acoustic events recorded during crystallization of iPP at various crystallization temperatures; data from [13].

6. Conclusions

The occlusion of portions of melt is inevitable during spherulitic crystallization. The pockets of melt between spherulites are called weak spots since further crystallization occurs without a supply of a fresh melt generating stresses. The lowest number of spherulites for a weak spot formation is three in thin films and four in bulk. The sizes of weak spots are closely related to sizes of spherulites, dimensionality of their growth and the mode of primary nucleation.

The buildup of a negative pressure resulting from uncompensated volume change due to crystallization leads to local decrease of supercooling which is reflected in the decreased spherulite growth rate and larger lamellae thickness. Both effects can serve as a measure of the level of negative pressure. Typical values of negative pressure, as determined by these two ways, are varying between -10 and -20 MPa. These values are unexpectedly high for liquids which were not purified and not de-aerated in any way. Instead these commercial polymers contain particles of stabilizer, antioxidants and residue of catalysts. Such high values of negative pressure for high molecular weight liquids differ from low molecular weight fluids for which special care must be taken to reduce and eliminate cavitation. This large difference can be attributed to the existence of entanglements of macromolecular chains in polymer melts which strengthen the high molecular weight liquid.

136

The buildup of a negative pressure in polymer melts leads also to cavitation which can be detected by monitoring of the acoustic emission. The cavitation occurs at a much higher negative pressure than for low molecular weight liquids. The buildup of negative pressure and cavitation phenomena depend on crystallization temperature; they are also influenced by sizes of weak spots. Both, a frozen-in negative pressure or cavities in weak spots weaken the material. The addition of a nucleating agent reduces the spherulite and weak spot sizes hence, limits the negative pressure buildup and suppresses the cavitation.

7. Acknowledgement

The authors wish to acknowledge the financial support from the State Committee for Scientific Research, Poland, through the Centre of Molecular and Macromolecular Studies, PAS, under Grant 4 T08E 055 22.

8. References

1. Galeski, A., Piorkowska, E. (1983) Localized Volume Deficiencies as the Effect of Spherulite Growth. I. The Two-Dimensional Case, *J.Polym.Sci.Polym..Phys.Ed.* **21**, 1299-1312.
2. Bartczak, Z., Galeski, A. (1986) Changes in interface shape during crystallization in two-polymer systems, *Polymer* **27**, 544-548.
3. Wool, R.P. (1995) Welding and fracture of polypropylene interfaces, in J.Karger-Kocsis (ed.), *Polypropylene, vol.1. Structure*, Chapmann & Hall, London, pp. 227-272.
4. Hoffman, J.D., Miller, R.L. (1997) Kinetics of crystallization from the melt and chain folding in polyethylene fractions revisited: Theory and experiment, *Polymer*, **38**, 3151-3212.
5. Phillips, P.J. (1990) Polymer crystals, *Rep.Prog.Phys.* **53**, 549-604.
6. Leute, U., Dollhopf, W. and Liska, E. (1976) Dilatometric measurements on some polymers: the pressure dependence of thermal properties, *Coll.Polym.Sci.* **254**, 237-246.
7. Pawlak.,A., Galeski, A. (1990) Stability of Spherulite Growth Rate, *J.Polym. Sci:Part B:Polym.Phys.* **28**, 1813-1821.
8. Nowacki, R., Kolasinska, J., Piorkowska, E. (2000) Cavitation during Isothermal Crystallization of Isotactic Polypropylene, *J.Appl.Polym.Sci.* **79**, 2439-2448.
9. Pawlak, A., Piorkowska, E. (1999) Effect of Negative Pressure on Melting Behavior of Spherulites in Thin Films of Several Crystalline Polymers, *J.Appl.Polym.Sci.* **74**, 1380-1385.
10. Piorkowska, E., Galeski, A. (1993) Crystallization of Isotactic Polypropylene and High-Density Polyethylene Resulting from Uncompensated Volume Change, *J.Polym.Sci:Part B:Polym.Phys.* **31**, 1285-1291.
11. Galeski, A., Piorkowska, E. (1983) Localized Volume Deficiencies as the Effect of Spherulite Growth. II. The Three-Dimensional Case, *J.Polym.Sci.Polym..Phys.Ed.* **21**, 1313-1322.
12. Galeski, A., Koenczoel, L., Piorkowska, E., Baer, E. (1987), Acoustic emission during crystallization of polymers, *Nature*, **327**, 40-41.
13. Galeski, A., Piorkowska, E., Koenczoel, L., Baer, E. (1990) Acoustic Emission During Crystallization of Polymers, *J.Polym.Sci:Part B:Polym.Phys.* **28**, 1171-1186.
14. Piorkowska, E., Galeski, A. (1985) Growth Sites in Space and Time, *J.Phys.Chem.*, **89**, 4700-4703.

CAVITATION DURING ISOTHERMAL CRYSTALLIZATION OF
ISOTACTIC POLYPROPYLENE AND POLY(METHYLENE OXIDE)

E.PIORKOWSKA AND R.NOWACKI
Centre of Molecular and Macromolecular Studies,
Polish Academy of Sciences
90 363 Lodz, Poland

1. Introduction

During polymer crystallization polycrystalline aggregates grow until the impingement with neighbours. Thus, in the course of crystallization they occlude pockets of the melt, blocking further inflow of the melt towards crystallization front [1,2]. Volume deficiency resulting from a lower specific volume of a solid results in a negative pressure buildup in bulk and in films confined between stiff plates [3-6].

The consequence of negative pressure for the polymers is a decrease of phase transition temperatures. At low and medium supercooling the decrease of the polymer equilibrium melting temperature T_m^0 is decisive; it results in the crystallization of thicker lamellae [3-5] and also in a slower growth rate of spherulites [3,6]. Both effects can be used to determine the level of negative pressure.

The negative pressure causes cavitation in the melt leading to an abrupt release of the pressure in melt [6]. Nevertheless, as a result the voids remain within the spherulitic structure. The gas pressure inside of freshly formed cavitation bubbles was found to be lower than 1 mm Hg [2]. It was discovered that the cavitation is associated with acoustic emission [7,8]; ultrasound emission depends on the crystallization temperature and can be suppressed by addition of a nucleating agent to a polymer. Since both the frozen stresses and the voids weaken a polymer, the occluded pockets of melt constitute weak spots in polymer supermolecular structure.

In this paper we describe the recent results of studies of cavitation in weak spots (i.e. occluded pockets of melt) during the isothermal crystallization of isotactic polypropylene (iPP) and poly(methylene oxide) (POM).

2. Cavitation during crystallization of iPP

The systematic study of the cavitation during isothermal crystallization of iPP films was described in [6]. IPP having M_w= 200 000 and M_n=17 000, a product of Polysciences, USA, was used. Fig.1 shows the micrographs of a weak spot formation and cavitation during crystallization. After occlusion of the pocket of melt the further crystallization proceeds with the buildup of negative pressure which may or may not cause the cavitation. Numerous cavities of various sizes appear simultaneously (i.e within 40 ms time interval) preferentially at the spherulite-melt interface and seldom within the melt. Some of them are poorly recognizable at the moment of cavitation but they become better visible after approximately 0.1 s. Very often the cavities appear in the vicinity of interspherulitic boundaries. At the moment when large bubbles are formed, dark lines appear where the crystallizing front adheres to the glass. Black lines were identified as rows of small voids of 1-2 μm diameter. Although from the atomic

A.R. Imre et al. (eds.), Liquids Under Negative Pressure, 137–144.
© 2002 *Kluwer Academic Publishers. Printed in the Netherlands.*

138

force microscopy investigations it followed that those small cavities are embedded in a polymer [9], a light microscopy of iPP films allowed to distinguish chains of cavities near the upper and support glasses confining the iPP film [6]. While at the lower crystallization temperature the chains of small voids are typical, they become less frequent with the increase of crystallization temperature. The cavitation occurs also in narrow channels up to 130°C. The voids appear simultaneously with the formation of bubbles inside the melt and along the crystallization front. In samples crystallizing isothermally above 130°C the radial fissures directed towards spherulite centers were observed; the detailed investigations revealed also in this case the cavitational mechanism.

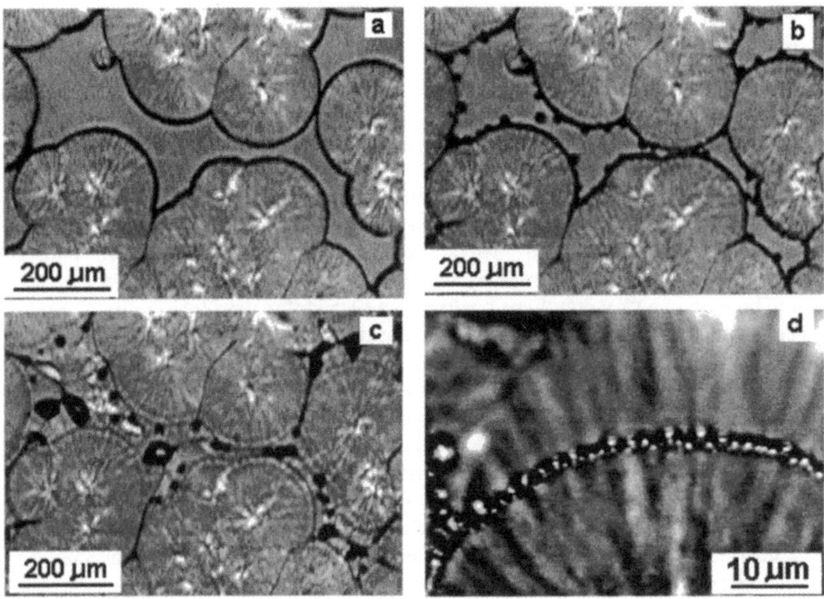

Figure 1. Crystallization in weak spot of iPP film between microscopic glasses at 128°C: (a) occlusion, (b) cavitation, (c) final structure, (d) details of black lines such as those visible in (c) [6].

The cavitation in weak spots ceases with the increase of crystallization temperature. In weak spots crystallized isothermally without cavitation at temperature higher than 130°C the voids appear along interspherulitic boundaries during cooling after isothermal crystallization (Fig.2b). At 140°C the cavitation in weak spots does not occur but numerous voids are formed along interspherulitic boundaries during cooling. Cavitation along the spherulites radii was also noticed. In the past Monasse and Haudin [10] and Varga and Ehrenstein [11] observed in iPP crystallized at elevated temperature the fractures along the interspherulitic boundaries instead of the cavitation. As it follows from our observation such fracture can involve also the cavitation step.

Scanning electron microscopy (SEM) of surfaces of iPP films after removal of the upper glass supports the conclusions drawn from light microscopy studies; chains of small cavities were found in the surface layer of the films and also larger voids between spherulites and along interspherulitic boundaries are present. The longitudinal voids along the spherulite radii were also observed (Fig.3).

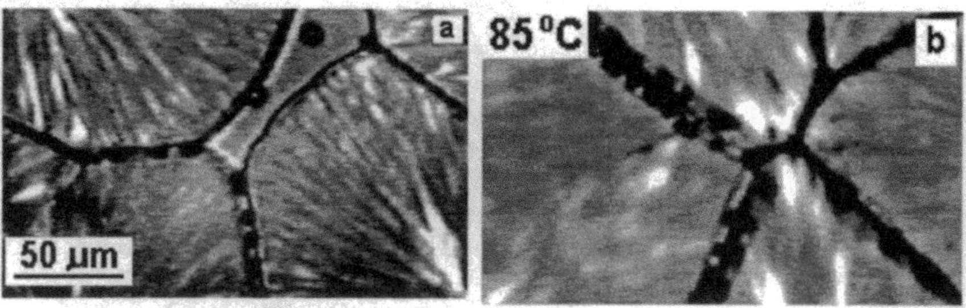

Figure 2. Light micrographs of cavitation along interspherulitic boundaries in iPP films: (a) during isothermal crystallization at 130°C, (b) at 85°C, during cooling after isothermal crystallization at 137°C [6].

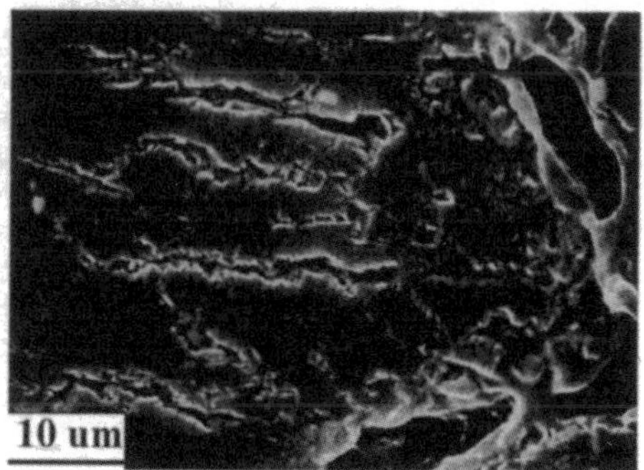

Figure 3. SEM micrograph of voids along interspherulitic boundaries and within spherulite in iPP film crystallized at 132°C [6].

The cavitation bubbles in weak spots are either overgrown by the crystallization front or enlarge during a weak spot crystallization (Fig.1c). The voids disappear very seldom. Often, the nucleation of spherulites on cavitation bubbles was observed, frequently of the β crystallographic form. The nucleation of β spherulites on cavitation bubbles was also reported in [11,12]

The growth rate of spherulites decreases continuously from the moment of weak spot formation due to the T_m^o depression and the decrease of an isothermal supercooling [5,6]. The growth rate of isolated spherulites and of spherulites in weak spots was measured in iPP studied [6] during isothermal crystallization. After the cavitation the growth rate increases rapidly (Fig.4); at temperature below 130°C the growth rate assumes usually its initial value. At higher temperature the growth rate recovery is incomplete after cavitation, which is attributed to fractionation during crystallization lowering the growth rate, since some decrease of the growth rate measured for isolated spherulites was also found. This corresponds also to the observations of cavitation occurring along interspherulitic boundaries in films crystallized at elevated temperature. Fractions having lower ability to crystallize are rejected, accumulate in interspherulitic boundaries which leads to a formation of residual layers of polymer melt undergoing cavitation either during isothermal crystallization or during cooling.

The growth rate of iPP spherulites at temperature T can be described by the equation [13]:

140

$$G = G_o \exp\{-U[R(T-T_g+30K)]^{-1}\} \exp\{-K_g[RT\Delta Tf]^{-1}\} \qquad (1)$$

where U=1500 cal mol^{-1}, is the activation energy of polymer chains transport, T_g=261.2K and T_m^o =458.2K are the glass transition temperature and the equlibrium melting temperature, respectively, ΔT denotes the supercooling : T_m^o- T. K_g and G_o and are coefficients depending on the crystallization regime; K_g is proportional to $\sigma\sigma_e T_m^o \Delta h_f^{-1}$, where σ and σ_e are the lateral and fold surface energies, while Δh_f is the heat of fusion per unit volume of crystal at T_m^o . F=2T (T+T_m^o)$^{-1}$ is a correction term originating from the temperature dependence of Δh_f.

Assuming that the decrease of isothermal supercooling is the main factor responsible for the depression of the growth rate from G at the atmospheric pressure to G_p under the negative pressure, one can calculate the shift of the equilibrium melting temperature, ΔT_m^o:

$$\Delta T_m^o = \Delta T - \{1/\Delta T + T \ln (G/G_p)/K_g\}^{-1} \qquad (2)$$

where ΔT is the supercooling at the atmospheric pressure.

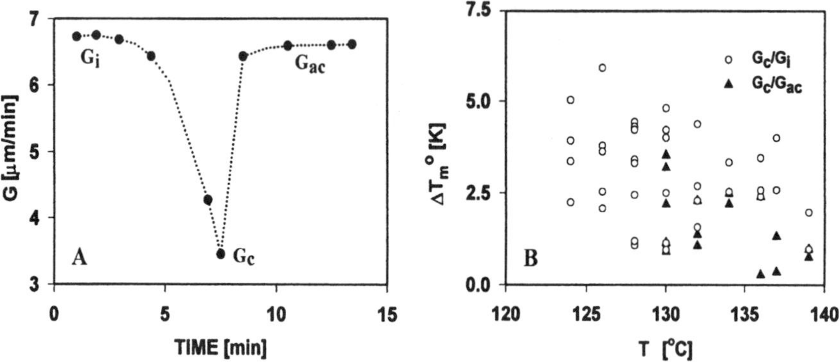

Figure 4. (a) time dependence of growth rate in weak spot of iPP film at 130°C; G_i, G_c and G_{ac} denote the spherulite growth rate: initial, right before cavitation and after cavitation, respectively, (b) temperature dependence of ΔT_m^o calculated from the decrease of growth rate to G_c. Data points from [6].

The values of K_g in the regime II and regime III of crystallization were found experimentally for iPP studied which allowed to determine ΔT_m^o from the value of growth rate under negative pressure in weak spots according to eq.(2). The possible influence of the T_g shift on the growth rate was estimated and found to be a marginal effect.

The negative pressure builds up due to the volume deficiency developing during crystallization in weak spots and therefore should be related to progression of crystallization characterized by the ratio of a volume of a polymer melt in a weak spot , S, to its initial value at the moment of the weak spot formation, S_i. It was found that the decrease of growth rate, hence the negative pressure buildup and the cavitation phenomena, depend on the initial sizes of weak spots and the crystallization temperature. In smaller weak spots the decrease of growth rate with the progression of crystallization is slower than in larger weak spots. In larger weak spots the cavitation occurs at earlier stages of crystallization than in smaller ones, while the smallest weak spots crystallize further without cavitation. This indicates that in planar smaller weak spots the relaxation of negative pressure by deformation of surrounding spherulites becomes more efficient due to larger ratio of surface to volume. The temperature increase acts like the decrease of the weak spot sizes and leads to slower negative pressure buildup, caused possibly by easier relaxation of pressure due to slower growth rate and easier deformation of surrounding spherulites. The latter might be also connected with the decreasing number of cilia and tie molecules which results in weaker bonding of the remaining amorphous phase to crystalline phase as well as between lamellae allowing easier displacements. The secondary crystallization phenomena are also delayed at elevated crystallization temperature.

The values of ΔT_m^o calculated based on the growth rate, G_c, right before cavitation measured in a range of crystallization temperatures are collected in Fig. 4b. The value of growth rate before the weak spot formation, G_i, was taken as the growth rate at atmospheric pressure. In addition, for temperatures above 130°C the values of ΔT_m^o were also calculated assuming that the growth rate measured after cavitation, G_{ac}, corresponds to the value at atmospheric pressure. The scatter of data points visible in Fig. 4b indicates that the cavitation occurs at a given temperature at different levels of negative pressure. We have to note that lower values of the growth rate measured before the cavitation correspond to higher S/S_i ratios; i.e. to the cavitation occurring at earlier stages of weak spot crystallization. With the increase of temperature the ΔT_m^o values decrease demonstrating the decrease of cavitation threshold. The highest values of ΔT_m^o correspond to negative pressure of 18 MPa. Similar results were obtained based on the measurements of the increase of melting temperature [4,5].

3. Cavitation during crystallization of POM

The cavitation phenomena were also studied during crystallization of poly(methylene oxide), Tarnoform 300, Polish product having MFI=9±1[g/10min] at 190°C /2.16 kg. [14]. The cavitation was studied during the isothermal crystallization of films in the temperature range of 152-162°C. The cavitation occurs in the weak spots preferably at the spherulite-melt interface, very seldom in the melt (Fig. 5). Similarly as in iPP, the voids appear in the vicinity of interspherulitic boundaries. Chains of small cavitation bubbles develop near cover glasses confining the polymer film (Fig.6a). With the increase of temperature they become less frequent. Some of the cavitation bubbles disappear after formation although this effect is rather seldom. The reason for bubble disappearance is not yet understood. Usually the holes enlarge during crystallization or become overgrown by crystallization fronts. They also nucleate new spherulites within weak spots. In addition to the cavitation at spherulite–melt interface and inside the melt the formation of numerous cavities within spherulites was observed appearing simultaneously with the cavitation in weak spots, as it is visible in Fig.5b, at later stages of crystallization and also during cooling after the completion of isothermal crystallization. The development of cavities within spherulites during cooling is visible in the micrograph in Fig.6a. The appearance of such voids around interspherulitic boundaries formed within weak spots crystallized without cavitation was also observed during cooling (Fig.6b). In the scanning electron micrograph in Fig.7 the large voids between spherulites as well as the smaller ones within spherulites are visible. In the weak spots crystallized isothermally without the cavitation the fracture along interspherulitic boundaries and within spherulites was noticed as well. With the increase of temperature the cavitation becomes less frequent and practically disappears above 157°C. The growth rate measurements show that the buildup of negative pressure is

142

faster in larger than in smaller weak spots. The growth rate depression in the weak spots right before cavitation differs at the same temperature indicating that the cavitation occurs at various levels of negative pressure.

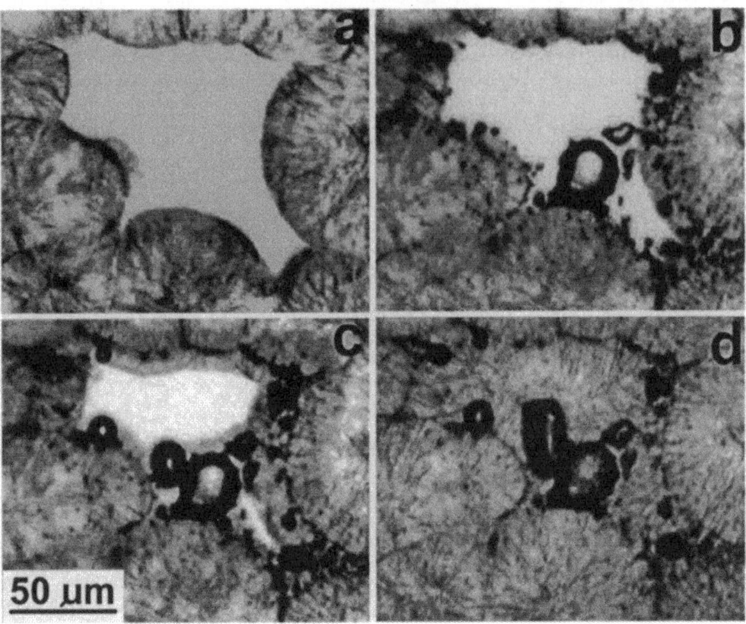

Figure 5. Light micrographs of crystallization in weak spot of POM film between microscopic glasses a 154°C: (a) occlusion, (b) cavitation, (c) crystallization after cavitation, (d) final structure.

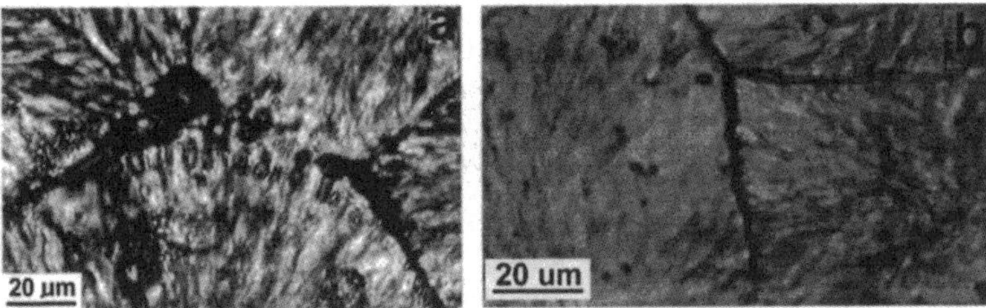

Figure 6. Light micrographs of weak spots in POM films: (a) chains of small cavitation bubbles formed near upper glass at 154°C, (b) voids within spherulites developed during cooling of film crystallized at 156°C .

4. Conclusions and discussion

The cavitation phenomena in weak spots during crystallization of iPP and POM have much in common - the voids appear predominantly at the spherulite-melt interface. Chains of small cavities are formed near the cover glasses. Similar features are recognized also in the light micrograph of weak spot in poly(ethylene oxide) demonstrated in Fig. 8 [1].

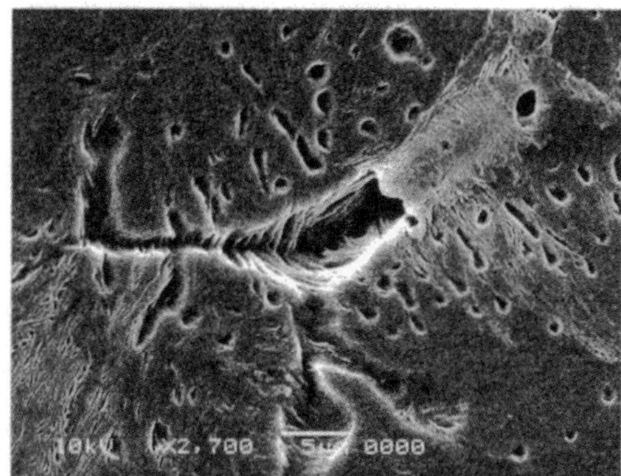

Figure 7. SEM micrograph of surface of POM film crystallized at 152°C.

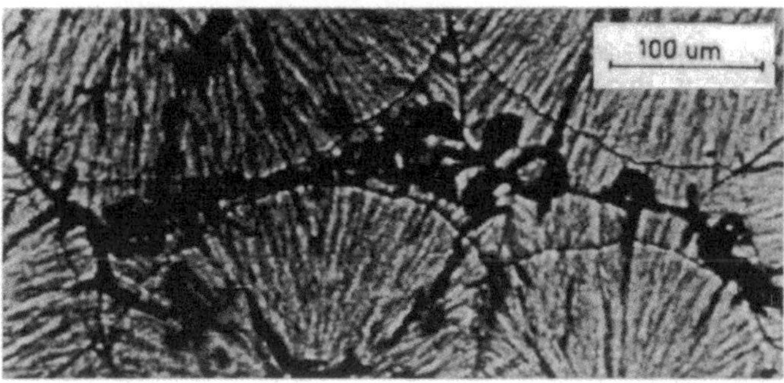

Figure 8. Light micrograph of weak spot in poly(ethylene oxide) film crystallized at 50°C [1].

The growth rate measurements show that the cavitation occurs in weak spots in the same polymer sample during isothermal crystallization at various level of negative pressure. Although the voids of different size appear simultaneously, predominantly at the spherulite-melt interface, there is a possibility that they are secondary events produced by an acoustic wave originating from a primary cavitation event. It was observed in the past that the mechanically induced negative pressure wave caused cavitation during crystal growth in supercooled organic liquids while no cavitation occurred in the same liquids during the melting of crystals [15].

The scatter of negative pressure value seems to indicate the heterogeneous nucleation of cavitation possibly related to impurities present in the polymer. We have to point out that the solubility of gases, although low in polymers, decreases suddenly during crystallization which can also affect the cavitation phenomena.

The direct observation of cavitation and growth rate measurements require thin films between stiff plates, so foreign surfaces confining the polymer are unavoidable. Different cleaning procedures of microscopic cover glasses reported in [6] did not introduce any changes to the obtained results. In addition, similar values of $\Delta T_m^{\,o}$ were determined from the elevation of melting temperature of weak spots in thin sections of bulk iPP samples which indicates that microscopic glasses do not affect significantly the cavitation threshold.

144

The negative pressure causing the cavitation in iPP decreases with the increase of crystallization temperature and the cavitation becomes less frequent. The latter is related to the slower negative pressure buildup in weak spots. The decrease of the cavitation threshold with the temperature can be connected to the negative temperature dependence of the cohesion and surface tension of polymer melts.

5. References

1. Galeski, A., Piorkowska, E. (1983) Localized Volume Deficiencies as the Effect of Spherulite Growth. I. The Two-Dimensional Case, *J.Polym.Sci.Polym..Phys.Ed.* **21**, 1299-1312.
2. Galeski, A., Piorkowska, E. (1983) Localized Volume Deficiencies as the Effect of Spherulite Growth. II. The Three-Dimensional Case, *J.Polym.Sci.Polym..Phys.Ed.* **21**, 1313-1322.
3. Pawlak.,A., Galeski, A. (1990) Stability of Spherulite Growth Rate, *J.Polym. Sci:Part B:Polym.Phys.* **28**, 1813-1821.
4. Piorkowska, E., Galeski, A. (1993) Crystallization of Isotactic Polypropylene and High-Density Polyethylene Under Negative Pressure Resulting from Uncompensated Volume Change, *J.Polym.Sci:Part B:Polym.Phys.* **31**, 1285-1291.
5. Pawlak, A., Piorkowska, E. (1999) Effect of Negative Pressure on Melting Behavior of Spherulites in Thin Films of Several Crystalline Polymers, *J.Appl.Polym.Sci.* **74**, 1380-1385.
6. Nowacki, R., Kolasinska, J., Piorkowska, E. (2000) Cavitation during Isothermal Crystallization of Isotactic Polypropylene, *J.Appl.Polym.Sci.* **79**, 2439-2448.
7. Galeski, A., Koenczoel, L., Piorkowska, E., Baer, E. (1987), Acoustic emission during crystallization of polymers, *Nature*, **327**, 40-41.
8. Galeski, A., Piorkowska, E., Koenczoel, L., Baer, E. (1990) Acoustic Emission During Crystallization of Polymers, *J.Polym.Sci:Part B:Polym.Phys.* **28**, 1171-1186.
9. Thomann, R., Wang, C., Kressler, J., Mulhaupt, R. (1996) Negative pressure effects during the isothermal crystallization of isotactic poly(propylene) studied by light and atomic force microscopy, *Macromol.Chem.Phys.* **197**, 1085-1091.
10. Monasse, B., Haudin, J.M. (1988) Growth transition and morphology change in polypropylene, *Coll.Polym.Sci.* **263**, 822-831.
11. Varga, J., Ehrenstein, G.W. (1996) Formation of β-modification of isotactic polypropylene in its late stage of crystallization, *Polymer*, **37**, 5959-5963.
12. Zadlo, A. (1986) M.Sc.Thessis: Conditions of β-spherulites formation in polypropylene (in Polish), Polish Academy of Sciences.
13. Clark, E., Hoffman, J.D. (1984) Regime III Crystallization in Polypropylene, *Macromolecules*, **17**, 878-885.
14. Nowacki, R., Piorkowska E., Cavitation during crystallization of poly(methylene oxide), in preparation.
15. Pease, D.C., Blinks, R.L. (1947) Cavitation from Solid Surfaces in the Absence of Gas Nuclei, *J.Phys.Coll.Chem*, **51**, 556-567.

THE LIMITS OF METASTABILITY OF LIQUID HELIUM

SÉBASTIEN BALIBAR AND FRÉDÉRIC CAUPIN
Laboratoire de Physique Statistique
de l'Ecole Normale Supérieure,
associé au CNRS et aux Universités Paris 6 et Paris 7,
24 rue Lhomond, 75231 Paris Cedex 05, France

Abstract. We review the present understanding of liquid helium outside its usual stability range. Most theoretical predictions and experiments concern the negative pressure region, where the phenomenon of cavitation has to be considered in relation with the existence of a liquid-gas spinodal limit. New experiments at high pressure also open perspectives in the study of the homogeneous nucleation of solid helium from the overpressurized liquid.

1. Introduction

During the last decade, the phase diagram of liquid helium has been considerably extended. One used to study liquid helium 4 in its stability region only, between its boiling line at low positive pressure and its melting line around 25 bar. Today, not only in their theoretical predictions but also in their experiments, physicists study liquid helium 4 down to about - 9 bar and also far above the melting pressure, possibly up to +200 bar. As for its light isotope helium 3, it was only between 0 and about 30 bar, and interesting results have now been obtained down to -3 bar, where an unexpected similarity was found with possible properties of metastable water. In our first section, we briefly summarize the main theoretical predictions concerning the extended phase diagram of liquid helium 4 and 3. We then describe the acoustic technique which allowed recent experiments on metastable liquid helium and new studies of the nucleation of either bubbles in stretched liquid helium or crystals when it is overpressurized. In our last section, we compare predictions with experimental results. We finally conclude with a list of open questions to be studied in the near future.

2. Theory: the extended phase diagram of liquid helium 4

We are interested in the limits of metastability of liquid helium. As shown in Fig.1, this metastability is possible below the saturated vapor pressure, i.e. below the liquid-gas equilibrium line, down to a liquid-gas spinodal (SP) line. It is also possible above the melting line, up to a high pressure instability (HPI) line. The SP

A.R. Imre et al. (eds.), Liquids Under Negative Pressure, 145–160.
© 2002 *Kluwer Academic Publishers. Printed in the Netherlands.*

146

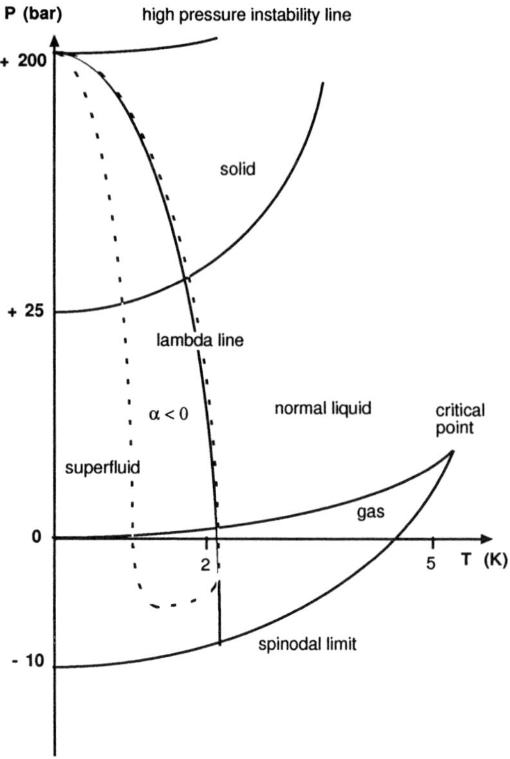

Figure 1. The phase diagram of helium 4: qualitative features. At T=0, liquid helium 4 is stable between 0 and 25 bar. Its metastable region extends down to about -10 bar and possibly up to +200 bar. The superfluid transition occurs when crossing the "lambda line". A region of negative expansion coefficient exists inside a loop which is bound by extremum density lines (broken lines)

line is now rather well established. It corresponds to a divergence of the isothermal compressibility, and H.J. Maris[1] first calculated its location from an extrapolation of the sound velocity at very low temperature. His extrapolation was based on the following form for the equation of state of liquid helium:

$$P - P_{sp} = \frac{b^2}{27}(\rho - \rho_{sp})^3 \qquad (1)$$

His most recent values are[2] P_{sp} = -9.6201 bar, ρ_{sp} = 0.094262 g/cm^3 and b = 1.4054 $\times 10^6$ g^{-1}cm^{4}s^{-1}. This form was criticized by Krotscheck[3] but recently confirmed by the Monte Carlo simulations of Bauer et al.[4]. In order to obtain the spinodal limit, Boronat[5] also used Monte Carlo simulations, as for the Barcelona group[10] and Dalfovo[6], they used density functional theories. The results of these various works and a few others are summarized in Fig.2.

As can be seen, all theoretical predictions agree that, at T = 0, P_{sp} = -9.4 $\pm$ 0.3 bar at T=0. The corresponding density is ρ_{sp} = 0.100 $\pm$ 0.005 g/cm^3. The temperature variation of the spinodal pressure was calculated by the Barcelona group[9]. Their results agree with the extrapolations by Hall and Maris[8] and

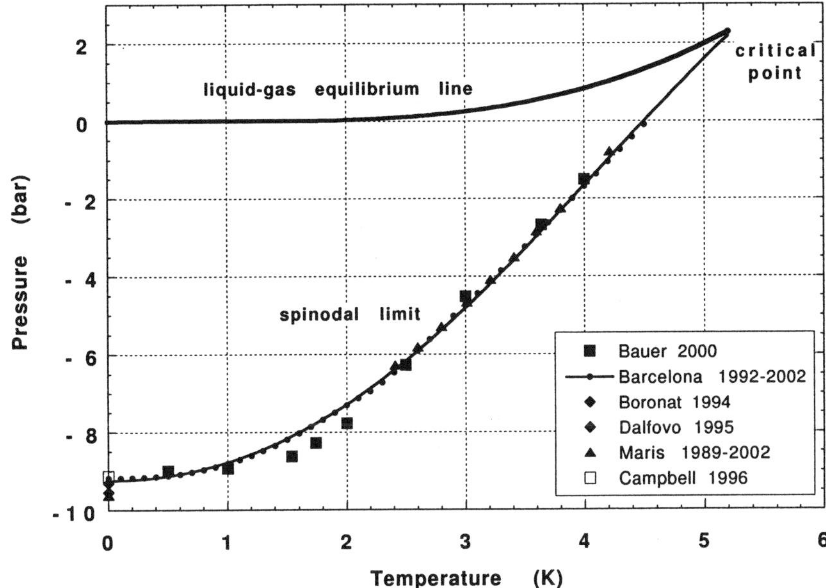

Figure 2. Successive theories and calculations lead to a rather well established prediction for the liquid-gas spinodal limit in helium 4. It is a line in the (P-T) plane extending from -9.4 bar at T=0 up to the critical point at +2.289 bar and 5.2K

the Monte Carlo calculations by Bauer[4]. The spinodal line has to end at the liquid-gas critical point ($P_c = 2.289$ bar, $T_c = 5.2$ K).

An interesting property of the spinodal line is its slope dP_{sp}/dT. It was noticed by Speedy[12] that the sign of this slope is the same as the sign of the isobaric expansion coefficient $\alpha_P = (1/V)(\partial V/\partial T)_P = -(1/V)(\partial S/\partial P)_T$. As shown in Fig.1, we now believe that the slope of the spinodal line in helium 4 is positive at all temperatures, meaning that α_P is also always positive near the spinodal line. This is far from obvious since, in the stable region at positive pressure, α_P changes sign twice. There is a line of maximum density (MD) in the phase diagram, but also a line of minimum density; they form a loop and do not reach the spinodal. This was first predicted by Skripov[13] and recently confirmed by the calculations of Edwards, Maris and Caupin[2, 15]. The physical origin of these extremum density lines in helium 4 is important to understand.

Since the work of Landau[14] and subsequent neutron scattering studies, it is known that the elementary excitations in superfluid helium are "phonons" and "rotons". The phonons are quantized sound waves; the rotons are special phonons with large wave-vectors whose density of states diverge near their energy minimum

Δ. Landau thought that rotons were kinds of quantized elementary vortices, but it is now generally accepted that they are free atoms dressed by the interaction with their neighbours. Since the roton energy gap Δ is of order 10 K in the stable region ($0 < P < 25$ bar), the thermodynamics of superfluid helium is controlled by rotons above about 1K and by phonons at low temperature.

The pressure dependence of the energy of phonons is the opposite of that of rotons. As P decreases towards the spinodal line, the phonon energy decreases (liquid helium becomes soft at large wavelength). On the contrary, the roton energy increases, and the roton minimum might change into an inflexion point around the spinodal limit[6, 15]. In the stable region, $\partial\Delta/\partial P$ is sufficiently negative for the roton contribution to α_P to overcome the phonon contribution and lead to a positive $\partial S/\partial P$ and a negative α_P. As one approaches the spinodal line, the phonon contribution increases and the roton contribution decreases. This leads to a positive expansion coefficient.

Another interesting feature of the phase diagram of helium 4 is the superfluid transition or "lambda" line. Superfluid helium becomes normal when it is invaded by excitations, and one can calculate the lambda line within Landau's "two fluids" model by calculating the "normal density" associated with thermal excitations and determining the temperature at which it equals the full density of liquid helium. It is again the particular pressure dependence of the roton gap which is responsible for the negative slope of the lambda line in the positive pressure region (see Fig. 1). Since the compressibility diverges at the spinodal limit, one could expect the superfluid to be invaded by an avalanche of phonons as one approaches the SP line. Some authors[3, 13] thus predicted that the lambda line bent back and reached the SP line at T=0. However, more recently, Skripov[13] considered the possibility that the lambda line meets the SP line at finite temperature. Bauer[4] and Apenko[16] found from their recent simulations that the lambda line does not bend back so much; it seems to be almost vertical at negative pressure in the phase diagram, reaching the SP line around $T = 2$K. This is now confirmed by the analytic calculations of Caupin[15], and by the numerical calculations of Edwards and Maris[2]. The reason for this is as follows. It is only the static compressibility which diverges at the spinodal limit. As a consequence, only the zero frequency phonons have zero energy there. The phonon branch in the dispersion curve goes from linear above the spinodal limit to quadratic at the spinodal limit[17]. As a result, there is no avalanche of phonons occuring at the spinodal limit. At finite temperature one needs to reach temperatures of order 2 K for the density of excitations to be high and destroy superfluidity.

Other interesting properties of the helium phase diagram are now considered at high pressure, far above the melting line, but they are mostly speculative at present. Some of them date back from the work of Schneider and Enz[18]. In 1971, they considered the possible instability associated with the vanishing of the roton energy gap. Indeed, since the roton gap decreases when the pressure increases, one could imagine that, at some critical pressure P_{HPI}, $\Delta =0$. Since the roton wavevector is finite, of order the inverse of the average distance between nearest neighbours in the liquid, it seems likely that the liquid would spontaneously organize in a periodic state which could be crystalline helium. At that time, one had no precise method to extrapolate $\Delta(P)$, no hope to make measurements up to the necessary pressure either. This has changed recently. Indeed, the density functional theory by Dalfovo et al.[6] allows to calculate the phonon-roton dispersion relation at any pressure, so that Maris and Edwards[15] recently predicted the roton gap to vanish around $P_{HPI} = + 200$ bar. Furthermore, as explained in Section 4, it might be possible to overpressurize liquid helium up to such a very high pressure by using fast acoustic pulses. The recent experiments by Chavanne et al.[19] have demonstrated that it is possible to pressurize liquid helium more than 4 bar above the liquid-solid equilibrium line. They were performed in the presence of a clean glass wall and the heterogeneous nucleation of solid helium was easily detected. They opened interesting perspectives for a much larger overpressurization when their glass plate will be removed in order to look for the homogeneous nucleation limit.

In the high pressure region also, the extrapolation of the lambda line is an interesting open problem. Very recent calculations [15] predict that the lambda line meets the HPI line at $T = 0$. This is first because superfluidity cannot be suppressed at $T = 0$ and high pressure except if the roton energy is zero. Secondly, on the HPI line the zero value of Δ induces an avalanche of rotons at any finite temperature, so that superfludity is destroyed at finite temperature. However, on the HPI line, the Landau theory is not valid and it is not easy to make rigorous predictions[15]. Furthermore, the slope of the lambda line close to $T = 0$ is not known and it has been drawn horizontal arbitrarily on Fig.1. Finally, it seems natural to expect that if the lambda line meets the HPI line at $T = 0$, so do the extremum density lines as well. Basically, all the above predictions at high pressure remain to be checked experimentally. The slope of the HPI line near $T = 0$ is uncertain at present.

3. Theory: the extended phase diagram of liquid helium 3

The case of liquid helium 3 is similar to the one of helium 4, with some important differences. Predictions have been limited to the negative pressure region and are shown in Fig. 3. The first difference to be noticed is that helium 3 is three times more fragile than helium 4: its spinodal pressure is predicted to be three times less negative. This is due to the larger quantum kinetic energy which makes the molar volume larger and the cohesion forces weaker in helium 3 than in helium 4. Another difference is the existence of one MD line only. The physical origin of this MD line has to do now with the Fermi liquid properties of liquid helium 3 at low temperature. As explained by Landau[14], the entropy of a Fermi liquid is proportionnal to the effective mass of quasi-particles which is an increasing function of pressure. As a result the expansion coefficient is negative at low temperature and changes sign at higher temperature when the Fermi liquid behavior is smeared out by thermal fluctuations. As summarized in Section 5, Caupin et al.[20] found experimental indications for the existence of a minimum in the spinodal line of helium 3. They also found strong theoretical arguments supporting this, as a consequence of the

150

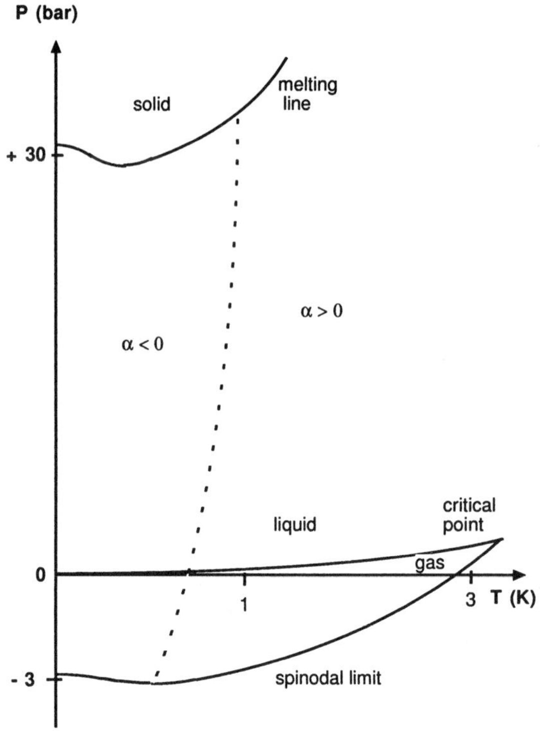

Figure 3. Phase diagram of helium 3: qualitative features. The spinodal line extends from -3 bar at T=0 to the critical point at + 1.165 bar and 3.324K. A line of maximum density (MD) exists which bounds a low temperature region where the expansion coefficient α is negative. This MD line meets the spinodal line at finite temperature so that the latter shows a shallow minimum around 0.4K

MD line meeting the spinodal line near 0.4 K. The case of helium 3 illustrates a property of spinodal lines which was first proposed by Speedy[12] for water.

4. Experiments on liquid helium at high pressure

In order to describe the acoustic technique which recently allowed to extend experiments far away from the stable region of liquid helium, let us start by describing what Chavanne et al.[19] measured at high pressure. As shown in Fig.4, they used a hemispherical piezo-electric transducer to emit and focus 1 MHz sound waves. At the acoustic focus, they were able to produce a very large acoustic pressure oscillation, up to about ± 20 bar despite the small acoustic impedance of liquid helium. In order to measure the instantaneous acoustic intensity at the acoustic focus, i.e. at the center of the transducer, they introduced a glass plate

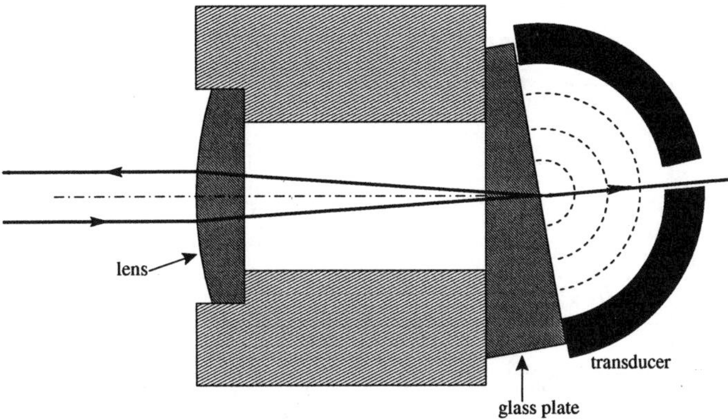

Figure 4. Chavanne et al.[19] studied metastable liquid helium 4 with an acoustic technique. A large pressure and density oscillation is produced at the center of a hemispherical piezo-electric transducer. This transducer is pressed against a glass plate and light is also focused at the acoustic focus. The local instantaneous density is measured from the amplitude of the light reflected at the interface between glass and liquid helium. The transmitted light is very sensitive to the possible nucleation of either crystals or bubbles at the center. The glass plate is useful to calibrate the transducer but it affects the nucleation. By removing the glass plate and recording the transmitted light, homogeneous nucleation can also be studied.

and measured the intensity of the light reflected at the glass/helium interface. Indeed, the reflectivity depends on the refractive index of liquid helium which is a known function of its density and is modulated by the wave.

Fig. 5 shows two pairs of recordings of sound intensity and nucleation events at the acoustic focus. In the lowest part, the "reflexion" traces are averages on 1000 acoustic bursts. The transducer is pulsed during 6 periods, and it has a finite quality factor Q of order 50. This is the reason for the envelope of the detected burst. As the amplitude is increased just above a certain threshold (about 10 Volt in the excitation voltage for this particular experiment), a positive peak is detected, superimposed to the sinusoidal wave. It is due to the nucleation of a small crystal whose density is 10% larger than that of the liquid. One trace corresponds to an excitation just below the nucleation threshold. The other one, which is superimposed on the previous one, corresponds to the threshold where the nucleation probability is 0.5. The reflected light is mainly used for the measurement of the instantaneous density. Since the equation of state of liquid helium is well established, it is easy to convert densities into pressures thanks to a slight extrapolation. In this particular experiment, the solid nucleated when the density reached 0.17543 g.cm^{-3}, i.e. 4.3 bar above the melting line at P=25.3 bar where the density is 0.17245 g.cm^{-3}. By also studying the light transmitted through the acoustic focus, Chavanne et al. showed that the observed nucleation was influenced by the presence of the wall. As shown by the same Fig.5, the transmitted light is very sensitive to phase changes in the acoustic focal region, so that single events could easily be detected. There is no averaging on the two "transmission" traces in the upper part fo this figure. They were recorded for the same excitation and showed that nucleation is stochastic: it occurs with some probability in a certain density region. This good

152

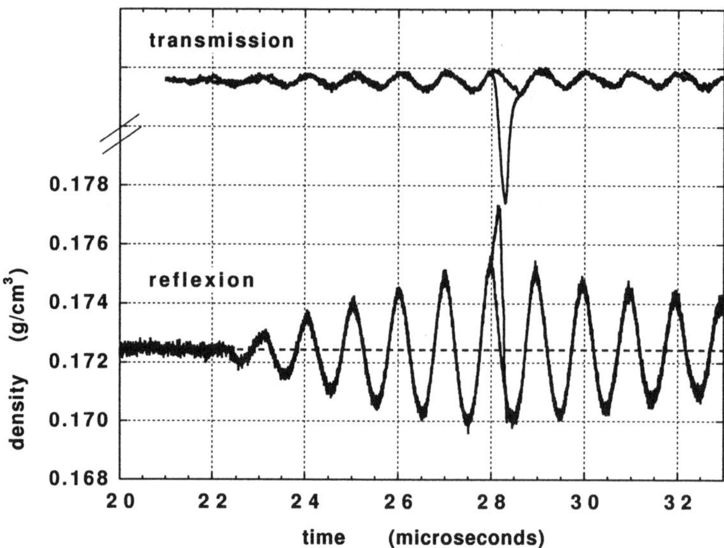

Figure 5. Two pairs of recordings in the experiment by Chavanne et al.[19]. The two in the lower part of the figure correspond to the reflected light. One recording is obtained for a sound amplitude just below the nucleation threshold. The sinusoidal oscillation is associated with the acoustic burst passing at the acoustic focus. Superimposed on it is another recording at slightly larger amplitude, where crystals nucleate in the sound wave with a probability of about 0.5. The crystal density is larger so that an extra peak appears in the reflected light. The two recordings in the upper part of the figure show recordings of the transmitted light, which were obtained at the same excitation level. One of them shows light scattering by the acoustic wave only. The other one shows additional scattering due to one nucleation event. Such recordings allowed measurements of the nucleation statistics

sensitivity allowed the measurement of the nucleation statistics as a function of sound amplitude and temperature. From all these measurements, Chavanne et al. were able to determine the activation energy E and the number of nucleation sites. They found $E/k_B T \approx 10$ along the nucleation curve, and a number of sites of order one. From this they infered that, in their experiment, there was one small defect on the glass surface which favored the nucleation by lowering the local energy barrier more than others.

Indeed, if one had homogeneous nucleation, the energy barrier could be estimated as about 3000K from the standard equation

$$E = \frac{16\pi}{3} \frac{\gamma^3}{\Delta G^2} \tag{2}$$

where $\Delta G = (\rho_L/(\rho_C - \rho_L))(P - P_m)$ is the difference in free energy per unit volume between the crystal and the liquid phase (P_m is the equilibrium melting

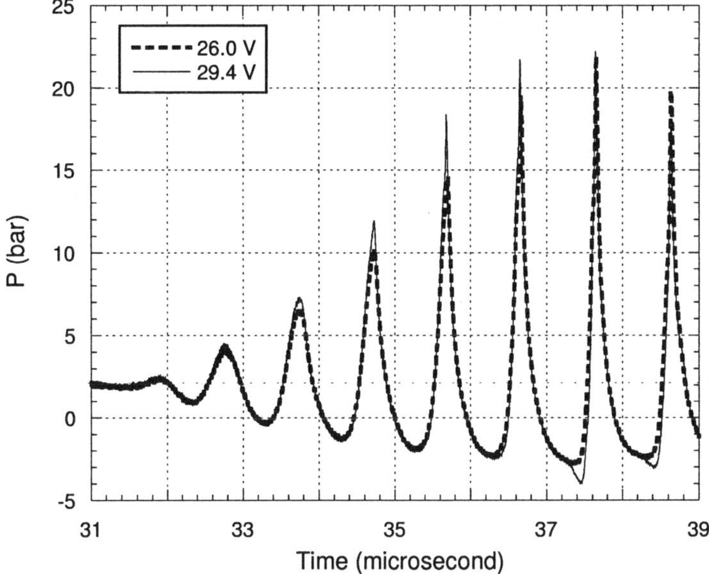

Figure 6. A recording by Chavanne et al.[19] of a large amplitude sound wave showing the existence of non-linear effects. The sinusoidal shape of the wave is distorted in a way consistent with the equation of state of liquid helium 4. The largest negative swings produce heterogeneous cavitation on the glass plate at -3 bar. The presence of such non-linear effects in the focusing of large amplitude acoustic waves makes the calibration of the density or pressure difficult when the glass plate is removed (see text).

pressure and ρ_C and ρ_L are the respective densities of the crystal and the liquid). Furthermore, the number of nucleation sites would be of order 10^7, the size ratio of the acoustic focal region ($\approx 10 \ \mu m^2$) to the critical nucleus ($\approx 1 \ nm^2$). Thanks to the cleanliness of their experiment, Chavanne et al. showed that it is possible to overpressurize liquid helium 100 to 1000 times more than had been previously achieved[21]. After removing their glass plate, they should be able to look for homogeneous nucleation and the possible existence of an instability around + 200 bar, as explained in Section 2. However, as we shall see now, the calibration of such experiments is not easy without a glass plate.

In the case shown in Fig.5 the acoustic intensity is moderate and the wave nearly sinusoidal. However, one can see on Fig.6 that, if the sound amplitude is very large, its shape is distorted by non-linear effects. Indeed, the sound velocity is much smaller for the minima than for the maxima of the wave. We have shown from both measurements and numerical integration of the propagation equations, that the wave minima have to be broader and smaller in amplitude than the maxima. This is due to the curvature of the equation of state, and this effect is enhanced by the focusing[22]. As a consequence, without precise knowledge of the non-linear effects, it is difficult to relate the magnitude of the positive swings or that of the negative swings to the excitation voltage. In the case shown in Fig.6, cavitation occurs if the density is low enough. In the next sections, we review experimental studies of cavitation. They were done without a glass plate to ensure that the

154

nucleation of bubbles was homogeneous. The calibration of the density or pressure
at the nucleation time was difficult.

5. Experiments on liquid helium at negative pressure

In its upper part, Fig.7 presents most of the measurements of the cavitation pressure in helium 4. The early data by Sinha et al.[23] were obtained at positive
pressure near the critical point, by heating a bismuth crystal. All the others
have been obtained with an acoustic method similar to the one described in
the previous section. It was first introduced by Nissen et al.[24]. A hemispherical
acoustic transducer focuses a high intensity sound wave and cavitation is detected
optically. In the absence of glass plate, the acoustic focus is far from any wall, so
that cavitation is homogeneous (it is an intrinsic property of the liquid). Nissen
et al.[24] calibrated the amplitude of the focused wave by two methods. They
first calculated the displacement of the inner wall of the transducer from the
characteristics of its electrical resonance. They also analyzed the light scattering
from the density oscillation at the focus. The two methods led to consistent results.
The recent experiments by Caupin et al.[7] are shown as vertical bars between an
upper bound and a lower bound. Indeed, they also used the characteristics of
the electrical resonance of their transducer to calibrate their amplitude, but they
noticed that, in the presence of non-linear effects, this only gives a lower bound for
the cavitation pressure. For the upper bound, they studied the cavitation voltage
as a function of the static pressure in their experimental cell. The last group
of data by Pettersen et al.[26] were not calibrated: the cavitation threshold was
given in Volt, not in bar. For this review, we have adjusted them to agree with
Nissen, and realized that this gave agreement also with the data by Caupin at low
temperature. All the data by Caupin, Pettersen, Nissen and Sinha look consistent
with each other. The data by Hall[27] appear to be slightly above all others. They
were calibrated from a comparison with the pressure at which "electron bubbles"
exploded, which can be calculated[28].

Fig.7 calls for several remarks. One first sees that it is only at low temperature
that the results by Caupin discriminate between the standard nucleation theory
and the more elaborate calculation done in Barcelona. Only the latter accounts for
the existence of a spinodal limit. However, above about 1.5K, nucleation occurs
far above the spinodal limit, and the size of the critical nucleus is rather large. Its
energy can be calculated as that of a bubble with a sharp surface whose energy per
unit area is nothing but the macroscopic surface tension. This is the "thin wall"
approximation of the "standard theory". Nissen used the following expression for
the nucleation rate per unit volume and per unit time from the work of Blander
and Katz[25]:

$$J = N \left(\frac{2\gamma}{\pi m} \right)^{1/2} \exp \left(\frac{-16\pi\gamma^3}{3kT(P_V - P_L)^2} \right), \tag{3}$$

where N is the particle number density of the liquid, m is the molecular mass, γ
is the surface tension, $P_{V,L}$ are the respective pressures in the vapor inside and in
the liquid outside the nucleating bubble. The quantity $P_V - P_L$ is related to the

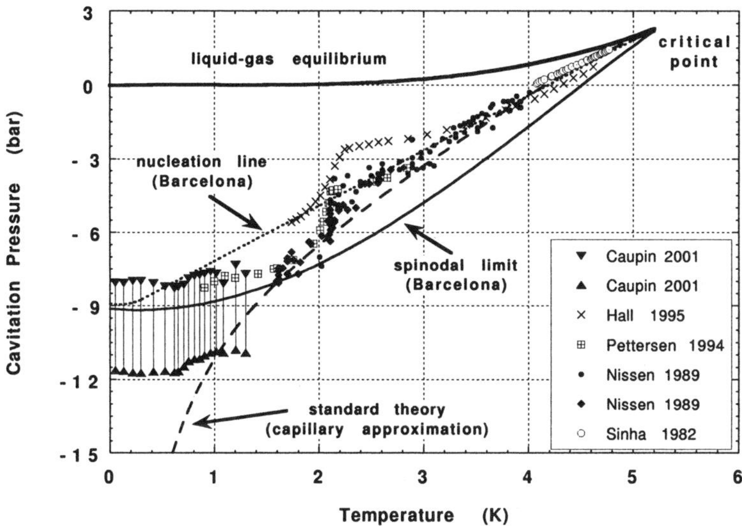

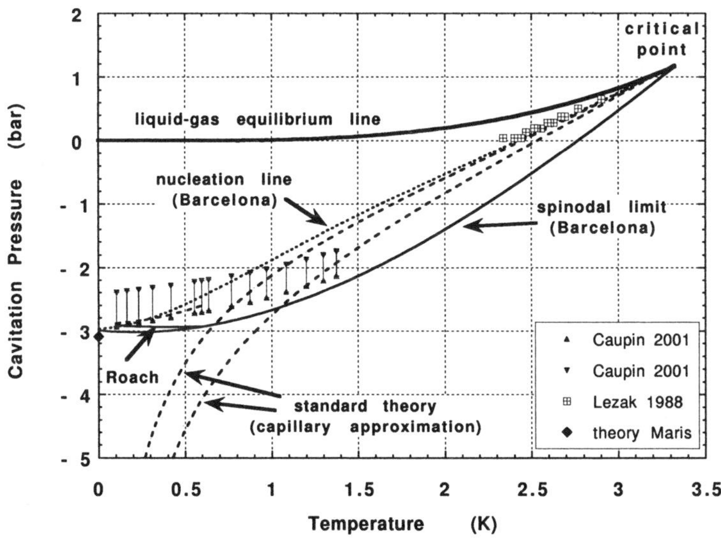

Figure 7. Cavitation in helium 4 (top) and in helium 3 (bottom). Comparison between experiments and theory. The standard theory models the nucleus as a bubble with a surface energy equal to the macroscopic surface tension. The low temperature measurements by Caupin et al. agree with the more elaborate calculations done in Barcelona[10], with a density functional theory which accounts for the existence of a spinodal line.

departure from the equilibrium pressure $P_L - P_{eq}$ by the relation

$$P_V - P_L = (P_{eq} - P_L)\delta, \qquad (4)$$

where $\delta = 1 - \rho_V/\rho_L$. The dashed curve in this figure is obtained with Eq.3. For this, we have written $JV\tau = 1$ and taken the product of the experimental volume V by the experimental time τ equal to 2×10^{-16} cm^{3}s, as in the experiments by Caupin and by Pettersen.

As can be seen from Eq.3, the cavitation pressure depends only logarithmically on the product $V\tau$. If we had chosen $V\tau = 4 \times 10^{-5}$ cm^{3}s as in the experiment by Sinha et al., the "standard theory" curve would have been slightly above, in better agreement with Sinha's data. This is illustrated in the lower part of Fig.7 which concerns helium 3. In this other case, the two dashed curves respectively correspond to $V\tau = 7.3 \times 10^{-16}$ (the lowest one) and to 4×10^{-5} cm^{3}s.

The curve labelled "nucleation line (Barcelona)" results from successive calculations [10] in the years 1992 to 2002. By using a density functional theory, the density profile of the nucleus was adjusted to minimize its energy E at all temperatures and pressures. This functional accounted for the existence of a spinodal limit whose temperature variation was calculated. The nucleation line was obtained from the equation:

$$E = kT \ln \left(\Gamma_0 V\tau \right) \qquad (5)$$

The "prefactor" Γ_0 was taken as a thermal frequency kT/h divided by the typical volume of a critical nucleus (1nm^3). This is not significantly different from the prefactor in Eq.3. As above, we have chosen the product $V\tau = 2 \times 10^{-16}$cm^{3}s. The Barcelona result is close to a straight line from T^* to the critical point. At the temperature T^* a crossover occurs from a thermally activated nucleation regime to a quantum one where nucleation takes place by quantum tunneling through the energy barrier. The discussion of this quantum regime is beyond the scope of this review article. As explained elswhere[29] experimental results confirm the theoretical prediction that T^* is about 0.2 K. A good agreement between theory and experiments is found after a temperature correction is applied to the data, which is due to the adiabatic cooling in the acoustic wave at low temperature[29].

Fig.7 shows that experiments had to be done at low temperature to test the existence of the spinodal limit and check its location in the phase diagram. Of course, the evidence is clearer below about 1.5K, where the nucleation line of the standard theory crosses the spinodal line. Despite the good agreement found, some disagreement remains in the region from 1.5 to 2.2 K, below the superfluid transition. Despite the large scatter in the experimental data, one sees a cusp in the temperature variation of the cavitation pressure. Furthermore, cavitation seems to occur at a pressure which is more negative than predicted by available theories. It was also noticed by Maris[28] that, in the presence of electon bubbles, the cavitation threshold shows no such cusp. In the latter case, cavitation results from the mechanical instability of already existing seeds (the electron bubbles). There seems to be here an effect which calls for new physical arguments. Is it possible that the spinodal line also has a cusp where the lambda line meets it?

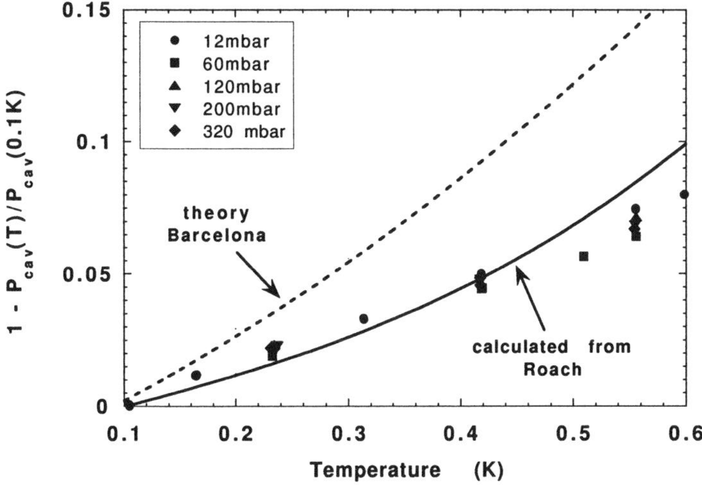

Figure 8. A comparison of experiments in helium 3 with the cavitation curve obtained in Barcelona[10]. The various data points correspond to the relative temperature variation of the cavitation threshold in successive experiments done at different static pressures in the experimental cell. A better agreement is found with a similar theory using a modified spinodal line. Indeed, according to extrapolations of the sound measurements by Roach[11], a shallow minimum should exist in the spinodal line. It is consistent with calculations of the expansion coefficient of liquid helium 3, which was found negative at low temperature by Caupin et al.[32].

Could it be that the prefactor has some anomaly when the temperature is increased through the lambda line? Could this be a consequence of the presence of quantized vortices in the superfluid? Since such vortices should favor cavitation[30], their existence should probably lead to an increase of the cavitation pressure, instead of the observed decrease. The observed cusp has no clear interpretation yet and obviously requires further studies.

Let us now come to the case of helium 3 (lower part of Fig.7). The high temperature data by Lezak[31] were obtained with the same method and in the same group as those by Sinha et al. for helium 4. They show good agreement with the standard nucleation theory if one takes 4×10^{-5} cm^{3}s for the product $V\tau$ (upper dashed curve). As in helium 4, one needs to study cavitation at low enough temperature (below 1K for $V\tau = 7 \times 10^{-16}$ cm^{3}s) if one looks for the existence of a spinodal limit. Caupin et al.[32] confirmed that, near T=0, the spinodal pressure is around -3 bar in helium 3 instead of -9 bar in helium 4. They carefully studied the temperature variation of the cavitation threshold in the range 0.1 to 0.6 K and noticed that it was found weaker than predicted by the Barcelona theory. As shown in Fig.8, they found a better agreement by using a spinodal line having a shallow minimum at 0.4K. In fact, the existence of such a minimum was also found by extrapolating the sound velocity measurements of Roach et al.[11]. It was further supported by considering the sign of the expansion coefficient in liquid helium 3, which is negative at low enough temperature where the Fermi liquid behavior is well established. Caupin et al.[32] thus concluded that liquid helium 3 shows a minimum in its spinodal line, as had been proposed for water by Speedy[12]. The reason is the same: both liquids expand instead of contracting when cooled down at low temperature, but the physical origin of the anomalous sign of α_P is very different in the two liquids. In the temperature range from 0.6 to 1.4 K, Caupin et al. also measured the cavitation threshold, and the agreement with the Barcelona calculation is not very good. As in helium 4, experiments indicate that cavitation occurs at a pressure which is more negative than predicted by theory. It is possible that the prefactor in the calculation is overestimated but this remaining problem also needs further study.

6. Conclusion: open questions

Liquid helium 4 and liquid helium 3 are model systems. They are extremely pure, and homogeneous cavitation can be studied in experiments using high intensity ultrasound. Furthermore, they are simple liquids and it is possible to calculate some features of their phase diagram rather accurately. During the last decade, all theoretical approaches have converged and the spinodal limits in both liquids are now well established from the theoretical point of view. Experiments at low temperature have confirmed most of the predictions and showed where the spinodal limit have to be considered to interpreat cavitation. As noticed in this review, several important questions remain to be answered. Why do experiments show a singularity in the cavitation threshold at the superfluid transition temperature? Moreover, could cavitation studies measure the location of the superfluid transition in the metastable parts of the phase diagram of helium 4? Is there an instability line for the liquid-solid transition at high pressure? Is it around +200 bar as recently suggested, and could it be checked in some future experiments? Eventually, what is the superfluid transition temperature of overpressurized liquid helium 4? Some of these questions are already under experimental and theoretical investigation.

Acknowledgements

We are very grateful to Manuel Barranco, Marti Pi and Montserrat Guilleumas for prividing us the last results of their calculations, which made the comparison with our experiments more accurate and easier to do.

References

1. Maris, H.J. (1991) Critical Phenomena in ^{3}He and ^{4}He at T=0, *Phys. Rev. Lett.* **66**, 45-47; Maris, H.J. (1995) Theory of quantum nucleation of bubbles in liquid helium, *J. Low Temp. Phys.* **98**, 403-424.
2. Edwards, D.O. and Maris, H.J. (unpublished).

3. Campbell, C.E., Folk, R., and Krotscheck, E. (1996) Critical behavior of liquid ^{4}He at negative pressures, *J. Low Temp. Phys.* **105**, 13-36.

4. Bauer, G.H., Ceperley, D.M., and Goldenfeld, N. (2000) Path-integral Monte Carlo simulation of helium at negative pressures, *Phys. Rev. B* **61**, 9055-9060.

5. Boronat, J., Casulleras, J. and Navarro, J. (1994) Monte Carlo calculations for liquid ^{4}He at negative pressure, *Phys. Rev. B* **50**, 3427-3430.

6. Dalfovo, F., Lastri, A., Pricaupenko, L., Stringari, S., and Treiner, J. (1995) Structural and dynamical properties of superfluid helium: a density-functional approach, *Phys. Rev. B* **52**, 1193-1209.

7. Caupin, F. and Balibar, S. (2001) Cavitation pressure in liquid helium, *Phys. Rev. B* **64**, 064507 (1-10).

8. Hall, S.C. and Maris, J. (1997) Thermodynamics and Nucleation of Bubbles in Normal and Superfluid Liquid Helium-4 at Negative Pressures, *J. Low Temp. Phys.*, **107**, 263-282.

9. Guilleumas, M., Pi, M., Barranco, M., Navarro, J., and Solís, M.A. (1993) Thermal nucleation of cavities in liquid helium at negative pressures, *Phys. Rev. B* **47**, 9116-9119.

10. Guilleumas, M., Barranco, M., Jezek, D.M., Lombard, R.J., and Pi, M. (1996) Quantum cavitation in liquid helium, *Phys. Rev. B* **54**, 16135-16138; Guilleumas, M., Barranco, M., Jezek, D.M., Lombard, R.J., and Pi, M. (1996) Thermally assisted quantum cavitation in liquid helium, *Proc. of LT21, Czech. Jour. Phys. Suppl. S1* **46**, 389-390.

11. Roach P.R., Eckstein, Y., Meisel, M.W., and Aniola-Jedrzejek, L. (1983) Thermal expansion, velocity of sound, and compressibility in liquid ^{3}He under pressure, *J. Low Temp. Phys.* **52**, 433-447

12. Speedy R.J. (1982) Stability-limit conjecture. An interpretation of the properties of water. *J. Phys. Chem.* **86**, 982-991.

13. Skripov, V.P. (1994) The metastability boundary in the ^{4}He diagram of state, *Zh. Fiz. Kh.* **68**, 1382-1385 [(1994) *Russ. J. Phys. Chem* **68**, 1252-1255]; Skripov, V.P. (2000) Extension of the λ curve of ^{4}He into the region of the metastable state of liquid helium, *Usp. Fiz. Nauk* **170**, 559-563 [*Phys. Usp.* **43**, 515-519].

14. For an introduction to Landau's theory of superfluidity, more generally to the physics of liquid and solid helium, see Wilks, J. (1967) *The properties of liquid and solid helium*, Clarendon Press, Oxford.

15. Caupin, F. and Balibar, S. (2002) Quantum statistics of metastable liquid helium, *this conference.*

16. Apenko, S.M. (1999) Critical temperature of the superfluid transition in Bose liquids, *Phys. Rev. B* **60**, 3052-3055.

17. Xiong, Q. and Maris, H.J. (1989) Liquid helium at negative pressure : nucleation of bubbles and anomalous phonon dispersion, *J. Low Temp. Phys.* **77**, 347-369.

18. Schneider, T. and Enz, C.P. (1971) Theory of the superfluid-solid transition of ^{4}He, *Phys. Rev. Lett.* **27**, 1186-1188.

19. Chavanne, X., Balibar, S., and Caupin, F. (2001) Acoustic nucleation of solid helium 4 on a clean glass plate, *J. Low Temp. Phys.* **125**, 155-164.

20. Caupin, F., Balibar, S., and Maris, H.J. (2001) Anomaly in the stability limit of liquid helium 3, *Phys. Rev. Lett.* **87**, 145302 (1–4).

21. Balibar, S., Castaing, B. and Laroche, C. (1980) Nucleation and orientation of ^{4}He crystals, *J. Phys. (Paris) Lett.* **41**, 283-285; Tsymbalenko, V.L. (1992) A possible observation of quantum nucleation in superfluid helium 4 near crystallization; Sasaki, Y. and Mizusaki, T. (1998) Temperature independent nucleation of solid helium 4 below 1K, *J. Low Temp. Phys.* **110**, 491-496; Ruutu, J.P., Hakonen, P.J., Penttila, J.S., Babkin, A.V., Saramaki, J.P. and Sonin, E.B. (1996) Evidence for helium 4 crystallization via quantum tunneling at mK temperatures, *Phys. Rev. Lett.* **77**, 2514-2517.

22. Appert, C., Tenaud, C., Chavanne, X., Balibar, S., Caupin, F. and d'Humières, D. (2002), Non-linear effects and shock formation in the focusing of a spherical acoustic wave, submitted to *Eur. Phys. J. B.*.

23. Sinha, D.N., Semura, J.S., and Brodie, L.C. (1982) Homogeneous nucleation in ^{4}He: A corresponding states analysis, *Phys. Rev. A* **26**, 1048-1061.

24. Nissen, J.A. Bodegom, E., Brodie, L.C. and Semura, J.S. (1989) Tensile strength of liquid ^{4}He, *Phys. Rev. B* **40**, 6617-6624.

25. Blander, M.A. and Katz, J.L. (1975) *J. Am. Inst. Chem. Eng.* **21**, 853.

26. Pettersen, M.S., Balibar, S., and Maris, H.J. (1994) Experimental investigation of cavitation in superfluid ^{4}He, *Phys. Rev. B* **49**, 12062-12070.

27. Hall, S.C., Classen, J., Su, C.K. and Maris, H.J. (1995) Experimental evidence for bubble nucleation on electrons in liquid ^{4}He, *J. Low Temp. Phys.* **101**, 793-797.

28. Classen, J., Su, C.K., Mohazzab, M. and Maris, H.J. (1998) Electrons and cavitation in liquid helium, *Phys. Rev. B* **57** 3000-3010; see also Maris, H.J. (2002) *this conference*;.

29. Balibar, S., Caupin, F., Roche, P., and Maris, H.J. (1998) Quantum cavitation: a comparison between superfluid helium-4 and normal liquid helium-3, *J. Low Temp. Phys.* **113**, 459-471.

30. Maris, H.J. (1994) Nucleation of bubbles on quantized vortices in helium-4, *J. Low Temp. Phys.* **94**, 125-144.

31. Lezak, D., Brodie, L.C., Semura, J.S. and Bodegom, E. (1988) Hopmogeneous nucleation temperature of liquid ^{3}He, *Phys. Rev. B* **37**, 150-154.

32. Caupin, F., Balibar, S., and Maris, H.J. (2001) Anomaly in the stability limit of liquid helium 3, *Phys. Rev. Lett.* **87**, 145302 (1-4).

CAVITATION IN ^{3}HE-^{4}HE LIQUID MIXTURES

M. BARRANCO, M. GUILLEUMAS and M. PI
Departament E.C.M., Facultat de Física.
Universitat de Barcelona. 08028 Barcelona, Spain

D.M. JEZEK
Departamento de Física, Universidad de Buenos Aires,
and CONICET. 1428 Buenos Aires, Argentina

J. NAVARRO
IFIC (CSIC and Universitat de València).
46071 València, Spain

1. Introduction

Phase transitions under equilibrium conditions are experimentally well determined and take place in the coexistence regime. However, phase transitions do not always occur under equilibrium conditions. As the new phase forms, the free energy of the system is lowered, but the original phase can be held in a metastable state close to the equilibrium transition point. Although they are internally stable, in each case there exists another configuration that has a lower thermodynamical potential. The metastable state is separated from the stable state by a thermodynamic barrier. Due to statistical fluctuations in density or concentration, this barrier can be overcome as the result of the formation and growth of small clusters of the new phase in the metastable state (bubbles in the liquid or droplets in the vapor; bubbles and droplets will be generically referred to as clusters). Nucleation is the process of the first localized appearance of a new stable phase in a metastable state. The nucleation process can proceed by thermal activation or by quantum tunnelling, and depending on the energy barrier the nucleation rate will be very slow or very fast.

Within the classical theory of nucleation [1], the grand potential of the growing cluster is evaluated in the capillarity approximation. In the case of cavitation for example, this means that the bubble is strictly empty and limited by a sharp surface. Such a macroscopic approximation is intuitively appealing and allows for simple and sometimes analytical estimates, but it has the obvious shortcoming that it cannot describe a surface region having a finite, and sometimes quite large thickness. Moreover, for mixtures made of liquids of limited miscibility, as it is the case of ^{3}He-^{4}He systems at low temperatures, it is impossible to guess which

A.R. Imre et al. (eds.), Liquids Under Negative Pressure, 161–174.
© 2002 *Kluwer Academic Publishers. Printed in the Netherlands.*

162

is the density profile of the 'bubble'. These limitations can be overcome using the density functional theory (DFT) to describe the thermodynamical properties of the system [1, 2, 3]. Density functional theory has been the most successful approach in addressing nucleation in liquid helium so far.

Liquid helium is especially appealing for nucleation studies due to its particular features at low temperatures. Since it does not wet the walls of the experimental cells and the samples can be prepared free of impurities, this avoids undesired heterogeneous nucleation and allows the study of *homogeneous nucleation*, a true property of the bulk liquid. Moreover, both helium isotopes remain liquid at zero temperature. This fact can be potentially exploited to study the transition from thermal to quantum nucleation regimes. As ^{4}He is superfluid below the lambda temperature $T_\lambda = 2.17$ K and ^{3}He is in the normal phase down to 3 mK, the comparative study of both isotopes is expected to shed light on the role played by superfluidity in the manifestation of quantum tunnelling.

In recent years, theoretical and experimental interest has been focused on the study of the negative pressure region of the phase diagram of pure ^{3}He and ^{4}He liquids. Obviously, liquid helium in this region can only be in a metastable state which is obtained by focusing sound waves generated by a hemispherical ultrasonic transducer into the bulk of the liquid. Negative pressures are produced during the negative part of the pressure swing at the acoustic focus [4]. The analysis of these experiments is complicated by the fact that only the static pressure and temperature of the experimental cell are known. Neither the pressure (P) nor the temperature (T) at the focus, where cavitation takes place, can be directly measured. This has caused gross errors in the past when trying to calibrate the potential voltage applied to the transducer in terms of the pressure generated at the focus.

Experiments on supersaturated helium mixtures at $P \geq 0$ [5, 6] are carried out under quite different conditions than those aiming to study cavitation in pure helium. In particular, the time scale of the supersaturation process is considerably large, of the order of 1-3 hours, whereas the ultrasound pulses used for cavitation last between 30 and 70 μsec. The kinetics of nucleation in dilute helium mixtures was first addressed by Lifshitz et al. [7]. We refer the interested reader to two recent works [8, 9] on nucleation in supersaturated ^{3}He-^{4}He mixtures.

In this review we will restrict ourselves to the case of negative pressures. Although no experimental information for mixtures is available up to now in this regime, its study constitutes a natural extension of that carried out in pure liquid helium. We will only address the low temperature regime below $\sim 0.15 - 0.2$ K, as this region can be sensibly studied by current density functionals for liquid helium mixtures [10, 11]. A comprehensive review on nucleation in helium can be found in [12].

2. Thermal cavitation

The formation of a new phase from a metastable phase proceeds through the formation of a 'critical cluster' of the novel phase. The nucleation rate J, i.e., the number of critical clusters formed in the homogeneous system per unit time and volume, is given by the expression

$$J = J_{0T} \exp(-\Delta\Omega_{max}/kT) \; , \tag{1}$$

where $\Delta\Omega_{max}$ is the free energy required to form the critical nucleus and therefore is the difference between the grand canonical potential of the critical cluster and that of the homogeneous metastable system, and k is the Boltzmann constant. The prefactor J_{0T} depends on the dynamics of the nucleation process, and there are many proposals of different degrees of complexity in the literature. It turns out that J_{0T} can be varied by several orders of magnitude without appreciably changing the physical results. A simple estimate of J_{0T} consists in writing it as an 'attempting frequency per unit volume', $J_{0T} = kT/(hV_{cl})$, where V_{cl} is the volume of the critical cluster roughly represented by a sphere of 10 Å radius [2, 3], and h is the Planck constant.

The application of DFT to the cavitation problem starts with the determination of the equilibrium phase diagram, and to delimit the metastable region where it may occur. Within DFT, at $T = 0$ one has a rather simple, analytical expression for the free energy density that allows one to obtain the phase diagram. Let ρ_i be the particle density of the ^{i}He isotope and ρ the total density $\rho = \rho_3 + \rho_4$. The boundaries of the different regions in the (P, x) plane, where x is the ^{3}He concentration $(x = \rho_3/\rho)$, can be determined as follows. Necessary and sufficient stability conditions for a binary system are given by the inequalities on the compressibility

$$K = \left(\frac{\partial P}{\partial \rho}\right)_x \geq 0 \tag{2}$$

and the chemical potentials

$$\text{either} \quad \left(\frac{\partial \mu_4}{\partial x}\right)_P \leq 0 \quad \text{or} \quad \left(\frac{\partial \mu_3}{\partial x}\right)_P \geq 0 \; . \tag{3}$$

A positive compressibility guarantees mechanical stability, whereas the condition on the chemical potentials (one inequality implies the other) ensures diffusive stability. Taken as equalities, the above equations determine two curves on the (P, x) plane which are shown in Fig. 2 as a dot-dashed line (Eq. (2)) and as a dashed line (Eq. (3)). This phase diagram was calculated [10] using a zero temperature DF proposed for the mixture [11]:

$$f(\rho_3, \rho_4) = f_{vol}(\rho_3, \rho_4) + f_{sur}(\rho_3, \nabla\rho_3, \rho_4, \nabla\rho_4) \; , \tag{4}$$

where $f_{vol}(\rho_3, \rho_4)$ only depends on the densities and $f_{sur}(\rho_3, \nabla\rho_3, \rho_4, \nabla\rho_4)$ is a surface term that vanishes in the homogeneous liquid. The explicit form of the density functional can be found in [10].

It can be seen from Fig. 2 that condition Eq. (3) is violated first and thus defines the spinodal line $P_{sp}(x)$, where the mixture becomes macroscopically unstable. At $P = 0$, it cuts the x axis at $x_{sp} \sim 30\%$. To draw the border between metastable

164

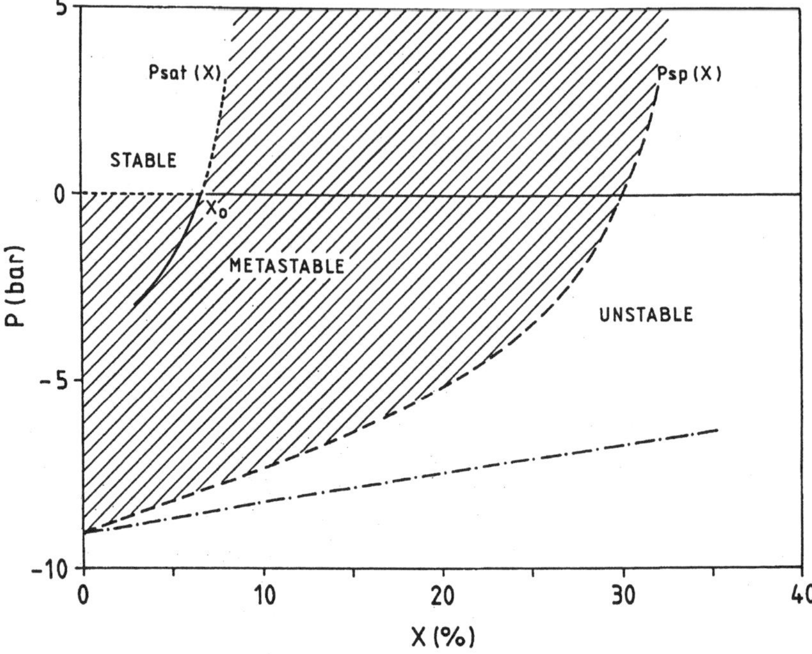

Figure 1. DF phase diagram of the ^{3}He-^{4}He liquid mixture at $T = 0$.

and stable regions one has to obtain the maximum concentration (or saturation) line by solving the two-phase equilibrium conditions

$$P(\rho, x) = P(\rho_{3p}, x = 1) \tag{5}$$
$$\mu_3(\rho, x) = \mu_3(\rho_{3p}, x = 1) \, ,$$

where ρ_{3p} is the density of segregated (pure) ^{3}He. These equations determine the curve denoted as $P_{sat}(x)$ in the (P, x) plane. It is interesting to see that Eqs. (5) have solutions at negative pressures down to the value corresponding to the spinodal point of pure ^{3}He.

Experimentally [13], the ^{3}He concentration in the mixture at saturation is $x_s \sim 6.6\%$ at $P = 0$, which reaches a maximum value of $\sim 9.4\%$ at $P \sim 10$ atm. However, supersaturated ^{3}He-^{4}He mixtures can be found in a metastable state for concentrations above the saturation value; at $P \geq 0$, the formation of critical ^{3}He drops is responsible for phase separation.

As in the pure case, another kind of metastability arises from the application of a tensile strength that drives the system into the negative pressure region. In the case of mixtures, the system either will develop a free surface for low x values, or will segregate ^{3}He, producing a mixture-pure ^{3}He interface as in the $P \geq 0$ regime. Phase separation will proceed in the first case by nucleating bubbles 'coated' with ^{3}He, and in the second case by nucleating ^{3}He-rich drops [10, 14].

Once the phase diagram has been established, the application of DFT to the cavitation problem proceeds in two steps. One first determines the critical cluster for P and x values that correspond to metastable states. This allows calculation of the nucleation barrier $\Delta\Omega_{max}(P,x)$. Next, for given x and T is determined the pressure at which the nucleation rate Eq. (1) times the experimental volume and time $(V\tau)_e$ equals a conventional number, say one $[(V\tau)_e \cdot J = 1]$, indicating that there is an appreciable probability of cluster formation which causes the onset of phase separation. This pressure is called homogeneous cavitation pressure (P_h), and it is one of the magnitudes (or equivalently the tensile strength defined as $-P_h$) that the experiments aim to determine.

At fixed P and x, the density profiles of the critical cluster ρ_3^0, ρ_4^0 are obtained by solving the coupled Euler-Lagrange (E-L) equations for the grand potential density $\omega(\rho_3,\rho_4) = f(\rho_3,\rho_4) - \mu_3\rho_3 - \mu_4\rho_4$:

$$\frac{\delta\omega}{\delta\rho_3} = 0 \qquad \frac{\delta\omega}{\delta\rho_4} = 0, \qquad (6)$$

imposing the physical conditions that $\rho_i'(0) = 0$, and $\rho_i(r \to \infty) = \rho_{im}$, where ρ_{im} is the particle density of isotope i in the metastable homogeneous liquid. The cavitation barrier $\Delta\Omega_{max}$ is

$$\Delta\Omega_{max} = \int d\mathbf{r}\,[f(\rho_3,\rho_4) - f_{vol}(\rho_{3m},\rho_{4m}) - \sum_i \mu_i(\rho_i - \rho_{im})] . \qquad (7)$$

This equation gives $\Delta\Omega_{max}$ as a function of x and $P = -f_{vol}(\rho_{3m},\rho_{4m})+\sum_i \mu_i\rho_{im}$. Several cavitation barriers are shown in Fig. 2 for $x = 0$ and $1\% - 6\%$. It can be seen that the barriers diverge near the saturation curve, and become negligible when the system approaches the spinodal line. The latter fact is missed in the capillarity approximation, which is completely unphysical in this region. This is crucial for liquid helium, since cavitation always takes place near the spinodal line.

In Fig. 2 we display some density profiles corresponding to the critical clusters for $x = 4\%$ and $P = -7, -5, -4,$ and -2 bar. As mentioned, the cavitation seed evolves from a ^{4}He bubble coated with ^{3}He to a ^{3}He-rich droplet. It is quite apparent from this figure that the flexibility of the DF approach is especially indicated for helium mixtures, where the limited miscibility of both isotopes at low T and the existence of surface Andreev states make the shape of the nucleation clusters hard to guess and mimic by means of simple-minded sharp-surface models.

Figure 2 shows P_h as a function of T and x. Thermal and quantum (see next Section) regimes are displayed. The dashed line is the extrapolation of the thermal regime to temperatures close to $T = 0$. One may see that even for the lower x, the differences with pure ^{4}He are appreciable. This is attributed to the sizeable dependence of the surface tension of the liquid mixture to the ^{3}He concentration [14]. Results for x values above 6% can be found in this reference.

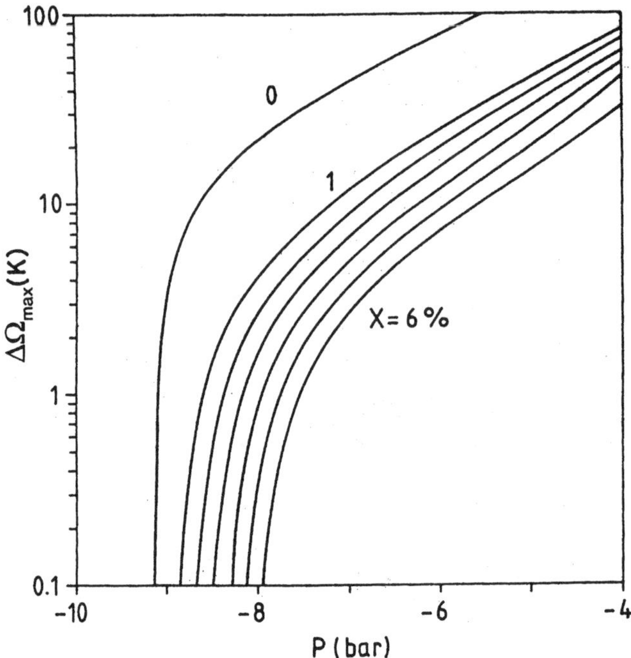

Figure 2. From top to bottom, cavitation barriers as a function of P for pure ^{4}He and for ^{3}He concentrations $x = 1\% - 6\%$.

3. Quantum cavitation

So far we have discussed that cavitation is a thermally activated process at high enough temperatures and the energy barrier is overcome by the energy provided to the system by a thermal bath. At low enough T this is no longer possible. However, cavitation may proceed by quantum tunnelling: the metastable state 'tunnels' through the energy barrier. The transition from one regime to the other is very abrupt, so that a thermal-to-quantum crossover temperature T^* may be defined by indicating whether nucleation takes place thermally ($T > T^*$) or quantically ($T < T^*$). In the limit of zero temperature, the transition is purely quantal, but for $T^* > T > 0$ thermally assisted quantum nucleation is the physical process.

In this section we present a determination of T^* for liquid helium mixtures based on the application of the functional-integral approach (FIA) in conjunction with the DFT description of liquid helium [15].

For $T < T^*$ the tunnelling rate is

$$J_Q = J_{0Q} \exp(-S^Q) , \qquad (8)$$

where $\mathcal{P} = \exp(-S^Q)$ is the tunnelling probability, and the prefactor J_{0Q} is of the order of the number of nucleation sites per unit volume times an attempting frequency. According to Feynman [16], the tunnelling probability can be written

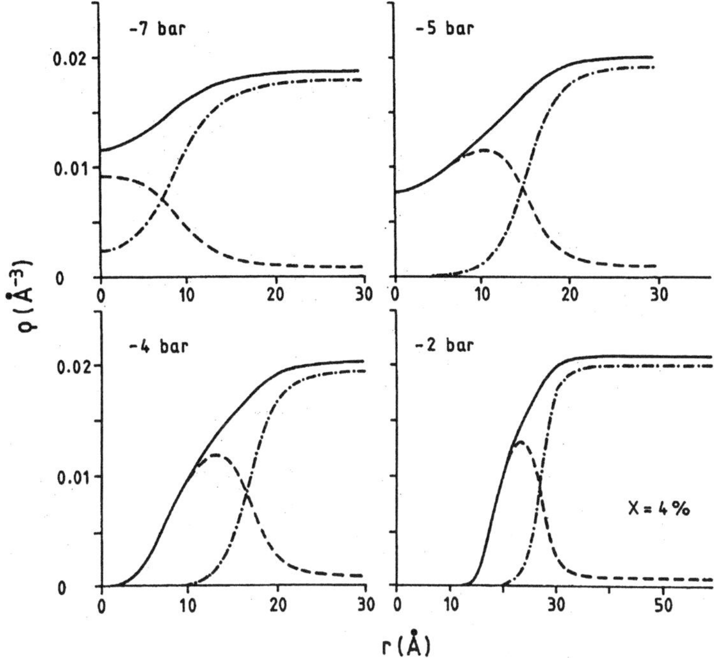

Figure 3. Density profiles corresponding to the critical bubbles for $x = 4\%$ and several values of P. The solid line represents the total density, and the dash-dotted (dashed) line, the ρ_4 (ρ_3) density.

as the functional integral

$$\int D[q(\tau)] \exp\left[-\frac{1}{\hbar}\oint d\tau \mathcal{L}[q(\tau)]\right] , \qquad (9)$$

where $\mathcal{L}[q(\tau)]$ is the imaginary-time ($\tau = it$) classical Lagrangian of the system and $D[q(\tau)]$ denotes integration over all periodic trajectories $q(\tau)$ with period $\tau_p = \hbar/kT$ in the potential well that results from inverting the energy barrier. The integral in the exponent of Eq. (9) is the imaginary-time action $S(T) = \oint d\tau \mathcal{L}$ evaluated over the period τ_p. In the semiclassical limit $S(T) \gg \hbar$, the trajectory that contributes the most from all possible periodic orbits is that which minimizes the action. It leads to Eq. (8) with $S^Q = S_{min}(T)/\hbar$, where S_{min} is the minimum action.

For helium mixtures, the method is extremely cumbersome to apply, but nevertheless it has been worked out [17]. Since T^* is expected to be small, one resorts to a zero temperature DF and first determines the critical cluster densities ρ_3^0, ρ_4^0 by solving Eqs. (6) and the barrier height $\Delta\Omega_{max}$ from Eq. (7). The next step consists in describing the dynamics of the cavitation process in the inverted barrier well, whose equilibrium configuration corresponds to (ρ_3^0, ρ_4^0) and has an energy $-\Delta\Omega_{max}$. The imaginary-time Lagrangian $\mathcal{L}$ can be easily written supposing that the collective velocities $\vec{u}_q(\vec{r}, t), q = 3, 4$ of both helium fluids associated with the

168

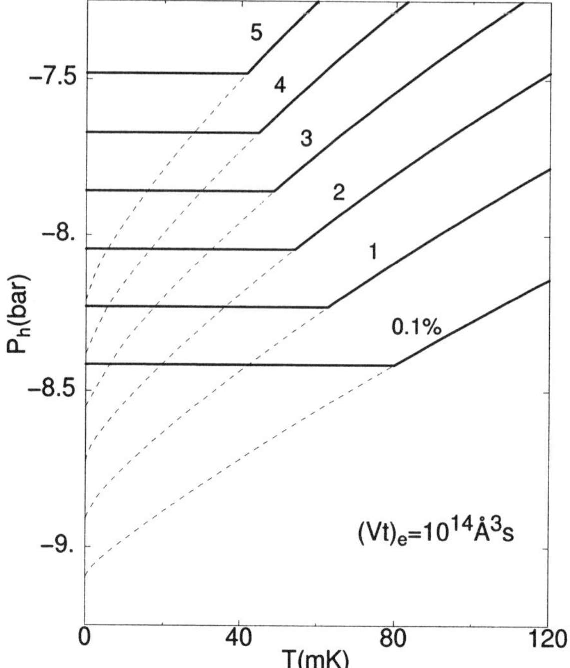

Figure 4. Homogeneous cavitation pressure P_h as a function of T for different ^{3}He concentrations. The dashed line is the extrapolation of the thermal regime to temperatures below T^*.

bubble growth are irrotational, which is not a severe restriction as one expects only radial displacements (spherically symmetric bubbles). One can then define a velocity potential field $s_q(\vec{r}, t)$ for each isotope such that $\vec{u}_q(\vec{r}, t) = \nabla s_q(\vec{r}, t)$. It follows that

$$\mathcal{L} = \sum_q m_q \dot{\rho}_q s_q - \mathcal{H}(\rho_3, \rho_4, s_3, s_4) \,, \tag{10}$$

where $\mathcal{H}(\rho_q, s_q)$ is the imaginary-time Hamiltonian density

$$\mathcal{H}(\rho_3, \rho_4, s_3, s_4) = \frac{1}{2} \sum_q m_q \rho_q \vec{u}_q{}^2 - [\omega(\rho_3, \rho_4) - \omega(\rho_{3m}, \rho_{4m})] \,. \tag{11}$$

Hamilton's equations yield the following four equations :

$$m_q \dot{\rho}_q = \frac{\delta \mathcal{H}}{\delta s_q} = -m_q \nabla(\rho_q \vec{u}_q) \tag{12}$$

$$m_q \dot{s}_q = -\frac{\delta \mathcal{H}}{\delta \rho_q} \,. \tag{13}$$

Eqs. (12) are the continuity equations. Taking the gradient of Eqs. (13) one gets the equations of motion

$$m_q \frac{d\vec{u}_q}{dt} = -\nabla \left\{ \frac{1}{2} m_q \vec{u}_q{}^2 - \frac{\delta\omega}{\delta\rho_q} \right\} . \tag{14}$$

To determine T^* one has to find the small amplitude, periodic solutions of Eqs. (12) and (14) linearized around ρ_3^0 and ρ_4^0. Defining the 'transition densities' $\rho_q^1(r)$ as

$$\rho_q(\vec{r}, t) \equiv \rho_q^0(r) + \rho_q^1(r) \cos(\omega_p t) , \tag{15}$$

and keeping only first-order terms in $\vec{u}_q(r, t)$ and in $\rho_q^1(r)$, one gets:

$$\omega_p^2 \rho_q^1(r) = \frac{1}{m_q} \nabla \left[\rho_q^0(r) \nabla \left(\sum_{q'=3,4} \frac{\delta^2\omega}{\delta\rho_q \delta\rho_{q'}} \bullet \rho_{q'}^1(r) \right) \right] , \quad q = 3, 4. \tag{16}$$

In this equation, $\frac{\delta^2\omega}{\delta\rho_q \delta\rho_{q'}} \bullet \rho_{q'}^1(r)$ means that $\delta\omega/\delta\rho_q$ has to be linearized, keeping only terms in ρ_3^1 and ρ_4^1, and their derivatives.

Eq. (16) is a fourth-order linear differential, eigenvalue equation for the 'vector' $(\rho_3^1(r), \rho_4^1(r))$. Physical solutions to Eqs. (16) have to fulfill $(\rho_q^1)'(0) = (\rho_q^1)'''(0) = 0$, and have to fall exponentially to zero at large distances. For a given pressure and ^{3}He-concentration, only a positive eigenvalue ω_p^2 has been found, from which we get $T^* = \hbar\omega_p/2\pi$.

Figure 3 shows T^* (mK) as a function of P (bar) for $x = 0.1, 1, 2, 3, 4$ and 5 %. Compared to the pure ^{4}He case [15], $T^*(P)$ has now a more complex structure. It is worth noting that the maximum of the $T^*(P)$ curve has decreased from $\sim$240 mK for pure ^{4}He down to $\sim$140 mK for ^{3}He-concentrations as small as 1%.

Figure 3 shows two different bubble configurations for $x = 1\%$. Configuration (a) corresponds to $P = -8$ bar and $T^* = 67.6$ mK, and configuration (b) to $P = -5$ bar and $T^* = 102.1$ mK. The solid lines represent the ^{3}He and ^{4}He critical bubble densities in Å^{-3}, and the dashed (dash-dotted) lines represent $\rho_3^1(r)$ $(\rho_4^1(r))$ in arbitrary units. Near the spinodal region, the 'bubble' configuration is filled with ^{3}He: the surface tension that matters for bubble formation is that of the ^{3}He-^{4}He interface. Away from the spinodal region (configuration (b)), the critical bubble is a true bubble covered with ^{3}He: the surface tension that matters now is that of the ^{3}He-^{4}He liquid free-surface, which is about ten times larger than the previous surface tension.

It is interesting to see that the transition densities ρ_q^1 evolve from those corresponding to 'volume oscillations' (Fig. 3, panel (a)) to 'surface oscillations' for ^{4}He, and a mixed surface-volume type for ^{3}He (panel (b)), to eventually become pure surface oscillations for both isotopes when we go from the spinodal towards the saturation line.

The different surface tensions involved in these processes, together with the existence of a ^{3}He-^{4}He segregation curve at negative pressures down to $x \sim 2.4\%$ (see Fig. 2) are the cause of the structures displayed in Fig. 3. Using the surface or volume character of the transitions densities as a useful guide, the figure can

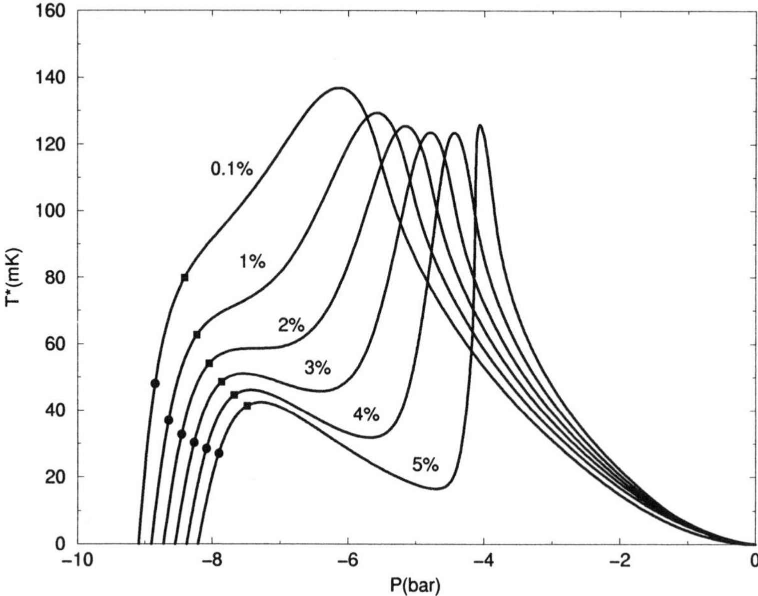

Figure 5. T^* as a function of P for the same x values as in Fig. 2. $P_h(T^*)$ is shown as circles(squares) for $(V\tau)_e = 10^4 \text{Å}^3 \sec (10^{14} \text{Å}^3 \sec)$, respectively.

be understood as follows. Below $x \sim 2.4\%$, no pure ^{3}He drop can 'co-exist' with the homogeneous mixture, and the critical configurations look as drawn in Fig. 3. Above $x \sim 2.4\%$, the situation changes, and the existence of a segregation line allows the system to develop critical configurations resembling pure ^{3}He drops 'co-existing' with the mixture. Thus, one first finds the kind of configurations that, as before, correspond to pressures close to the spinodal line and originate a rise in T^* (left-hand side of all curves in Fig. 3). Next, one finds ^{3}He-rich droplets embedded in the mixture, whose interface vibration originates the first decrease of $T^*(P)$, followed by a ^{3}He volume vibration which causes the rise at the second maximum. At pressures closer to zero, the ^{3}He drop is reabsorbed, the critical configuration is that of a ^{4}He bubble covered with ^{3}He, and the second decrease of the $T^*(P)$ curve is eventually associated with surface vibrations of the mixture free-surface. Here we do not give any further detail since, as in the pure ^{4}He case, only the part of the $T^*(P)$ curve near the spinodal region is relevant for the cavitation problem.

$P_h(T^*)$ is shown as circles (squares) on the curves in Fig. 3. The circles correspond to $(V\tau)_e = 10^4 \text{Å}^3$ sec, and the squares to 10^{14}Å^3 sec. Compared to the pure ^{4}He case, and depending on the $(V\tau)_e$ value, for $x = 1\%$ T^* has been reduced by a factor of 4 or 5, respectively. P_h has been shown in Fig. 2.

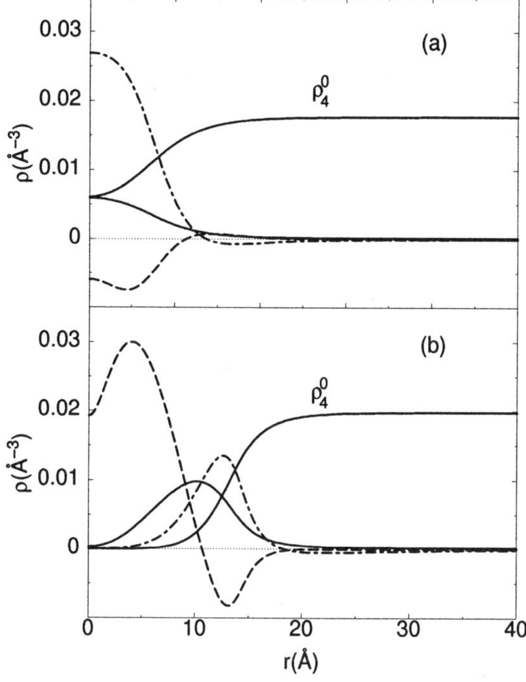

Figure 6.   Particle densities $\rho_4^0(r)$ and $\rho_3^0(r)$ of the critical bubbles (solid lines), and the $\rho_4^1(r)$ (dash-dotted lines) and $\rho_3^1(r)$ (dashed lines) transition densities for $x = 1\%$, corresponding to: (a) $P = -8$ bar and $T^* = 67.6$ mK. (b) $P = -5$ bar and $T^* = 102.1$ mK. $\rho_q^1(r)$ are drawn in arbitrary units, and $\rho_q^0(r)$ in Å^{-3}.

4. Cavitation on vortex lines

An interesting case of heterogeneous cavitation in superfluid liquid helium is that caused by the existence of quantized vortices. The fact that the presence of quantized vortices in liquid ^{4}He below T_λ results in a decrease of the tensile strength has been known experimentally for quite a long time [18]. Detailed DFT calculations [19] have indeed reproduced this effect.

On the other hand, the degree of critical supersaturation reached in the experiments on supersaturated helium mixtures at $P \geq 0$ is very small, of the order of 1% [5, 6], whereas the calculations yield a value of $\sim 10\%$ [7, 8]. A possible explanation of this discrepancy is that phase separation may be caused by heterogeneous nucleation on vortices [8]. To conclude this review, let us see how the presence of vortex lines in the mixture affects the cavitation process [20].

To this end, the energy density for the mixture $f(\rho_3, \rho_4)$, Eq. (4), has been sup-

172

plemented with a centrifugal energy term associated with the vortex velocity field
in the Feynman-Onsager approximation, namely $\hbar^2 \rho_4/(2m_4 r^2)$, where r is now the
radial distance to the vortex line placed along the z axis (cylindrical symmetry).
The E-L equations are solved again for given x and P, and in the metastability
region two different solutions for the *same* P and x conditions are obtained. The
lower energy configuration corresponds to a metastable vortex, and the higher
energy configuration corresponds to a metastable, critical configuration[21]. The
barrier height per unit vortex length is

$$\Delta\Omega_{max} = 2\pi \int r dr \left[\omega(\rho_3^c, \rho_4^c) - \omega(\rho_3^m, \rho_4^m) \right] , \tag{17}$$

where ρ_q^c and ρ_q^m are the particle densities of the critical and metastable vortices,
respectively.

In Fig. 4 we have plotted the critical and metastable density profiles corresponding to a configuration with $P = -1.66$ bar, $x = 1\%$. This figure illustrates
that metastable and critical vortex line configurations can indeed be found in the
$P < 0$ metastability region.

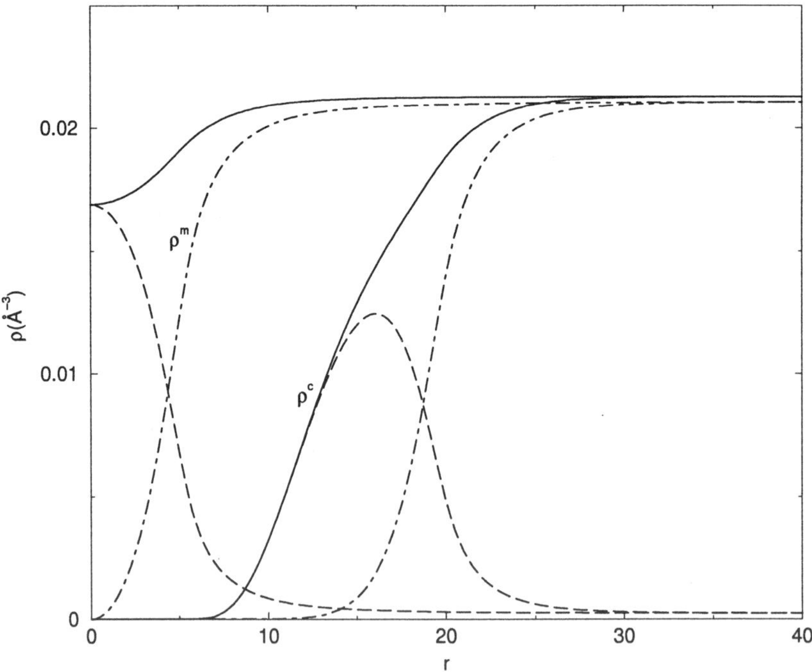

Figure 7. Vortex density profiles for $x = 1\%$ and $P = -1.66$ bar as a function of r (Å). The
solid lines represent the total particle density, and the dash-dotted (dashed) lines, the ρ_4 (ρ_3)
densities. The critical (metastable) configuration is denoted as ρ^c (ρ^m).

Finally, Fig. 4 displays several barrier heights per unit vortex length. A comparison with the results displayed in Fig. 2 shows that the presence of vortex

lines decreases $|P_{sp}|$ by several bars, even for small x's. This is a manifestation of the well-known general result that heterogeneous cavitation takes place at smaller tensile strengths.

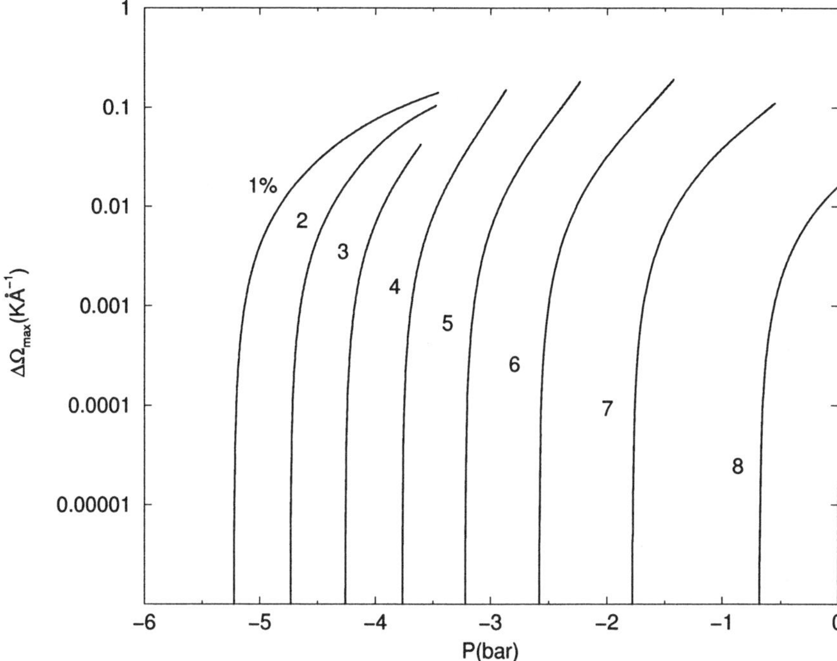

Figure 8. Barrier height per unit vortex length as a function of P for the indicated ^{3}He concentrations.

Acknowledgments

This work has been performed under grant 2000SGR00024 from Generalitat de Catalunya. D. M. J. acknowledges the CONICET (Argentina) and the Generalitat de Catalunya ACI program for financial support.

References

1. Oxtoby, D.W. (1992) Homogeneous nucleation: theory and experiment, *J. Phys.: Condens. Matter* **Vol. no. 4**, pp 7627-7650.
2. Q. Xiong, Q. and Maris, H.J. (1989) Liquid helium at negative pressures: nucleation of bubbles and anomalous phonon dispersion, *J. Low Temp. Phys.* **Vol. no. 77**, pp. 347-369.
3. Jezek, D.M., Guilleumas, M., Pi, M., Barranco, M., and Navarro, J. (1993) Thermal nucleation and cavitation in ^{3}He and ^{4}He, *Phys. Rev. B* **Vol. no. 48**, pp. 16582-16588.
4. Lambaré, H., Roche, P., Balibar, S., Maris, H.J., Andreeva, O.A., Guthmann, C., Keshishev, K.O. and Rolley, E. (1998) Cavitation in superfluid helium-4 at low temperature, *Eur. Phys. J. B* **Vol no. 2**, pp. 381-391.

5. Satoh, T., Morishita, M., Ogata, M. and Katoh, S. (1992) Critical supersaturation of ^{3}He-^{4}He liquid mixtures: Decay of metastable states at ultralow temperatures, *Phys. Rev. Lett.* **Vol no. 69**, pp. 335-338.

6. Maidanov, V.A., Mikheev, V.A., Mikhin, N.P., Omelaenko, N.F., Rudavskii, E.Ya., Chagovets, V.K. and Sheshin, G.A. (1992) Supersaturation of ^{3}He-^{4}He superfluid solutions in the phase separation region, *Sov. J. Low Temp. Phys.* **Vol. no. 18**, pp. 663-669.

7. Lifshitz, I.M., Polesskii, V.M. and Khokhlov, V.A. (1978) Kinetics of nucleation and stratification of dilute ^{3}He-^{4}He solutions under pressure at low temperatures, *Sov. Phys. JETP* **Vol no. 47**, pp. 137-140.

8. Barranco, M., Guilleumas, M., Jezek, D.M., Lombard, R.J., Navarro, J., and Pi, M. (1999) Nucleation in dilute ^{3}He-^{4}He liquid mixtures at low temperatures, *J. Low Temp. Phys.* **Vol. no. 117**, pp. 81-100.

9. Tanaka, E., Hatakeyama, K., Noma, S., Burmistrov, S.N., and Satoh, T. (2002) Critical supersaturation of superfluid ^{3}He-^{4}He mixtures, *J. Low Temp. Phys.* **Vol. no. 127**, in press.

10. Guilleumas, M., Jezek, D.M., Pi, M., Barranco, M. and Navarro, J. (1995) Cavitation in ^{3}He-^{4}He liquid mixtures at low temperatures, *Phys. Rev. B* **Vol. no. 51**, pp. 1140-1146.

11. Dalfovo, F. and Stringari, S. (1985) Hartree-Fock calculations for ^{3}He-^{4}He mixtures at zero temperature, *Phys. Lett. A* **Vol. no. 112**, pp. 171-174.

12. Barranco, M., Guilleumas, M., Pi, M. and Jezek, D.M. (2002) Cavitation in liquid helium, *Quantum liquids in Confined Geometries*, Krotscheck, E., and Navarro, J. Eds., World Scientific, Singapore.

13. Edwards, D.O. and Pettersen, M.S. (1992) Lectures on the properties of liquid and solid ^{3}He-^{4}He mixtures at low temperatures, *J. Low Temp. Phys.* **Vol. no. 87**, pp. 473-523.

14. Guilleumas, M., Pi, M., Barranco, M., Jezek, D.M. and Navarro, J. (1995) Nucleation in supersaturated solutions of ^{3}He in ^{4}He at negative pressures, *Phys. Rev. B* **Vol. no. 52**, pp. 1210-1214.

15. Guilleumas, M., Barranco, M., Jezek, D.M., Lombard, R.J. and Pi, M. (1996) Quantum cavitation in liquid helium, *Phys. Rev. B* **Vol. no 54**, pp. 16135-16138.

16. Feynman, R.P. (1972) *Statistical Mechanics*. Benjamin, New York.

17. Jezek, D.M., Guilleumas, M., Pi, M. and Barranco, M. (1997) Thermally assisted quantum cavitation in solutions of ^{3}He in ^{4}He, *Europhys. Lett.* **Vol. no 38**, pp. 601-606.

18. McConnell, P.M., Chu, M.L., and Finch, R.D. (1970) Mechanism of ultrasonic cavitation nucleation in liquid helium by quantized vortices, *Phys. Rev. A* **Vol. no. 1**, pp. 411-418.

19. Maris, H.J. (1994) Nucleation of bubbles on quantized vortices in helium-4, *J. Low Temp. Phys.* **Vol. no. 94**, pp. 125-144.

20. Jezek, D.M., Guilleumas, M., Pi, M. and Barranco, M. (1997) Stability of vortex lines in liquid ^{3}He-^{4}He mixtures at zero temperature, *Phys. Rev. B* **Vol. no. 55**, pp. 11092-11095.

21. Jezek, D.M., Guilleumas, M., Pi, M. and Barranco, M. (1995) Critical supersaturation of ^{3}He-^{4}He liquid mixtures at low temperature, *Phys. Rev. B* **Vol. no. 51**, pp. 11981-11983.

A QUANTUM MONTE CARLO STUDY OF THE NEGATIVE PRESSURE REGIME IN QUANTUM LIQUIDS

J. BORONAT and J. CASULLERAS
Departament de Física i Enginyeria Nuclear,
Universitat Politècnica de Catalunya,
Campus Nord B4-B5, E-08034 Barcelona, Spain

1. Introduction

In the last years, several theoretical studies of negative pressures [1, 2, 3] and cavitation [4, 5] in liquid helium have been published. Apart from its intrinsic theoretical interest, the progress achieved has also been stimulated by the successful experimental work in this area [6, 7]. From long time, liquid helium has been chosen as an ideal benchmark for thermal and quantum cavitations and for the negative pressure regime [8]. There are several relevant reasons for that interest. The first one is its *simple* character, i.e., its lack of internal structure (electronic activation energies are ten thousands larger than binding energies of the liquid), which permits avoiding the difficulties associated to chemical bondings, dipolar moments, internal degrees of freedom, and so on. Secondly, bulk helium is an extremely pure liquid with essentially no impurities other than isotopic. This unusual feature is a fundamental aspect in order to achieve experimental homogeneous nucleation and it is hardly reachable in other liquids. Finally, helium remains liquid down to zero temperature. This special circumstance makes possible to strongly reduce thermal effects and, more importantly, to approach the necessary conditions for observing quantum cavitation. These properties, together with the superfluid character of helium isotopes below the critical temperature, give raise to a really unique system.

A relevant concern for theory has been the determination of the spinodal line in order to *complete* the equation of state. First attempts to determine the spinodal point, in the limit of zero temperature, were based on extrapolations to negative pressures of the experimental equation of state [9]. These extrapolations used either expected/predicted behaviors of the speed of sound with pressure or functional forms coming from density functional theories [1]. Both approaches arrived to similar predictions but the uncertainty introduced by the extrapolation could not be determined. Microscopic approaches allowed for a cleaner determination since no more extrapolations were required. Both variational theory, based on the HNC framework [3], and quantum Monte Carlo methods [10] have been applied to characterize the physics of helium at negative pressure. In particular, diffusion

A.R. Imre et al. (eds.), Liquids Under Negative Pressure, 175–186.

Monte Carlo calculations, that reproduce accurately the experimental equations of state of liquid ^{4}He [2, 11] and liquid ^{3}He [12], provide the most accurate results for that regime. The continuation of the spinodal line when the temperature increases has also been recently studied using the path integral Monte Carlo method [13].

In the following sections, we report diffusion Monte Carlo results for both liquid ^{4}He and liquid ^{3}He obtained in our group in the last years. Previously, in Sec. II we briefly introduce the basics on the DMC method with some remarks on the difficulties fermion statistics introduces.

2. The diffusion Monte Carlo method

The diffusion Monte Carlo (DMC) method solves in a stochastic way the imaginary-time Schrödinger equation for the wave function

$$f(\boldsymbol{R}, t) \equiv \psi(\boldsymbol{R})\, \Psi(\boldsymbol{R}, t) \ , \tag{1}$$

where $\psi(\boldsymbol{R})$ is a time-independent trial wave function, that describes approximately the ground state of the system at the variational level, and $\Psi(\boldsymbol{R}, t)$ is the exact wave function. The introduction of $\psi(\boldsymbol{R})$ corresponds to the use of importance sampling, a standard Monte Carlo technique that helps to reduce the variance and make the problem attainable. Considering a Hamiltonian of the form

$$H = -\frac{\hbar^2}{2\,m}\, \boldsymbol{\nabla}_{\boldsymbol{R}}^2 + V(\boldsymbol{R}) \ , \tag{2}$$

the Schrödinger equation that DMC solves is

$$-\frac{\partial f(\mathbf{R}, t)}{\partial t} = -D\boldsymbol{\nabla}_{\boldsymbol{R}}^2 f(\boldsymbol{R}, t) + D\boldsymbol{\nabla}_{\boldsymbol{R}} \left(F(\boldsymbol{R}) f(\boldsymbol{R}, t)\right) + \left(E_L(\boldsymbol{R}) - E\right) f(\boldsymbol{R}, t) \ . \tag{3}$$

In Eq. (3), $D = \hbar^2/(2m)$, $E_L(\boldsymbol{R}) = \psi(\boldsymbol{R})^{-1} H \psi(\boldsymbol{R})$ is the local energy, and $F(\boldsymbol{R}) = 2\psi(\boldsymbol{R})^{-1}\boldsymbol{\nabla}_{\boldsymbol{R}}\psi(\boldsymbol{R})$ is a quantum drift force which guides the diffusion process to regions where Ψ is expected to be large.

The r.h.s. of Eq. (3) can be written as the action of three operators A_i acting on the wave function $f(\boldsymbol{R}, t)$,

$$-\frac{\partial f(\boldsymbol{R}, t)}{\partial t} = \left(A_1 + A_2 + A_3\right) f(\boldsymbol{R}, t) \equiv A\, f(\boldsymbol{R}, t) \tag{4}$$

The three terms A_i may be interpreted by similarity with classical differential equations. The first one, A_1, corresponds to a free diffusion with a a diffusion coefficient D; A_2 acts as a driving force due to an external potential, and finally A_3 looks like a birth/death term. In Monte Carlo, the Schrödinger equation (4) is best suited when it is written in a integral form by introducing the Green function $G(\boldsymbol{R}', \boldsymbol{R}, t)$, which gives the transition probability from an initial state $\boldsymbol{R}$ to a final one $\boldsymbol{R}'$ during a time t,

$$f(\boldsymbol{R}', t + \Delta t) = \int G(\boldsymbol{R}', \boldsymbol{R}, \Delta t)\, f(\boldsymbol{R}, t)\, d\boldsymbol{R} \ . \tag{5}$$

More explicitly, the Green function is given in terms of the operator A by

$$G(\boldsymbol{R}', \boldsymbol{R}, \Delta t) = \langle \boldsymbol{R}' | \exp(-A\Delta t) | \boldsymbol{R} \rangle \ . \tag{6}$$

The DMC algorithm relies on reasonable approximations of $G(\boldsymbol{R}', \boldsymbol{R}, \Delta t)$ for values of the time-step $\Delta t \to 0$ [10, 14]. Then, Eq. (6) is iterated repeatedly by using that short-time approximation until reaching the asymptotic regime $f(\boldsymbol{R}, t \to \infty)$, a limit in which the ground state can be effectively sampled.

The trial wave function $\psi(\boldsymbol{R})$ is chosen accordingly to the interatomic potential, the quantum statistics, the density and the geometry. For a bosonic liquid like ^{4}He a Jastrow model

$$\psi(\boldsymbol{R}) = \prod_{i<j}^{N} f(r_{ij}) \ , \tag{7}$$

with $f(r) = \exp(-0.5(b/r)^5)$, describes approximately the dynamical correlations induced by $V(r)$. The wave function is more involved if the liquid under study is ^{3}He due to its fermionic nature. In this case, the standard model is the Jastrow-Slater wave function which incorporates the required antisymmetry by means of a Slater determinant $D_{\downarrow(\uparrow)}$,

$$\psi(\boldsymbol{R}) = \prod_{i<j}^{N} f(r_{ij}) \, D_{\downarrow} \, D_{\uparrow} \ . \tag{8}$$

In the simplest approximation, the Slater determinant in Eq. (8) corresponds to the exact wave function for the ideal Fermi gas with orbitals $\varphi_\alpha(r_i) = \exp(i\boldsymbol{k}_\alpha \cdot \boldsymbol{r}_i)$. The interplay with the dynamical correlations contained in the Jastrow part drives to more realistic orbitals containing the so-called backflow correlations [15, 16]. Now, the single-particle position $\boldsymbol{r}_i$ is substituted by

$$\tilde{r}_i = r_i + \lambda_{\rm B} \sum_{j \neq i}^{N} \eta(r_{ij}) r_{ij} \ , \tag{9}$$

with $\eta(r) = \exp[-((r-r_{\rm B})/\omega_{\rm B})^2]$, $\lambda_{\rm B}$, $r_{\rm B}$ and $\omega_{\rm B}$ being parameters to be optimized variationally.

In order to allow for an exact solution of the Schrödinger equation, the DMC method requires that the wave function $f(\boldsymbol{R}, t)$ (1) be positive everywhere, since in the MC calculation it is considered a probability distribution function. That is the case for liquid ^{4}He but not for liquid ^{3}He. For the latter, the requirement of antisymmetry introduces the well-known sign problem that hinders the probabilistic interpretation of the wave function. In order to circumvent this handicap, the usual approach is to introduce a boundary condition in the problem that forces Ψ to have the nodal surface of the trial wave function ψ. This approximation, known as fixed-node (FN) method, generates an upper bound to the eigenvalue whose quality depends on the accuracy of the nodal surface of ψ. In our study [12], we have used both the FN method and the released-node (RN) technique. The latter is used to study the difference between the eigenvalue and the FN upper bound,

178

and allows for establishing the requirement that they match at the level of the statistical errors. If it were not the case, an analytical method is used to improve ψ until the above requirement is met. This combination of FN and RN has made possible a very accurate calculation of the ground state properties of liquid ^{3}He.

The main external ingredient in the DMC calculations is the interatomic potential. The high accuracy of the results obtained is very much related to the excellent potentials nowadays available for helium. In all the results presented in the following sections, we have used the HFD-B(HE) potential proposed by Aziz *et al.* [17]. The equation of state obtained with this potential reproduces the experimental data for both helium isotopes.

3. Liquid ^{4}He

The study of the negative pressure regime and the calculation of the spinodal points have been carried out both in bulk and in reduced (one and two) dimensions. We first present results for the bulk phase and then we briefly discuss the 1D and 2D liquids.

3.1. BULK

The equation of state of bulk liquid ^{4}He has been calculated for a wide range of densities, ranging from the spinodal point up to freezing [2, 11]. From this direct output of the calculation, and performing a numerical fit to the data, one can extract information on derived quantities like the pressure and the speed of sound through the thermodynamic relations

$$P(\rho) = \rho^2 \frac{\partial e}{\partial \rho} \qquad c(\rho) = \left(\frac{1}{m} \frac{\partial P}{\partial \rho} \right)^{1/2} , \qquad (10)$$

$e = E/N$ being the energy per particle.

In Fig. 1, results for the density dependence of the pressure are reported in comparison with experimental data [18]. In the positive pressure zone, where the comparison is possible, the agreement is excellent. The spinodal point is thermodynamically defined as the locus in which the isothermal compressibility diverges, or equivalently, the point in which the speed of sound becomes zero. Results for the speed of sound as a function of the pressure are shown in Fig. 2. For positive P, the agreement with experimental data [19] is again remarkable, a feature that gives us confidence in the prediction for negative pressures. In Fig. 2, it can be seen that c drops to zero very rapidly when approaching the spinodal point. According to our results, the spinodal point of liquid ^{4}He is located at a density $\rho_s = 0.264(2)\,\sigma^{-3}$ ($\sigma = 2.556\,\text{Å}$) and pressure $P_s = -9.30(15)$ atm.

The behavior of c near P_s is expected to be of the form $c \propto (P-P_s)^\nu$, with ν the critical exponent. Under quite general assumptions, Solís and Navarro [1] proved that if one can follow analytically the equation of state of the metastable homogeneous phase down to the spinodal point then the exponent would be $\nu = 1/4$.

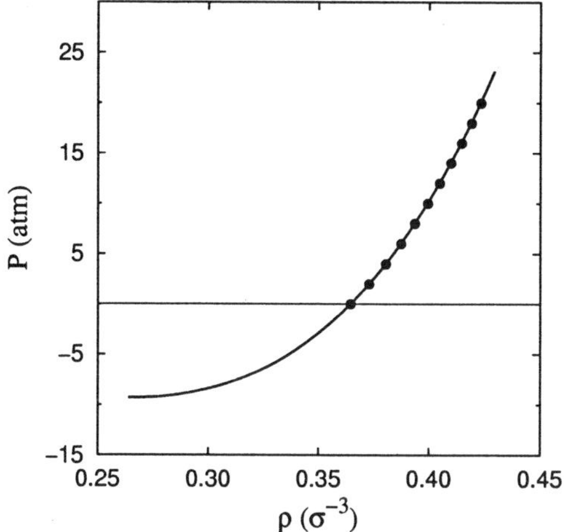

Figure 1. Liquid ^{4}He pressure as a function of density. The line is the DMC result and the points are experimental data from Ref. [18].

When the pressure increases, and in particular, in the positive pressure regime the experimental values are very well reproduced by $\nu = 1/3$ [9]. We have found that for pressures only slightly larger that $P_{\rm s}$ and up to almost solidification pressure, the behavior is also $c \propto (P - P_{\rm s})^\nu$ with exponent [1]

$$\nu = \nu_0 \equiv -P_{\rm s}\kappa_0 u_0 \ , \tag{11}$$

with κ_0 the isothermal compressibility and $u_0 = (\rho/c)\,(\partial c/\partial \rho)$ the Grüneisen constant, both evaluated at the equilibrium density ρ_0. Considering the DMC estimations for these quantities $\nu_0 = 0.322(7)$, a value which is very close to the reported $\nu = 1/3$.

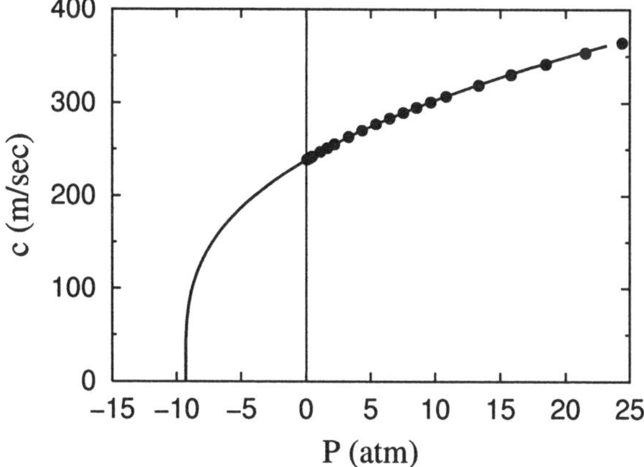

Figure 2. Speed of sound of liquid ^{4}He as a function of density. The line is the DMC result and the points are experimental data from Ref. [19].

180

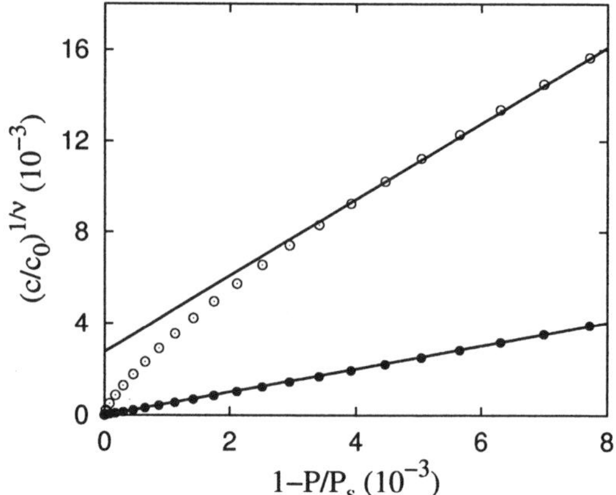

Figure 3. Behavior of the ^{4}He speed of sound close to the spinodal point. Full and empty circles correspond to $\nu = 1/4$ and $\nu = 1/3$, respectively. The straight lines are fits to the data in the respective linear regions.

In Fig. 3, the behavior of the speed of sound near the spinodal point is shown for $\nu = 1/4$ and $\nu = 1/3$. Our DMC data clearly show that in the vicinity of P_s a linear behavior is only observed when $\nu = 1/4$, and this is therefore our prediction. Campbell *et al.* [3] have performed an accurate analysis of the equation of state in the vicinity of the spinodal point summing up the ring-diagrams, which are the Feynman diagrams dominated by long-wavelength fluctuations. With this analysis, they predict a different value $\nu = 1/5$ that is expected to hold only in an extremely small area above ρ_s. Our DMC calculations have not been able to see this predicted regime.

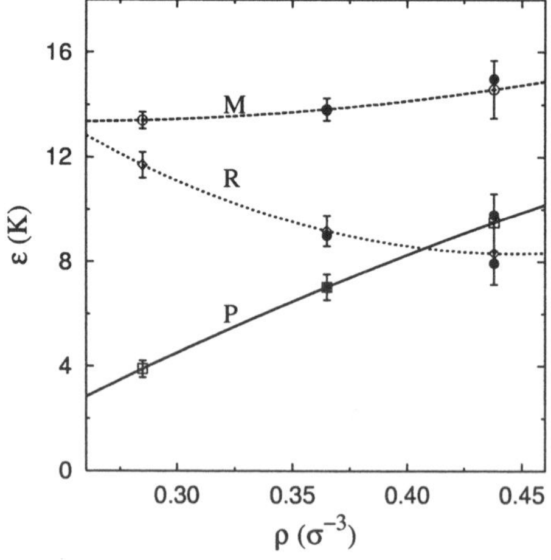

Figure 4. Phonon (P), roton (R), and maxon (M) excitation energies as a function of the density. The lines are polynomial fits to the data. Solid circles are available experimental data [21].

The DMC method can also be applied to the calculation of the excitation spectrum. In this case, ψ is eigenstate of the total momentum operator and then orthogonal to the ground state. The problem is then mapped onto a fermion-like calculation in which special care has to be taken with the sign problem. We have verified that the combination of FN and RN methods, together with the inclusion of backflow correlations in ψ, allows for a very accurate calculation of $\varepsilon(q)$ [20]. In Fig. 4, results for the characteristic phonon, maxon, and roton energies are shown in comparison with experimental data [21]. The agreement between theory and experiment at the equilibrium and near-freezing densities is remarkable. It has been predicted [22] that the roton mode would soften and even disappear when approaching the spinodal point. In Fig. 4, results for $\varepsilon(q)$ at a density close to ρ_s ($\rho = 0.285\,\sigma^{-3}$) are reported [23]. The lines on top of the calculated points show the dependence on pressure of the several modes. As one can see, the roton energy approaches the maxon curve near the spinodal point, in agreement with the predicted softening of the roton in the negative pressure regime.

3.2. RESTRICTED GEOMETRIES

^{4}He adsorbed on a planar substrate or into the interstitial and internal channels of carbon nanotube bundles are nowadays current realizations of two- and one-dimensional quantum fluids, respectively. The most studied system corresponds to helium adsorbed on top of the surface of substrates with deep and narrow potential wells like graphite or solid H_2. In this case, the degree of freedom perpendicular to the surface is practically frozen out and low coverage films are stabilized forming nearly perfect two-dimensional (2D) systems. We have studied [24] using the DMC method the equation of state of two-dimensional ^{4}He which serves as a limit of a thin film. In fact, if the binding energy to the substrate is subtracted from the total energy of the film, the remaining energy per particle is very close to the one of the idealized 2D geometry [25].

Two-dimensional ^{4}He is a self-bound system with a binding energy $E/N = -0.89706(61)\,$K at an equilibrium surface density $\rho_0 = 0.043440(23)\,\text{Å}^{-2}$. This result depends slightly on the interatomic potential used in the calculation. The results reported in this work have been obtained by using the same Aziz potential that in bulk has allowed for a very accurate calculation of its equation of state. From the curve $E/N(\rho)$, the pressure and speed of sound can be obtained. The speed of sound at equilibrium density is $c^{2D}(\rho_0) = 92.8(6)\,$m/sec, nearly three times smaller than the speed of sound in bulk ^{4}He at its corresponding ρ_0, $c^{3D}(\rho_0) = 238.3(5)\,$m/sec. The spinodal point is located where $c = 0$. The behavior of c with the pressure is shown in Fig. 5 from the spinodal point up to higher positive pressures. The spinodal density is $\rho_s = 0.03490(31)\,\text{Å}^{-2}$ which corresponds to a negative pressure $P_s = -0.0148(2)\,\text{KÅ}^{-2}$. In relative terms, the negative pressure regime in 2D is *shorter* than in bulk, a feature that can be observed by comparing Figs. 2 and 5.

One-dimensional quantum fluids have been recently the focus of several experimental and theoretical works [26, 27, 28]. From the experimental side, the current fabrication of carbon nanotubes has provided a perfect environment to

182

achieve physical realizations of nearly one-dimensional systems. Apart from its highly interesting application as possible fuel cells, the adsorption of helium or hydrogen into the interstitial channels, or directly inside the nanotubes, provides a unique opportunity to deal with 1D quantum fluids. Using the same methodology and potential than in bulk and in 2D we have calculated the equation of state of 1D ^{4}He [28]. This study is useful as a limiting situation for the narrowest channels.

The DMC equation of state of 1D ^{4}He shows the existence of a stable liquid phase with a small binding energy $E/N(\rho_0) = -0.0036(2)\,$K at an equilibrium density $\rho_0 = 0.062(1)\,$Å^{-1}. The total energy is the result of a big cancellation between the potential and kinetic energies; at ρ_0, $T/N = 0.2706(4)\,$K and $V/N = -0.2742(4)\,$K. This feature explains the high influence of the interatomic potential in this geometry; using the HFDHE2 Aziz potential [29] the energy is two times smaller $(-0.0018(3)\,$K$)$. Results for the speed of sound are shown in Fig. 6. The speed of sound drops to zero at the spinodal point, located at a density $\rho_s = 0.047(1)\,$Å^{-1} and pressure $P_s = -1.92(2) \cdot 10^{-4}\,$KÅ^{-1}. The metastable negative pressure part of the equation of state is very small in comparison with 2D and bulk, ρ_s being quite close to ρ_0. It is worth noticing that near equilibrium, the mean distance between atoms is really huge $(\sim 16\,$Å$)$ in comparison with bulk, *i.e.*, the negative pressure regime in 1D corresponds to a very dilute liquid.

4. Liquid ^{3}He

For many years, liquid ^{3}He has been studied using microscopic quantum-many body theories. The variational method based either on Fermi hypernetted chain theory (FHNC) [30] or on VMC [31], has allowed for a reasonable description of its ground-state properties. The best description has been achieved by generalizing the Jastrow-Slater wave function with the introduction of three-body correlations

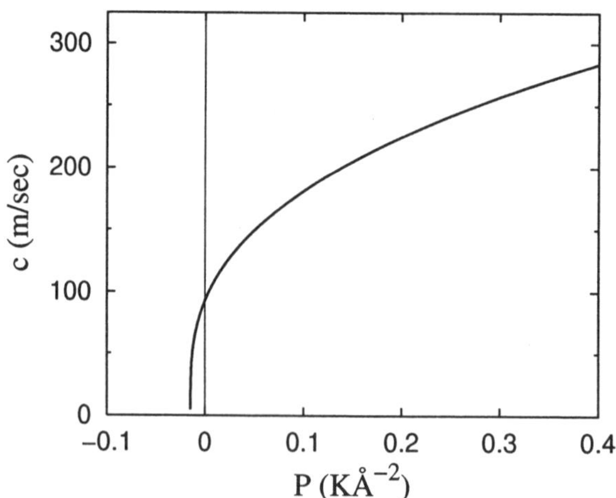

Figure 5.   DMC results for the speed of sound of 2D liquid ^{4}He as a function of the pressure.

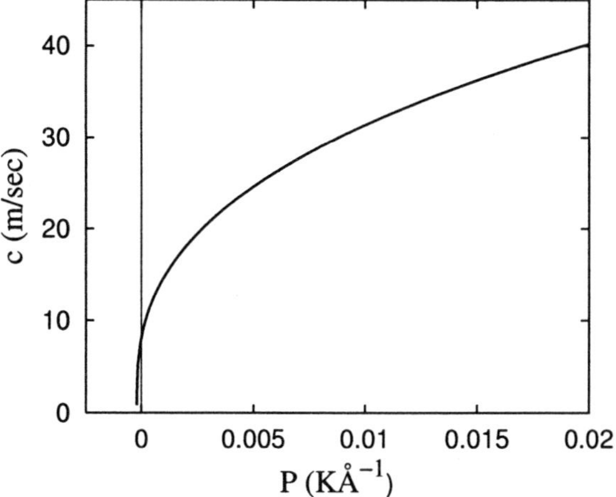

Figure 6.   DMC results for the speed of sound of 1D liquid ^{4}He as a function of the pressure.

in the dynamical part and momentum-dependent (backflow) correlations in the Slater determinant. However, the interplay between Fermi statistics and dynamical correlations causes the quality of the variational models to be poorer than in the description of its boson counterpart, liquid ^{4}He.

The necessary requirement of the antisymmetry of the wave function translates into the well-known sign problem that impedes a straightforward application of the DMC or GFMC methods. Recently, the equation of state of bulk liquid ^{3}He has been calculated using a DMC method that combines the FN approximation, the RN technique and an analytical method to improve the trial wave function [12]. This work concludes the high quality of the nodal surface provided by optimized

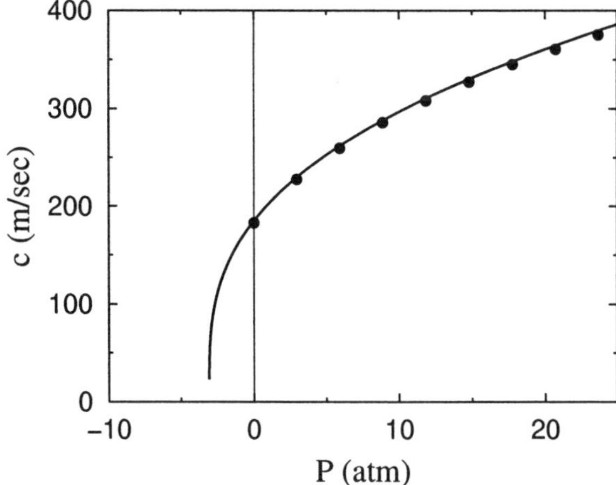

Figure 7.   Speed of sound of liquid ^{3}He as a function of the pressure (solid line). Points are experimental data from Ref. [32].

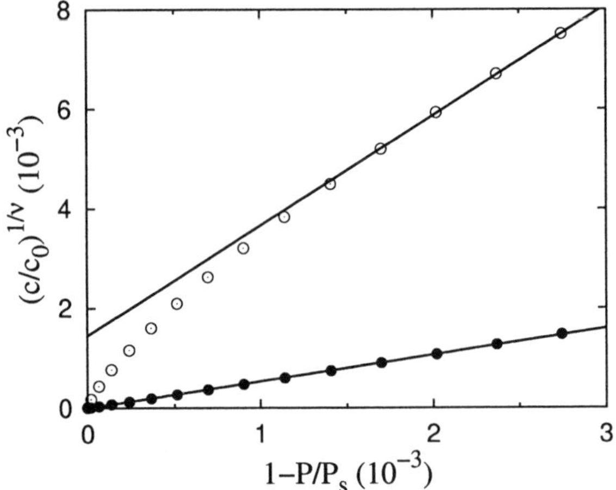

Figure 8. Critical behavior of the speed of sound of liquid ^{3}He near the spinodal point. Full and empty circles correspond to $\nu = 1/4$ and $\nu = 1/3$, respectively. The straight lines are fits to the data in the respective linear regions.

backflow correlations. The bias in the resulting energies is analyzed by means of the RN method and by the inclusion of the next-order terms in the antisymmetric trial wave function. Both calculations show that the bias introduced by the backflow wave function is reduced to the level of the typical statistical noise, and thus a practically exact result is obtained.

The DMC calculations generate a very accurate equation of state. The equilibrium point is predicted to be at density $\rho_0 = 0.274(1)\,\sigma^{-3}$ with a binding energy $E/N(\rho_0) = -2.464(7)\,\text{K}$, to be compared with the experimental data [18] $\rho_0^{\text{expt}} = 0.273\,\sigma^{-3}$ and $E/N(\rho_0)^{\text{expt}} = -2.473\,\text{K}$. The pressure and speed of sound are readily obtained from a polynomial fit to the energies. Both quantities are stringent tests to the equation of state. The results obtained for $P(\rho)$ and $c(\rho)$ show a perfect agreement with experimental data; for example, the speed of sound at ρ_0 is $182.2(6)\,\text{m/sec}$ to be compared with the experimental value $182.9\,\text{m/sec}$ [32].

Fig. 7 contains the DMC results for the speed of sound as a function of pressure. In the positive pressure region, $c(P)$ reproduces accurately the experimental data [32] with the same quality achieved in liquid ^{4}He (Fig. 2). The spinodal point is located at a density $\rho_s = 0.202(2)\,\sigma^{-3}$, corresponding to a negative pressure $P_s = -3.09(20)\,\text{atm}$. This estimation is free from possible extrapolation errors resulting from the extension of the positive regime down to negative pressures. Nevertheless, the extrapolations predict nearly the same value for the spinodal pressure, $P_s = -3.11\,\text{atm}$ [7]. The spinodal point has also been calculated in the density functional framework [1]; the results ($\rho_s = 0.199\,\sigma^{-3}$, $P_s = -3.10\,\text{atm}$) are also close to the DMC data.

The behavior of $c(P)$ near the spinodal point is shown in Fig. 8. The results are plotted considering two possible critical exponents: $\nu = 1/3$ and $\nu = 1/4$. The exponent $\nu = 1/3$ is the one used in extrapolations to negative pressure from the experimental data on $c(P)$. Our results confirm that this exponent matches accurately the behavior of the speed of sound from high pressures to values near P_s. However, as it is shown in Fig. 8, the critical exponent compatible with our results at pressures close to the spinodal value turns to $\nu = 1/4$. As in the ^{4}He case, it seems to be a consequence of following analytically the metastable curve.

The situation for ^{3}He in reduced geometries is completely different to the one of the liquid ^{4}He. A recent DMC calculation of the equation of state of two-dimensional ^{3}He has proven that its ground state corresponds to a gas phase [33]. Therefore, no spinodal point exists in 2D ^{3}He. The 1D case has not been studied yet. However, it is expected that also this system will not be self-bound.

Acknowledgements

We thank the collaboration of Stefano Giorgini, M. Carmen Gordillo and Jesús Navarro in some parts of this work. Partial financial support from DGES (Spain) Grant No. PB98-0922 and Generalitat de Catalunya Grant No. 2001SGR-00222 are acknowledged.

References

1. Solís, M.A. and Navarro, J. (1992) Liquid ^{4}He and ^{3}He at negative pressure, *Phys. Rev. B*, Vol. no. **45**, pp. 13080-13083.
2. Boronat, J., Casulleras, J. and Navarro, J. (1994) Monte Carlo calculations for liquid ^{4}He at negative pressure, *Phys. Rev. B*, Vol. no. **50**, pp. 3427-3430.
3. Campbell, C.E., Folk, R. and Krotscheck, E. (1996) Critical behavior of liquid ^{4}He at negative pressures, *J. Low Temp. Phys.*, Vol. no. **105**, pp. 13-16.
4. Maris, H.J. (1995) Theory of quantum nucleation of bubbles in liquid helium, *J. Low Temp. Phys.*, Vol. no. **98**, pp. 403-424.
5. Guilleumas, M., Barranco, M., Jezek, D.M., Lombard, R.J. and Pi, M. (1996) Quantum cavitation in liquid helium, *Phys. Rev. B*, Vol. no. **54**, pp. 16135-16138.
6. Caupin, F., Balibar, S. and Maris, H.J. (2001) Anomaly in the stability limit of liquid ^{3}He, *Phys. Rev. Lett.*, Vol. no. **87**, pp. 145302-1–145302-4.
7. Caupin, F., and Balibar, S. (2001) Cavitation pressure in liquid helium, *Phys. Rev. B*, Vol. no. **64**, pp. 064507-1–064507-10.
8. Maris, H.J. and Balibar, S. (2000) Negative pressures and cavitation in liquid helium, *Physics Today*, Vol. no. **98**, pp. 403-424.
9. Maris, H.J. and Xiong, Q. (1989) Nucleation of bubbles in liquid helium at negative pressure, *Phys. Rev. Lett.*, Vol. no. **63**, pp. 1078-1081.
10. Boronat, J. (2002) Monte Carlo simulations at zero temperature: helium in one, two, and three dimensions, *Quantum liquids in Confined Geometries*, Krotscheck, E., and Navarro, J. Eds., World Scientific, Singapore.
11. Boronat, J. and Casulleras, J. (1994) Monte Carlo analysis of an interatomic potential for He, *Phys. Rev. B*, Vol. no. **49**, pp. 8920-8930.
12. Casulleras, J. and Boronat, J. (2000) Progress in Monte Carlo calculations of Fermi systems: normal liquid ^{3}He, *Phys. Rev. Lett.*, Vol. no. **84**, pp. 3121-3124.
13. Bauer, G.H., Ceperley, D.M. and Goldenfeld, N. (2000) Path integral Monte Carlo simulation of helium at negative pressures, *Phys. Rev. B*, Vol. no. **61**, pp. 9055-9060.
14. Guardiola, R. (1998) Monte Carlo methods in quantum many-body theories, *Microscopic quantum many-body theories and their applications*, Navarro, J. and Polls, A. Eds., Springer, Berlin, pp. 269-336.

15. Schmidt, K.E., Lee, M.A., Kalos, M.H. and Chester, G.V. (1981) Structure of the ground state of a fermion fluid, *Phys. Rev. Lett.*, Vol. no. **47**, pp. 807-810.

16. Panoff, R.M. and Carlson, J. (1989) Fermion Monte Carlo algorithms and liquid ^{3}He, *Phys. Rev. Lett.*, Vol. no. **62**, pp. 1130-1134.

17. Aziz, R.A., McCourt, F.R.W. and Wong, C.C.K. (1987) A new determination of the ground state interatomic potential for He_2, *Mol. Phys.*, Vol. no. **61**, pp. 1487-1511.

18. De Bruyn Ouboter, R. and Yang, C.N. (1987) The thermodynamic properties of liquid ^{3}He-^{4}He mixtures between 0 and 20 atm in the limit of absolute zero temperature, *Physica B*, Vol. no. **144**, pp. 127-144.

19. Abraham, B.M., Eckstein, V., Ketterson, J.B., Kuchnir, M. and Roach, P.R. (1970) Velocity of sound, density and Grüneisen constant in liquid ^{4}He, *Phys. Rev. A*, Vol. no. **1**, pp. 250-257.

20. Boronat, J. and Casulleras, J. (1997) Microscopic calculation of the phonon-roton branch in superfluid ^{4}He, *Europhys. Lett.*, Vol. no. **38**, pp. 291-296.

21. See Glyde, H.R. (1994), *Excitations in liquid and solid helium*, Clarendon Press, Oxford.

22. Maris, H.J., private communication.

23. Boronat, J. and Casulleras, J. (1998) Diffusion Monte Carlo for excited states: phonons and rotons in superfluid ^{4}He, *J. Low Temp. Phys.*, Vol. no. **110**, pp. 443-448.

24. Giorgini, S., Boronat, J. and Casulleras, J. (1996) Diffusion Monte Carlo study of two-dimensional liquid ^{4}He, *Phys. Rev. B*, Vol. no. **54**, pp. 6099-6102.

25. Clements, B.E., Epstein, J.L., Krotscheck, E. and Saarela, M. (1993) Structure of boson quantum films, *Phys. Rev. B*, Vol. no. **48**, pp. 7450-7470.

26. Krotscheck, E. and Miller, M.D. (1999) Properties of ^{4}He in one dimension, *Phys. Rev. B*, Vol. no. **60**, pp. 13038-13050.

27. Boninsegni, M. and Moroni, S. (2000) Ground state of ^{4}He in one dimension, *J. Low Temp. Phys.*, Vol. no. **118**, pp. 1-6.

28. Gordillo, M.C., Boronat, J. and Casulleras, J. (2000) Quasi-one-dimensional ^{4}He inside carbon nanotubes, *Phys. Rev. B*, Vol. no. **61**, pp. R878-R881.

29. Aziz, R.A., Nain, V.P.S., Carley, J.S., Taylor, W.L. and McConville, G.T. (1979) An accurate interatomic potential for helium, *J. Chem. Phys.*, Vol. no. **70**, pp. 4330-4342.

30. Viviani, M., Buendía,E., Fantoni, S. and Rosati, S. (1988) Spin-dependent correlations in the ground state of liquid ^{3}He, *Phys. Rev. B*, Vol. no. **38**, pp. 4523-4538.

31. Schmidt, K.E., Lee, M.A., Kalos, M.H. and Chester, G.V. (1981) Structure of the ground state of a Fermion fluid, *Phys. Rev. Lett.*, Vol. no. **47**, pp. 807-810.

32. Wheatley, J.C. (1975) Experimental properties of superfluid ^{3}He, *Rev. Mod. Phys.*, Vol. no. **47**, pp. 415-470.

33. Grau, V., Boronat, J. and Casulleras, J. (2002) Zero-temperature equation of state of two-dimensional ^{3}He, to be published.

NUCLEATION OF BUBBLES ON ELECTRONS IN LIQUID HELIUM

HUMPHREY J. MARIS and DENIS KONSTANTINOV
Department of Physics, Brown University
Providence, RI 02912, USA

1. Introduction

Cavitation in liquid helium has been studied for over forty years, beginning with the work of Misener and Hebert [1] and Beams [2]. The interest in helium stemmed from the expectation that because all other elements freeze at higher temperatures, liquid helium should have exceptionally high purity. It was expected that this would greatly reduce the chance of heterogeneous nucleation, and thus enable the study of homogenous nucleation under controlled conditions. However, the early experiments gave values of the cavitation strength that varied significantly from one measurement to the next [3-11], and were in all cases considerably less than the strength that was anticipated theoretically. Most of these experiments were conducted with large volumes of liquid, typically of the order of 1 cm^3, and used liquid that was part of the main bath of a helium cryostat which could easily contain small particles of frozen air. In 1989, Nissen *et al.* [12] used a hemispherical ultrasonic transducer to focus 566 kHz sound into a small volume of liquid helium (volume ~10^{-5} cm^3). Their measurements gave a cavitation strength that increased rapidly as the temperature decreased. At their lowest temperature of 1.6 K, they estimated that vapor bubbles did not appear in the liquid until the pressure reached about − 8 bars. The difference between the results of Nissen *et al.* and the earlier workers is presumably due to the presence of some centers for heterogeneous nucleation in the liquid. Suppose, for example, that the density of these objects is 100 cm^{-3} and that one such object will result in bubble nucleation when the pressure becomes more negative than -0.1 bar. A measurement on a liquid sample of volume 1 cm^3 will then almost certainly give a cavitation strength of 0.1 bars, whereas a measurement on a sample of volume 10^{-5} cm^3 has a high probability of giving a result that is characteristic of the pure liquid.

Since the work of Nissen *et al.*, there have been many studies of bubble nucleation in helium at negative pressures [13], and much of this work is summarized in the articles by F. Caupin and S. Balibar in this book. Measurements have been made in both helium-3 [14-18] and helium-4 [14,17,19,20]. In helium-4, the results indicate that at high temperatures, the barrier against bubble nucleation is overcome as a result of thermal fluctuations, whereas below about 0.2 K quantum tunneling through the barrier is the dominant process [14,19,20]. In helium-3, nucleation by quantum tunneling has not yet been clearly demonstrated [16,18]. In these studies, it is believed that bubbles are forming as a result of homogeneous nucleation.

187

A.R. Imre et al. (eds.), Liquids Under Negative Pressure, 187–199.
© 2002 *Kluwer Academic Publishers. Printed in the Netherlands.*

188

Study of heterogeneous nucleation in most liquids is difficult because it is not usually easy to determine the type and characteristics of the centers responsible for the nucleation. In liquid helium, however, there is one type of "defect", the injected electron, that is ideally suited for a study of heterogeneous nucleation [21]. In this article, we describe studies that have been made of bubble nucleation by electrons and discuss some of the remaining open questions in this field.

2. Electrons in Liquid Helium

At large distances an electron is attracted to a helium atom because the helium atom is polarizable. However, at short distance there is a strong repulsion. This arises because when the electron is within the atom it is forced by the Pauli principle to go into a higher energy state with principal quantum number n of 2 or greater. As a result, an electron entering helium has to overcome a potential barrier of approximately 1 eV [22]. If an electron is injected into helium and comes to rest in the liquid, it forces open a spherical cavity which is almost free of helium atoms. The energy of this so called "electron bubble" can be approximated by the expression

$$E = \frac{h^2}{8m_e R^2} + 4\pi R^2 \alpha + \frac{4\pi}{3} R^3 P, \tag{1}$$

where R is the bubble radius, m_e the electron mass, α the surface tension, and P the pressure. The first term is the zero point energy of the electron confined in a spherical cavity. The second and third terms represent the surface energy of the bubble and the work done to form the bubble against the applied pressure, respectively. For zero pressure and low temperature, the bubble radius R_0 that gives the minimum total energy is $(h^2 / 32\pi m_e \alpha)^{1/4}$. This is 19 Å for helium-4 and 24 Å for helium-3.

In the derivation of Eq. 1 the following assumptions are made:
1) The wave function of the electron is taken to go to zero at the bubble wall, i.e., the potential barrier provided by the helium is assumed to be large.
2) The long range attraction of the electron to the helium atoms is neglected, i.e., the potential energy for the electron inside the bubble is taken to be zero.
3) The width of the bubble wall, i.e., the interface between the liquid and vapor, is assumed to be small compared to the bubble radius.
4) It is assumed that the effect of the helium vapor inside the bubble can be neglected.

Effects 1)-3) are all fairly small corrections [23], and their inclusion is estimated to change the size of the electron bubble by only 5 to 10%. Effect 4) becomes important at temperatures above about 3 K, and is discussed in more detail in the following section.

An electron forms a bubble state because the energy of this state is less than the 1 eV energy the electron would have if it were moving through the uniform liquid. In the absence of applied pressure, the radius of the bubble is $(h^2 / 32\pi\, m_e\alpha)^{1/4}$, and the energy of the electron bubble calculated from Eq. 1 is

$$E_0 = (2\pi\, h^2\alpha / m_e)^{1/2}, \tag{2}$$

which for helium-4 is 0.2 eV. It can be seen that the existence of electron bubbles is in large measure a consequence of the extremely small value of the surface tension of helium (0.375 erg cm^{-2}). Since the 1960's these bubbles have been studied by a number of techniques and their basic properties are well established [21].

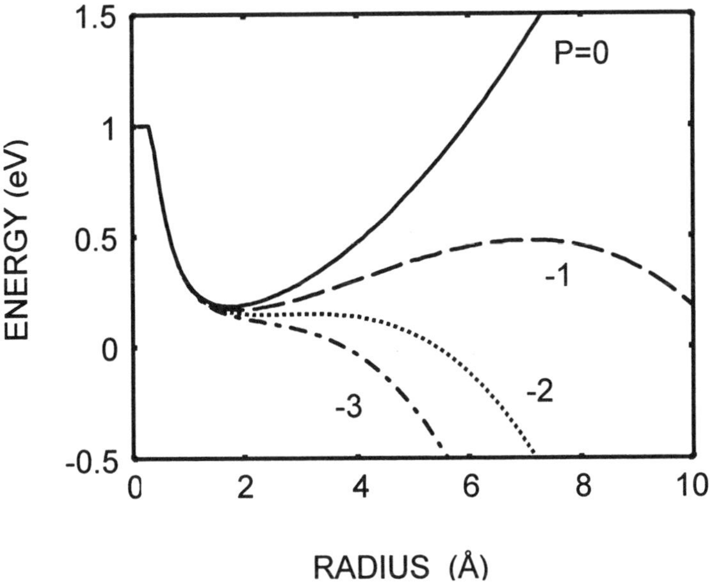

Figure. 1. Energy of an electron bubble as a function of the radius. The different curves are labeled by the pressure in bars.

3. Explosion of Bubbles at Negative Pressure

When a positive pressure is applied to the liquid, the radius at which the energy of the bubble is a minimum decreases. It is easy to show from Eq. 1 that for small pressures the variation of the equilibrium radius with pressure is

$$\frac{dR}{dP} = -\frac{R_0^2}{8\alpha}, \tag{3}$$

which for helium-4 is - 1.2 Å bar^{-1}. A negative pressure makes the bubble expand, and at a critical negative pressure P_c, the bubble becomes unstable and begins to grow without limit. As shown in Fig.1, this occurs because when the pressure is negative with respect to P_c, the energy of the bubble decreases monotonically with increasing radius. Then if a sample of liquid helium contains even a single electron bubble, that bubble will grow to a macroscopic size. The fact that at negative pressure an electron bubble will explode was first recognized by Akulichev and Boguslavskii in 1972 [24]. At low temperatures where the presence of helium atoms inside the bubble can be neglected, the critical pressure as calculated from Eq. 1 is

$$P_c = -\frac{16}{5}\left(\frac{2\pi m_e}{5h^2}\right)^{1/4}\alpha^{5/4} \quad . \tag{4}$$

For helium-4, this gives P_c = -1.9 bars [25]. One can consider that the explosion results from the combination of the liquid pressure and the outward pressure exerted by the electron overcoming the effect of the surface tension.

The calculation of the critical pressure at temperatures above 3 K is complicated by the presence of an increasing amount of helium vapor inside the bubble. Detailed calculations of the configuration of the bubble as a function of temperature and applied pressure have been performed by Classen $et\ al.$ [23]. As the pressure is varied, the bubble evolves as follows. When the pressure equals the saturated vapor pressure, the form of the wave function of the electron is only slightly modified from the form that it has at low temperatures. However, the bubble contains a finite number density of helium atoms, primarily in the region close to the bubble wall where the magnitude of the electron wave function is small and hence the interaction between the electron and the atoms is not large. As the pressure is lowered, the density of the helium atoms increases. The interaction between these atoms and the electron results in a decrease in the amplitude of the wave function in the region near to the wall. When the pressure is close to the instability pressure P_c, the electron is confined to a sphere of radius R_e significantly less than the bubble radius R. There is a layer of vapor filling the region between the volume occupied by the electron and the bubble wall. The calculated explosion pressure as a function of temperature is shown in Fig. 2, along with the results of experimental measurements [23]. Similar results and agreement with theory have been obtained for helium-3 [26].

It is important to note that to a very good approximation this cavitation resulting from electrons in helium is a purely "mechanical" process and is only slightly affected by thermal fluctuations or quantum tunneling. This distinguishes it from homogeneous nucleation. Let us elaborate on this point. In the experiments to study cavitation from electrons, sound of frequency in the range 100 kHz to 1 MHz is used to produce a transient negative pressure. Thus, the time scale of the pressure variation is of the order of 1 to 10 μs. The frequencies of the normal modes of vibration of the bubble, for example, the breathing mode, are of the order of 10^{10} Hz. As a result, it is correct to consider the sound as producing a quasi-static pressure field which changes very slowly in time and is able to pull the bubble apart. In principle, thermal fluctuations can result in bubble nucleation when the applied pressure P has not yet reached P_c. However, for

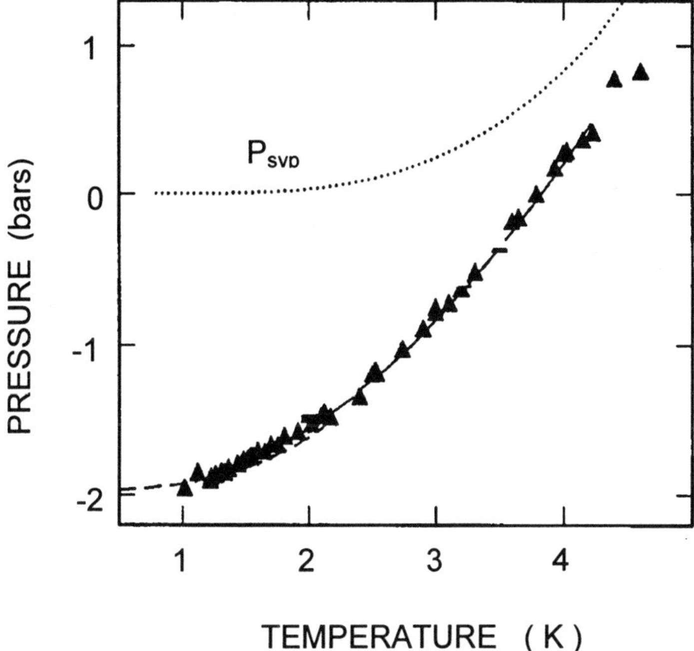

Figure 2. Negative pressure needed to explode an electron bubble in helium-4 as a function of temperature from ref. 23. The solid curve is from theory and the triangles are the measured values. The dotted line indicates the saturated vapor pressure.

this to happen, the energy barrier must have decreased to a value that is comparable to kT, e.g., $20kT$. From Eq. 1, it can be shown that for P close to P_c the energy barrier is

$$\Delta E = \frac{32}{15\sqrt{2}} E_0 \left(\frac{P - P_c}{P_c} \right)^{3/2} .$$ (5)

Since the energy E_0 in temperature units is 2300 K, this means that $\Delta E >> kT$ unless $(P\text{-}P_c) / P_c$ is extremely small. Thus, thermal fluctuations can help the bubble explode only when P is very close to P_c. A detailed calculation [27] shows that quantum tunneling is also unimportant except very close to P_c.

The experimental configuration is shown schematically in Fig. 3. Sound is generated by a hemi-spherical transducer immersed in the liquid, and a large oscillating pressure is produced at the acoustic focus. To a reasonable approximation, the maximum negative pressure produced at the focus is proportional to the driving voltage V_{tran} applied to the transducer. A laser beam is passed through this focus and when a bubble is formed, light is scattered and detected by a photomultiplier. In a typical experiment, the probability that a sound pulse results in cavitation is measured as a

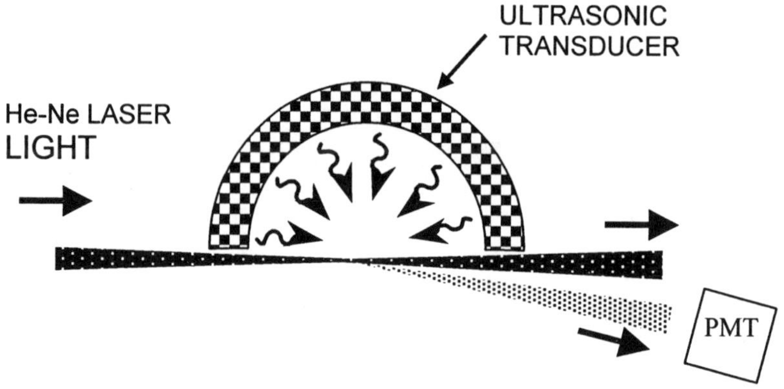

Figure 3. Schematic diagram of the experimental setup to study cavitation in liquid helium. Scattered laser light is detected by the photomultiplier (PMT).

function of the driving voltage applied to the transducer. The pressure swing resulting from unit voltage applied to the transducer can be determined through measurements of the cavitation threshold as a function of static pressure in the helium cell [23].

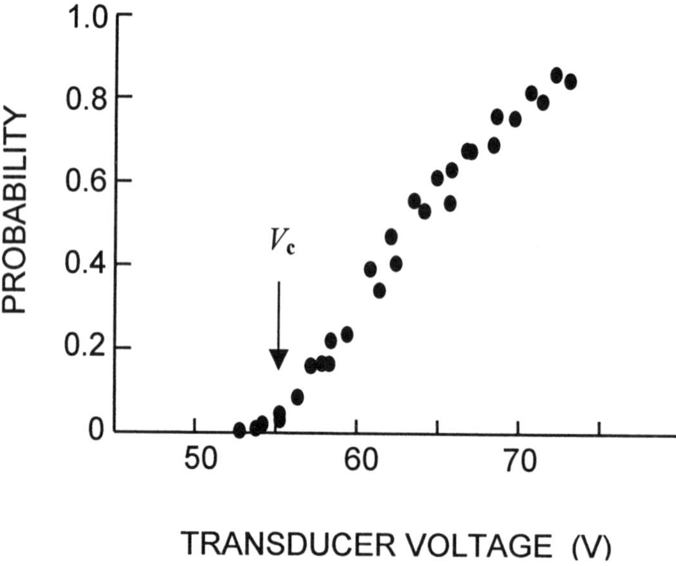

Figure 4. Example of the results from a measurement of the probability of cavitation as a function of applied transducer voltage.

Electrons can be introduced into the liquid either through the use of a radioactive source (β-emitter) or through an electrical discharge at a sharp metal tip in the liquid. Although the radioactive source emits electrons at a constant rate, the number density of electrons in the vicinity of the acoustic focus can be varied by applying a dc voltage to the transducer wall to change the amount of time that each electron spends in the liquid. When a tip is used, the density of the electrons at the focus can again be controlled by varying the applied voltage.

Some typical results are shown in Fig. 4. It can be seen that at a critical transducer voltage V_c of 55 V, the probability of cavitation begins to increase rapidly. At V_c the negative pressure swing at the center of the acoustic focus reaches the magnitude required to explode a bubble. However, the probability of cavitation is still small because nucleation will occur only if there is an electron bubble very close to the center of the acoustic focus. As the transducer voltage is increased, nucleation begins to be possible for electrons further removed from the center of the focus and hence the probability of cavitation increases. From an estimate of the variation of the pressure swing with position, it is possible to make a fit to the measured cavitation probability as a function of V_{tran} and from this to deduce the number density of electron bubbles.

4. Some Unsettled Questions

Now we discuss some extra effects that have been seen in the cavitation studies of helium containing electrons.

4.1. RARE EVENTS

In a sample of liquid helium containing electrons, the threshold at which electron bubbles begin to explode is readily identifiable, as was just discussed for Fig. 4. However, it has been found [23] that there may be a small probability of detecting cavitation below this threshold, even for pressure swings much too small to cause the explosion of normal electron bubbles. These rare events are only seen when a radioactive source is used to inject electrons into the helium; they do not appear when electrons are injected from a tip. The probability of these rare events is unaffected by the application of a dc voltage to the transducer, and increases with the duration of the sound pulse. Furthermore, the rare events disappear when the line of sight between the radioactive source and the acoustic focus is blocked.

These observations suggest that the rare events arise from electrons that pass through the acoustic focus at the same time that the sound oscillation is present. These electrons ionize helium atoms. The secondary electrons then find themselves traveling through liquid that is at a negative pressure. They quickly lose their energy and then begin to form a bubble which grows rapidly. The bubble soon grows to a size such that the energy given by Eq. 1 is a minimum. However, at this point the bubble is surrounded by liquid that is flowing away from it at high velocity. The inertia of this liquid can enable the bubble to keep growing, to overcome the energy barrier, and to continue on to become a bubble of macroscopic size [28].

194

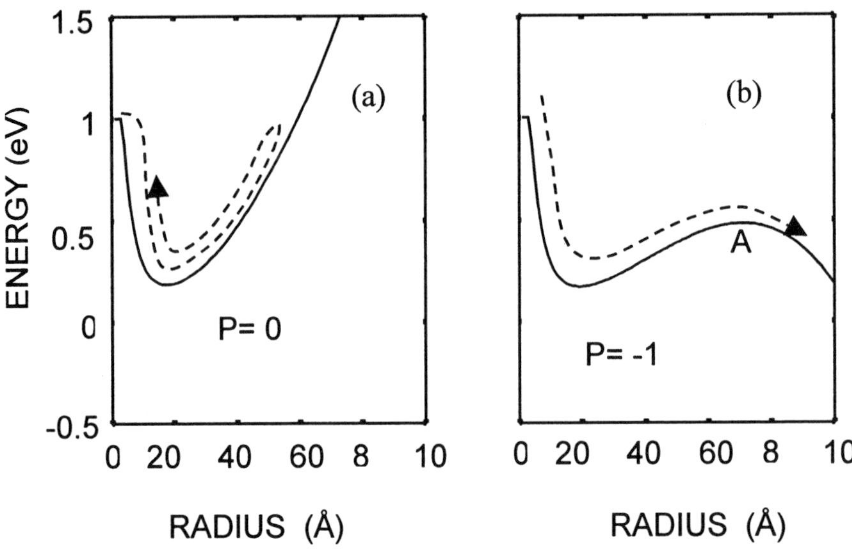

Figure 5. Origin of rare events. See the text for discussion.

Figure 5 shows the energetics of this process. When the radius of the bubble is zero, the electron is in uniform liquid helium and has an energy of approximately 1 eV. If the pressure is zero, the bubble will grow to a maximum radius, and then oscillate back and forth until it finally settles at the radius R_0 of minimum energy (Fig. 5a). At a sufficiently large negative pressure, the bubble may continue to grow past the energy barrier (Fig. 5b). How large a negative pressure is required for this to happen? The simplest approach is to ignore any form of dissipation, and to assume that the pressure must be such that the energy at the maximum (point A in Fig. 5b) is less than the starting energy of 1eV. This condition gives good agreement with experiment [28], but is probably a substantial oversimplification. In the first place, account needs to be taken of the compressibility of the liquid. The rapidly expanding bubble will radiate sound into the surrounding liquid, and this will give a loss in energy. A second correction arises from the energy deposited into the liquid *before* the bubble begins to form. Secondary electrons produced by fast particles in helium have an initial kinetic energy of the order of 50 eV and travel a distance of only about 100 Å before losing this energy. Hence, the 50 eV is deposited into a very small volume and there may be a significant local temperature rise. This temperature rise could affect the nucleation process in several ways. For example, the excitations produced in the liquid may further damp the motion of the bubble wall, thereby increasing the magnitude of the negative pressure that is required. On the other hand, an increase in temperature will give a decrease in surface tension and lower the cavitation threshold. The magnitude of these corrections is hard to determine, and has not yet been established.

4.2 ELECTRONS ON VORTICES

The number of electron bubbles per unit volume of the liquid in these experiments is determined by the balance between the rate at which electrons are injected by the source that is used and the rate at which they escape to the walls of the experimental cell. The escape rate is dependent on the space charge field and the voltage that is applied between the source and the transducer. The electron density also varies rapidly with temperature because of the variation of the bubble mobility with T. In the range 1 to 2 K, the mobility varies with temperature approximately as $\exp(\Delta/T)$ where Δ is the roton energy gap (~8.7 K). As the temperature approaches 1 K, the mobility becomes so large that the number of electrons in the liquid is very small and the probability that cavitation will occur is much less than unity, even when the pressure at the acoustic focus is significantly more negative than the explosion pressure P_c.

However, when the temperature is lowered to around 0.9 K, it is found that cavitation again becomes possible. In this temperature range, the mobility of the electron bubbles is sufficiently large that their velocity exceeds the critical velocity for nucleation of quantized vortices [21]. These are singularities in the liquid that may be in the form of lines ending on the container walls or closed loops entirely within the liquid. The flow of liquid around a vortex is quantized so that

$$\oint \vec{v}.\vec{dl} = \frac{h}{m_4}, \tag{6}$$

where the integral is along a path enclosing the vortex line, $\vec{v}$ is the liquid velocity, and m_4 is the mass of a helium atom. When an electron bubble is close to a vortex line, it displaces liquid that has a high velocity and the total energy is reduced. As a consequence, the electron bubble is attracted to the vortex line, and can become bound to it. When electrons become attached to vortices, they move very slowly through the liquid and the electron density then becomes very large.

It is found that when the electron bubbles are attached to vortices the magnitude of the negative pressure at which they explode is reduced by about 12 %. This is shown in Fig. 6. A reduction in the magnitude of the pressure is expected since the circulation of the liquid around the vortex line means that the pressure at the electron bubble will be negative with respect to the pressure in the bulk of the liquid. However, an apparently straightforward calculation [29] of the magnitude of this effect based on a simple model gives a shift of only 4 %, rather than the 12 % found experimentally. The reason for this difference is currently not known. It would be interesting to perform a more sophisticated calculation, using a density-functional method, for example.

4.3 ELECTRONS IN EXCITED STATES

The discussion so far has been limited to the consideration of bubbles with the electron in the lowest energy quantum state, i.e., the 1S state. Using light of the appropriate wavelength, it is possible to excite the electron to higher energy states. The simplest such transition is from the 1S ground state to the 1P state; at zero applied pressure this

196

transition takes place for a wavelength of approximately 10 μm (energy ~ 0.1 eV). The outward pressure exerted by the electron on the bubble wall is $\hbar^2 \, |\nabla \psi|^2 \, / 2m_e$; for any state in which the wave function does not have spherical symmetry this pressure varies with direction and hence the shape of the electron bubble that minimizes the total

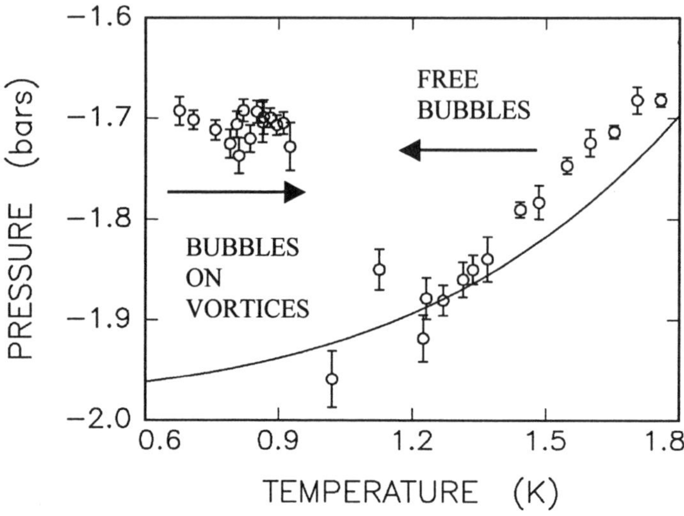

Figure 6. Critical pressure at which electron bubbles explode as a function of temperature from ref. [29]. The solid curve is the theory for a bubble in bulk liquid.

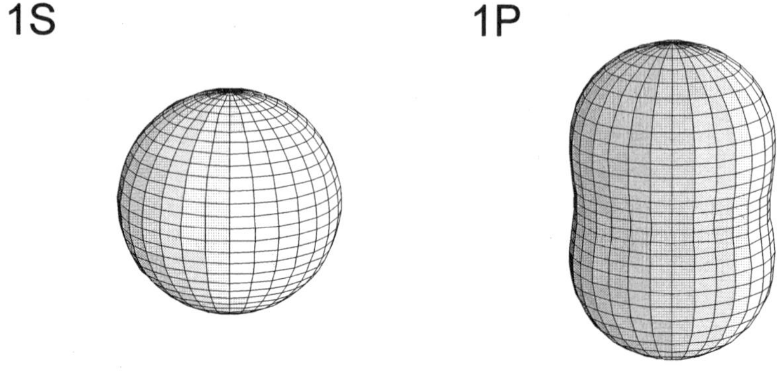

Figure 7. Equilibrium shape of electron bubbles with the electron in the 1S and the 1P states.

energy is no longer spherical. For the 1P state with azimuthal quantum number $m = 0$, the electron pressure is zero in the x-y plane, and the bubble has the shape shown in Fig. 7. The pressure at which the electron bubble explodes is different for each quantum state; the magnitude of the negative pressure required to explode the 1P state is calculated to be about 14 % less than for the 1S state [30].

In recent work, a CO_2 laser has been used to excite electron bubbles to the 1P state [31]. Results obtained at 1.99 K are shown in Fig. 8. It can be seen that the effect of the light is to create some new objects that break at a smaller negative pressure. These are the 1P electron bubbles. The transducer voltage required to explode them is about 16 % less than the voltage needed for the 1P bubble, and is thus in good agreement with theory. As expected, the number of these bubbles that are present is proportional to the laser intensity. From measurements of this type we are currently trying to determine an accurate value for the lifetime of the 1P excited state. As a result of re-emission of a photon, the 1P state should have a lifetime of the order of 10 μs [32]; our preliminary measurements indicate a lifetime much shorter than this, presumably due to some form of non-radiative transition.

Finally, it is of interest to consider in more detail what happens when an electron bubble is excited by light. According to the Franck-Condon principle, one should consider that the transition of the electron state takes place before the shape of the bubble changes. After the transition, the pressure on the bubble wall is changed and the wall then begins to move in response to this change in pressure. The motion that ensues is dependent on the magnitude of the dissipative forces acting on the bubble wall. If these are large, the shape of the bubble will change slowly from spherical to the peanut shape shown in Fig. 7. However, if the dissipation is very small, the bubble will quickly reach the equilibrium shape, and the inertia of the liquid around the bubble will then keep the bubble wall moving so that the bubble shape becomes even more distorted.

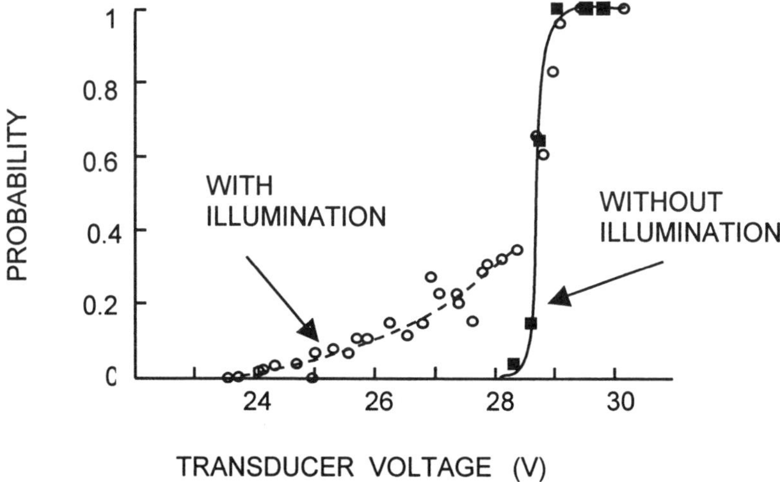

Figure 8. Probability of cavitation with and without illumination. Light produces bubbles containing 1P electrons which are more easily broken.

At 1.99 K, the temperature at which the data of Fig. 8 were obtained, the dissipation in liquid helium is large and so the bubble should slowly relax to the equilibrium 1P shape. However, as the temperature is lowered below about 1.5 K the dissipation becomes very small. At these temperatures, it is possible [33] that the inertial effects are sufficient to make the radius of the waist of the bubble shrink to zero, thus dividing the bubble into two daughter bubbles. What happens after this is not yet established. Experimental measurements in this temperature range give results that have a complicated dependence on the temperature and the ambient pressure in the experimental cell [31]. For example, when the pressure is zero some 1P bubbles are produced but at 1 bar, where the energy difference between the 1S and the 1P states is better matched to the energy of the photons from the CO_2 laser, no 1P bubbles are detected. Instead, there appears to be a reduction in the density of "normal" electron bubbles, together with the production of a large number of new bubbles that explode only when a larger negative pressure is applied. An additional intriguing result is that at temperatures below 1.5 K, the new bubbles that are produced have a long lifetime, at least one second. Thus these new objects that are produced by the light must be unable to relax back to ordinary electron bubbles by radiative decay. We hope to report on these experiments in more detail in the near future.

5. Acknowledgements

We thank F. Caupin, S. Balibar and D.O. Edwards for helpful discussions. This work was supported in part by the National Science Foundation through Grant No. DMR-0071507.

References

[1]   Misener, A.D. and Hebert, G.R. (1956) Tensile strength of liquid helium II, *Nature* **177**, 946-947.
[2]   Beams, J.W. (1956) Tensile strength of liquid helium II, *Phys. Rev.* **104**, 880-882.
[3]   Finch, R.D., Kagiwada, R., Barmatz, M. and Rudnick, I. (1964) Cavitation in liquid helium, *Phys. Rev.* **134**, A1425-A1428.
[4]   Finch, R.D. and Wang, T.G. (1966) Visible cavitation in liquid helium *J. Acoust. Soc. Am.* **39**, 511-514.
[5]   Finch, R.D., Wang, T.G., Kagiwada, R., Barmatz, M. and Rudnick, I. (1966) Studies of the threshold of cavitation noise in liquid helium, *J. Acoust. Soc. Am.* **40**, 211-218.
[6]   Edwards, M.H., Cleary, R.M. and Fairbank, W.M. (1966) Bubble formation on vortices in a liquid helium bubble chamber, in D.F. Brewer (ed.) *Quantum Fluids*, North-Holland, Amsterdam, pp. 140-145.
[7]   Finch, R.D. and Chu, M.L. (1967) Production and detection of solitary macroscopic quantized vortices in helium II, *Phys. Rev.* **161**, 202-206.
[8]   Jarman, P.D. and Taylor, K.J. (1970) The sonically induced cavitation of liquid helium, *J. Low Temp. Phys.* **2**, 389-402.
[9]   McConnell, P.M., Chu, M.L. and Finch, R.D. (1970) Mechanism of ultrasonic cavitation nucleation in liquid helium by quantized vortices, *Phys. Rev.* **A1**, 411-418.
[10] Marston, P.L. (1976) Tensile strength and visible ultrasonic cavitation of superfluid ⁴He, *J. Low Temp. Phys.* **25**, 383-407.
[11] Dhingra, H.C. and Finch, R.D. (1976) Experiments on ultrasonic cavitation in liquid helium in the presence of second sound, *J. Acoust. Soc. Am.* **59**, 19-23.
[12] Nissen, J.A., Bodegom, E., Brodie, L.C. and Semura, J.S. (1989) Tensile strength of liquid ⁴He, *Phys. Rev* **B40**, 6617-6624.

[13] Balibar, S. and Maris, H.J. (2000) Negative pressures and cavitation in liquid helium, *Physics Today* **53**, 29-34.

[14] Balibar, S., Caupin, F., Roche, P. and Maris, H.J. (1998) Quantum cavitation: a comparison between superfluid helium-4 and normal liquid helium-3, *J. Low Temp. Phys.* **113**, 459-471.

[15] Caupin, F., Roche, P., Marchand, S. and Balibar, S. (1998) Cavitation in normal liquid helium-3, *J. Low Temp. Phys.* **113**, 473-478.

[16] Caupin, F., Balibar, S. and Maris, H.J. (2001) Anomaly in the stability limit of liquid He-3, *Phys. Rev. Lett.* **87**, 145302.

[17] Caupin, F. and Balibar, S. (2001) Cavitation pressure in liquid helium, *Phys. Rev.* **B64**, 064507.

[18] Caupin, F., Balibar, S. and Maris, H.J. (2002) Nucleation in a Fermi liquid at negative pressure, *J. Low Temp. Phys.* **126**, 91-96.

[19] Balibar, S., Guthmann, C., Lambaré, H., Roche, P., Rolley, E. and Maris, H.J. (1995) Quantum cavitation in superfluid helium-4?, *J. Low Temp. Phys.* **101**, 271-7.

[20] Lambaré, H., Roche, P., Balibar, S., Maris, H.J., Andreeva, O.A., Guthmann, C., Keshishev, K.O. and Rolley, E. (1998) Cavitation in superfluid helium-4 at low temperature, *Eur. Phys J.* **2**, 381-391.

[21] For a review, see A.L. Fetter (1976) Vortices and ions in helium, in K.H. Benneman and J.B. Ketterson (eds.), *The Physics of Liquid and Solid Helium*, Wiley, New York, pp. 207-305.

[22] Sommer, W.T. (1964) Liquid helium as a barrier to electrons, *Phys. Rev. Lett.* **12**, 271-273.

[23] Classen, J., Su, C.-K., Mohazzab, M. and Maris, H.J. (1998) Electrons and cavitation in liquid helium, *Phys. Rev.* **B57**, 3000-3010.

[24] Akulichev, V.A. and Boguslavskii, Y.Y. (1972) Cavitation stability of liquid helium due to "electron bubbles", *Sov. Phys. JETP* **35**, 1012-1013.

[25] The radius and the critical pressure depend on the value that is used for the surface tension. There has been some disagreement about the correct value for this quantity. See Iino, M., Suzuki, M. and Ikushima, A. (1985) Surface tension of liquid ^{4}He: surface energy of the Bose-Einstein condensate, *J. Low Temp. Phys.* **61**, 155-169, and Roche, P., Deville, G., Appleyard, N.J. and Williams, F.I.B. (1997) Measurement of the surface tension of superfluid ^{4}He at low temperature by capillary wave resonances, *J. Low Temp. Phys.* **106**, 565-573.

[26] Su, C.-K., Cramer, C.E. and Maris, H.J. (1998) Electrons and cavitation in liquid helium-3, *J. Low Temp. Phys.* **113**, 479-484.

[27] Su, C.-K. and Maris, H.J. (1998) Quantum nucleation of bubbles from electrons in liquid helium at negative pressure, *J. Low Temp. Phys.* **110**, 485-490.

[28] Konstantinov, D., Homsi, W., Luzuriaga, J., Su, C.-K., Weilert, M.A. and Maris, H.J. (1998) How does a bubble chamber work?, *J. Low Temp. Phys.* **113**, 485-490.

[29] Classen, J., Su, C.-K., Mohazzab, M. and Maris, H.J. (1998) Explosion of electron bubbles trapped on vortices in He-II, *J. Low Temp. Phys.* **110**, 431-436.

[30] Maris, H.J. and Konstantinov, D. (2000) Bubbles in liquid helium containing electrons in excited states, *J. Low Temp. Phys.* **121**, 615-620.

[31] Konstantinov, D. and Maris, H.J. unpublished.

[32] Fowler, W.B. and Dexter, D.L. (1968) Electronic bubble states in liquid helium, *Phys. Rev.* **176**, 337-343.

[33] Maris, H.J. (2000) On the fission of elementary particles and electrons in liquid helium, *J. Low Temp. Phys.* **120**, 173-204.

QUANTUM STATISTICS OF METASTABLE LIQUID HELIUM

FRÉDÉRIC CAUPIN AND SÉBASTIEN BALIBAR
Laboratoire de Physique Statistique
de l'Ecole Normale Supérieure
associé aux Universités Paris 6 et Paris 7 et au CNRS
24 rue Lhomond 75231 Paris Cedex 05, France

Abstract. Experimental studies of homogeneous cavitation give information on the limit of stability of liquids, the "spinodal limit". The slope of the spinodal line in the $(P - T)$ plane has been related to the sign of the isobaric expansion coefficient of the liquid. Using quantum statistics, we give theoretical arguments to explain the existence of a minimum in the spinodal line of liquid helium 3, for which experimental evidence has been reported. The calculations involve the shape of the dispersion curve of collective excitations. We present a similar analysis in the case of superfluid helium 4; we also consider the overpressurized region, and the location of the superfluid transition line in the metastable regions.

1. Introduction

We have investigated homogeneous cavitation in liquid helium by using a high amplitude ultrasonic wave generated by a hemispherical transducer [1, 2, 3]. This brings the liquid in a metastable state, because at pressures below the saturated vapour pressure (svp), an energy barrier needs to be overcome for the gas phase to nucleate. This barrier vanishes at a negative pressure called the spinodal pressure P_s, where the compressibility of the liquid diverges (the sound velocity vanishes). Cavitation becomes likely when the wave produces a sufficiently large negative pressure swing, corresponding to the cavitation pressure P_{cav}.

In liquid helium 4, we had previously found a crossover from a high temperature regime where cavitation is a thermally activated process, to a low temperature regime where it occurs by quantum tunneling [1, 2, 3]. In liquid helium 3, where this crossover was predicted to take place at 120 mK [4, 5], measurements performed down to 35 mK have shown a different behaviour [2, 6]. To explain this, we have proposed to consider the Fermi degeneracy of helium 3 [7]. Here we will focus on the thermally activated regime.

In liquid helium 3, the experimental temperature dependence of P_{cav} supports the existence of a minimum at 0.4 K in the liquid-gas spinodal line $P_s(T)$. R. J. Speedy previously proposed this peculiar behaviour for the spinodal of water and related it to a change in sign of the expansion coefficient α, i. e. a line of density maxima (LDM) [8]. In Sec. 2 we show that this feature is consistent with

A.R. Imre et al. (eds.), Liquids Under Negative Pressure, 201–214.
© 2002 *Kluwer Academic Publishers. Printed in the Netherlands.*

202

extrapolations of sound velocity measurements and calculate α near the spinodal line.

We consider stretched liquid helium 4 in Sec. 3. After giving a short review of previous work, we use quantum statistics to derive α and the shape of the spinodal, and to describe the superfluid transition line at negative pressure.

In Sec. 4 we finally consider overpressurized helium 4 and we predict some properties of the phase diagram in this other metastable region.

2. Liquid helium 3 at negative pressure

In this section we summarize a recent article written with H.J. Maris [9] before giving more details on the calculations involving quantum statistics.

2.1. EXPERIMENTAL EVIDENCE FOR A MINIMUM IN THE SPINODAL

We have performed experiments in a volume V and during a time τ and measured the cavitation probability Σ [2]. At a pressure P and a temperature T, the energy barrier that must be overcome for liquid-gas separation to occur is $E_b(P,T)$, and we have:

$$\Sigma(P,T) = 1 - \exp\left[-\Gamma_0 V\tau \exp\left(-\frac{E_b(P,T)}{k_B T}\right)\right] \tag{1}$$

where Γ_0 has the dimensions of frequency times an inverse volume. It is thus natural to estimate Γ_0 as an attempt frequency ν at which the fluctuations try to overcome the nucleation barrier multiplied by the density of the critical nuclei which can be taken to be spheres of radius R_c [10, 11, 12]. Typically, R_c is around $1\,\mathrm{nm}$ and the attempt frequency varies from $k_B T/h$ to E_b/h; all the different estimates thus lie between $5 \times 10^{36}\,T$ and $1.5 \times 10^{38}\,T\,\mathrm{m^{-3}\,s^{-1}\,K^{-1}}$. Pettersen $et\ al.$ [12] have calculated V and τ for the experimental method which uses an acoustic wave to produce a negative pressure swing in the liquid. For ^{3}He and for a 1 MHz acoustic wave as in Ref. [2], this gives $V\tau = 1.2 \times 10^{-22}\,\mathrm{m^3\,s}$ [13]. The theoretical estimates of the factor $\Gamma_0 V\tau$ thus vary from $6 \times 10^{14}\,T$ to $1.8 \times 10^{16}\,T\,\mathrm{K^{-1}}$. Although this range extends over two orders of magnitude, it does not significantly affect the value of the energy barrier: for $\Sigma = 0.5$, all estimates give $E_b = (34 \pm 3)k_B T$.

Maris [4] has calculated $E_b(P)$ at low temperature by a density functional method; close to the spinodal pressure P_s, his results are well represented by a power law:

$$\frac{E_b}{k_B} = \beta(P - P_s)^\delta \tag{2}$$

with $\beta = 47.13\,\mathrm{K\,bar^{-3/4}}$ and $\delta = 3/4$. However, to calculate the cavitation pressure up to $0.6\,\mathrm{K}$, we need to know the temperature dependence of E_b. The strongest source of this dependence is that the spinodal pressure varies with temperature; therefore we write $E_b(P,T) = E_b(P - P_s(T))$ and assume that Eq. 2 remains valid at higher temperature with parameters β and δ held constant. The temperature

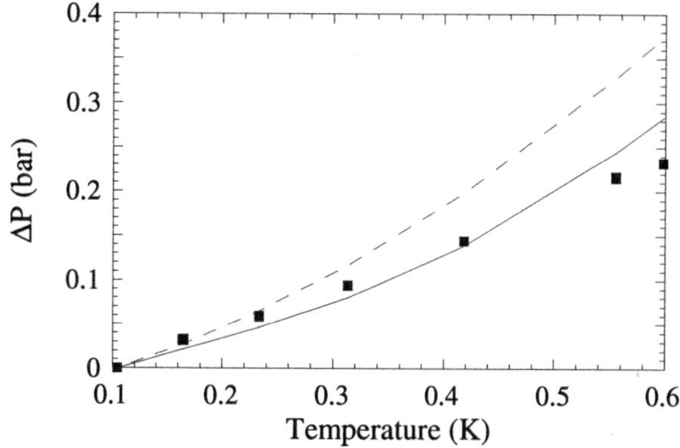

Figure 1. Temperature variation $\Delta P = P_{\text{cav}}(T) - P_{\text{cav}}(0.1\,\text{K})$ of the cavitation pressure. Experimental ΔP is given by solid squares. Other lines are calculated ΔP according to Guilleumas *et al.* [11] (dashed line) and this work (solid line).

dependence of the cavitation pressure follows from Eq. 1:

$$P_{\text{cav}}(T) = P_{\text{s}}(T) + \left[\frac{T}{\beta} \ln \left(\frac{\Gamma_0 V \tau}{\ln 2} \right) \right]^{1/\delta} \tag{3}$$

To compare the experimental and theoretical temperature variations of P_{cav}, we have plotted the quantity $\Delta P = P_{\text{cav}}(T) - P_{\text{cav}}(0.1\,\text{K})$ in Fig. 1, using two different spinodal lines $P_{\text{s}}(T)$ and the lowest estimate of the prefactor, namely $\Gamma_0 V \tau = 6 \times 10^{14}\, T\,\text{K}^{-1}$. The dashed line was deduced from Eq. 3 using the spinodal line calculated by Guilleumas *et al.* [11], which is displayed on Fig. 2. It does not agree with the experimental results; to do so, it would require $\Gamma_0 V \tau$ to be at least 3 orders of magnitude smaller than expected. We do not see any reasons to support this hypothesis. and we think that the discrepancy comes from the shape of the spinodal limit.

Before proceeding further, we need to recall how the spinodal pressure P_{s} can be obtained: Maris' method [15, 4] consists in extrapolating measurements of the sound velocity c at positive pressure with a law of the form $c = [b\,(P - P_{\text{s}})]^{1/3}$. Maris used for c the measurements of Abraham *et al.* at low temperature [16]. We used the same method with a set of data from Roach *et al.* [17] between 0.01 and 0.6 K [18] . The spinodal line we obtained is shown in Fig. 2: the spinodal pressure reaches a minimum of -2.9 bar around $T = 0.4\,\text{K}$. The new shape of the spinodal curve we propose is sufficient to remove the discrepancy stated above: using again Eq. 3 with the value $\Gamma_0 V \tau = 6 \times 10^{14}\, T\,\text{K}^{-1}$, we find a cavitation line which has a temperature dependence consistent with the experimental results (solid line in Fig. 1).

Let us now turn to the physical origin of such a minimum in the spinodal. A similar behavior was first proposed by Speedy in the case of water [8]. A review of related topics also describing alternative theories can be found in Ref. [19]. Following a thermodynamical analysis first developed in the case of helium [20], Speedy shows that close to the spinodal the sign of the isobaric thermal expansion

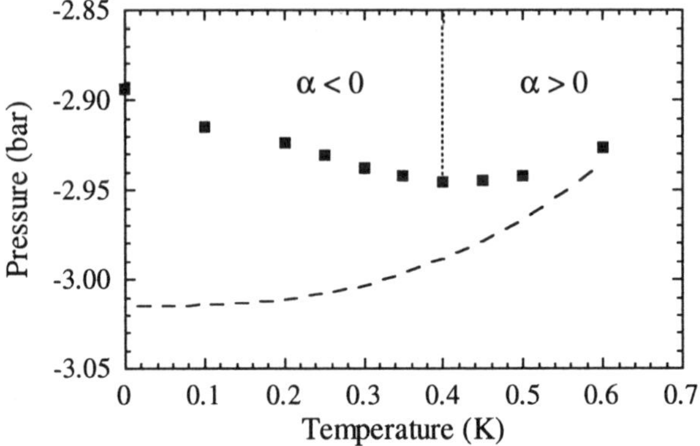

Figure 2. Comparison between two theoretical estimates of the spinodal line: Guilleumas *et al.* [11] (dashed line) and this work (solid squares). The spinodal found in this work exhibits a minimum at 0.4 K. The dotted line is a linear extrapolation of the LDM as measured by Boghosian *et al.* [14] between 0 and 11 bar (see Fig. 3).

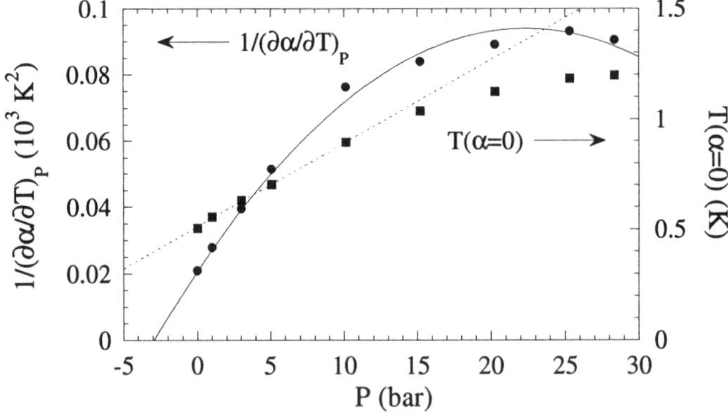

Figure 3. Temperature of density maximum (solid squares) and inverse of $(\partial\alpha/\partial T)_P$ around $\alpha = 0$ (solid circles) as functions of pressure, derived from measurements by Boghosian *et al.* [14]. The dotted line shows the extrapolation of the LDM used in Fig. 2. The solid line is a parabolic fit to $1/(\partial\alpha/\partial T)_P$ forced to vanish at the pressure of the minimum in the spinodal.

coefficient α of the liquid is the same as the sign of dP_s/dT. Therefore, if the locus of points such that $\alpha(P,T) = 0$ intersects the spinodal, this results in an extremum in the curve $P_s(T)$. Water and ^{3}He have in common that both liquids exhibit a LDM: in some temperature range, they expand upon cooling. Therefore they may exhibit such a minimum in the spinodal. To obtain the LDM in ^{3}He, we used measurements by Boghosian *et al.* [14], which extend to higher temperatures and agree well with Roach values in the region where both sets overlap; the result is shown in Fig. 3. A simple linear extrapolation of the LDM for pressures below 11 bar extends down to the minimum in the spinodal as shown in Fig. 2. In his original paper [8], Speedy shows that the expansion coefficient at the spinodal undergoes a jump from $-\infty$ to $+\infty$ at the temperature at which the LDM meets the spinodal. To find some evidence to support this prediction, we follow Speedy's analysis for water and derive the slope $(\partial\alpha/\partial T)_{\mathrm{P}}$ around $\alpha = 0$ for each isobar in the measurements by Boghosian *et al.* This slope should diverge when the pressure reaches the spinodal. Fig. 3 shows that the experimental values are consistent with this prediction.

2.2. INTERPRETATION

We now give theoretical arguments to explain why α should be negative in helium 3 at low temperature and near the spinodal. The negative sign of α in ^{3}He was first observed experimentally in 1958 by Lee *et al.* [21]. The same year, Brueckner and Atkins [22] pointed out how this behaviour was related to the variation of the effective mass m^* with the density ρ. Indeed, using a Maxwell relation, we can write:

$$\alpha = -\frac{1}{V}\left(\frac{\partial S}{\partial P}\right)_T = -\frac{\rho^2}{M}\chi_T\left(\frac{\partial S}{\partial \rho}\right)_T \tag{4}$$

At low temperature, the heat capacity of a Fermi gas is:

$$C_{\mathrm{F}} = \left(\frac{k_{\mathrm{B}}}{\hbar}\right)^2\left(\frac{\pi m}{3\rho}\right)^{2/3}T \tag{5}$$

Because of interactions, helium 3 must be described by the Landau theory of a Fermi liquid [23]. This leads to a corrective factor in the heat capacity $C_{\mathrm{V}} = (m^*/m)\,C_{\mathrm{F}}$, so that we have $S = C_{\mathrm{V}} = (m^*/m)\,C_{\mathrm{F}}$. Using Greywall's measurements of the effective mass [24] and extrapolating them at negative pressure as we did before [25], we find that the corresponding α_{F} given by Eq. 4 remains negative down to the spinodal; when $P \to P_{\mathrm{s}}$, it diverges as $-T\chi_T$.

We now consider the corrections to the linear regime of the heat capacity and their evolution close to the spinodal. We see two sources of corrections. At positive pressure, the main one comes from the coupling of the Landau quasiparticles to the incoherent spin fluctuations and varies as $T^3 \ln T$. This effect has been studied by Greywall [26], who has shown that its amplitude decreases when pressure decreases; we assume that this is the case until the spinodal is reached, so that α_{spin} remains negligible compared to α_{F}. The second correction α_{ph} comes from the contribution of phonons to the heat capacity. The phonons obey Bose statistics and can be described with the grand canonical formalism, using the grand potential

$$J = \frac{Mk_{\mathrm{B}}T}{2\pi^2\rho}\int_0^{+\infty} k^2\ln\left[1 - \exp\left(-\frac{\varepsilon(k)}{k_{\mathrm{B}}T}\right)\right]\mathrm{d}k \tag{6}$$

where $\varepsilon(k)$ is the energy of a phonon of momentum k. The usual linear dispersion relation writes as:

$$\varepsilon(k) = \hbar c k \tag{7}$$

where c is the sound velocity. The entropy is deduced by

$$S = -\left(\frac{\partial J}{\partial T}\right)_P \tag{8}$$

206

Using Eqs. 4, 6, 8, and 7, one obtains

$$\alpha_{\mathrm{ph}} = \frac{2\pi^2 {k_{\mathrm{B}}}^4}{15\hbar^3} \left(\frac{T}{c}\right)^3 \chi_T \left[\frac{1}{3} + \frac{\rho}{c}\left(\frac{\partial c}{\partial \rho}\right)_T\right] \tag{9}$$

This term is therefore usually neglected compared to α_{F} (linear in T), but it could become large near the spinodal where the isothermal sound velocity vanishes. However, this is relevant only for the long wavelength phonons: as stated by Lifshitz and Kagan [27], the first correction to the linear dispersion gives for the energy $\varepsilon(k)$:

$$\varepsilon(k) = \hbar k \sqrt{c^2 + 2\rho\lambda k^2} \tag{10}$$

where λ is a constant. As the spinodal is approached, the dispersion relation thus becomes quadratic, and we find near P_{s}

$$\alpha_{\mathrm{ph}} = \frac{35\,\rho\,{k_{\mathrm{B}}}^{5/2}\zeta(5/2)}{64\,\pi^{3/2}\hbar^{3/2}\,(2\rho\lambda)^{3/4}}\,T^{3/2}\chi_T \tag{11}$$

which diverges but remains negligible compared to α_{F} at sufficiently low temperature. We also note that, if the sound remains adiabatic at small k close to the spinodal, the use of the adiabatic (instead of isothermal) sound velocity, which does not vanish at P_{s}, would further reduce the phonon contribution.

We have therefore shown that, in helium 3, α is negative at low temperature near the spinodal, so that the spinodal pressure starts by decreasing with increasing temperature. In Sec. 2.1 we have reported experimental support for this behaviour; it also indicates that the spinodal pressure reaches a minimum value around $0.4\,\mathrm{K}$ and increases monotonically at higher temperature.

3. Superfluid helium 4 at negative pressure

3.1. LANDAU MODEL AND THE DISPERSION CURVE

Before discussing the case of negative pressure, we need to recall Landau's theory [28] which gives a qualitatively correct description of superfluid helium 4 at positive pressure. It is based on the dispersion relation of elementary excitations displayed in Fig. 4 (a); this shape was proposed by Landau [28] and measured by neutron scattering (see Refs. [29] and [30] for instance). There is a phonon branch at small momentum. A peculiarity of helium 4 is that it exhibits a structure at larger momentum, known as the maxon peak, and the roton minimum at k_0; the energy in the roton region can be approximated by:

$$\varepsilon(k) = \Delta + \frac{\hbar^2(k - k_0)^2}{2\mu} \tag{12}$$

where Δ is the roton gap and μ the roton effective mass. Typical svp values for these parameters are: $\Delta = 8.62\,\mathrm{K}$, $k_0 = 19.29\,\mathrm{nm}^{-1}$ and $\mu = 0.161\,m_4$ where

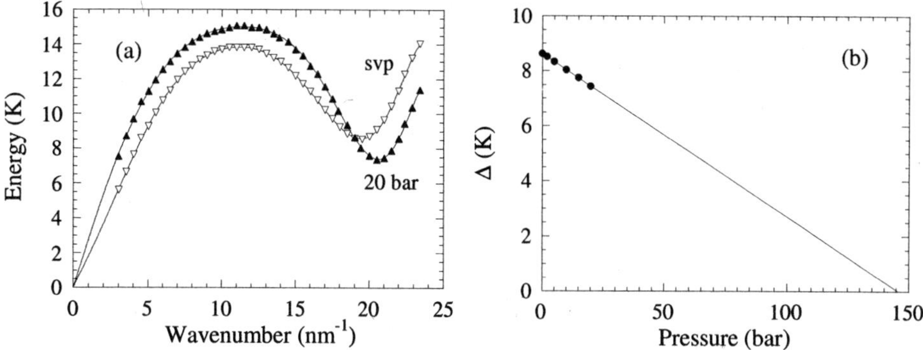

Figure 4. (a) Dispersion curve of elementary excitations in superfluid helium 4 at 0.5 K, as obtained by neutron scattering [30]. Open triangles correspond to svp data and solid triangles to 20 bar; the solid lines are guides to the eye, and their slopes at the origin are deduced from the experimental sound velocity [31]. (b) Pressure variation of the roton gap Δ; solid circles are experimental values from Ref. [30], and the solid line is a linear extrapolation at high pressure (see Sec. 4).

$m_4 = 6.65 \times 10^{-27}$ kg is the mass of an atom of helium 4 [30]. The energy levels are populated by thermal activation, starting by the phonon branch at low temperature, and then filling the roton region at higher temperature. This model allows us to calculate the thermodynamic functions of the superfluid. It also gives an estimate of the superfluid transition temperature T_λ, when equating the total density to the density of normal fluid ρ_n coming from the excitations [29]:

$$\rho_n = -\frac{\hbar^2}{6\pi^2 k_B T} \int_0^{+\infty} \frac{k^4 \exp\left[\varepsilon(k)/(k_B T)\right]}{\exp\left[\varepsilon(k)/k_B T\right] - 1} \, dk \tag{13}$$

Unfortunately, this leads to an overestimate of T_λ: for example, one would find $T_\lambda = 2.76$ K at svp, instead of the measured 2.17 K. The discrepancy arises from the interaction between excitations; however, the model gives quantitatively correct results at low temperature (up to 1.6 K at svp) where there are few excitations. It also accounts for the experimentally observed decrease of T_λ with increasing pressure. This comes from the decrease of the roton gap [32]. It has been measured by neutron scattering (see Ref. [30] for instance): the dispersion curves at low and high pressures are displayed on Fig. 4 (a), and $\Delta(P)$ in Fig. 4 (b). The shape of the dispersion curve has also been calculated using a density functional theory [33]. In the negative pressure region, Quantum Monte-Carlo calculations [34] show a trend of the maxon peak and the roton gap to merge together. This is also obtained in a recent extension of the density functional results to negative pressure [35].

We will therefore use Landau's simple model in the following, because it is valid at low temperatures, it allows analytical calculations, and captures the physics of superfluid helium 4.

3.2. EXPANSION COEFFICIENT AND THE SPINODAL LINE

The spinodal pressure of helium 4 at zero temperature has been estimated by several methods [3]; all of them give $P_{\rm s}$ around $-10\,{\rm bar}$. Density functional theory [11] and Quantum Monte-Carlo calculations [34] give a monotonic spinodal line with positive slope. Hall and Maris [36] expect a change in slope of the spinodal line if a roton minimum still exists down to $P_{\rm s}$, and a monotonic spinodal if it disappears. In this section we will try to clarify this point by calculating the expansion coefficient near the spinodal.

Liquid helium 4 exhibits two lines of density extrema: a line of density minima (around $1\,{\rm K}$ at svp) and a line of density maxima just above T_λ. The line of density minima was measured by Atkins and Edwards [37]; they found good agreement with Landau's model. Indeed, at positive pressure the phonon contribution to the expansion coefficient is given by Eq. 9, and the roton contribution is derived from Eqs. 4, 6, 8, and 12:

$$
\begin{aligned}
\alpha_{\rm rot} = {} & -\frac{2\,k_{\rm B}{}^{1/2}\mu^{1/2}k_0{}^2\Delta}{(2\pi)^{3/2}\hbar\,T^{1/2}}\left(1+\frac{3k_{\rm B}T}{2\Delta}\right)\exp\left(-\frac{\Delta}{k_{\rm B}T}\right)\\
& \times\chi_T\left\{\frac{\rho}{2\mu}\left(\frac{\partial\mu}{\partial\rho}\right)_T+\frac{2\rho}{k_{\rm B}}\left(\frac{\partial k_0}{\partial\rho}\right)_T-1\right.\\
& \left.-\frac{\rho}{\Delta}\left(\frac{\partial\Delta}{\partial\rho}\right)_T\left[\frac{\Delta}{k_{\rm B}T}\left(1+\frac{\frac{3}{2}\left(\frac{k_{\rm B}T}{\Delta}\right)^2}{1+\frac{3k_{\rm B}T}{2\Delta}}\right)-1\right]\right\}
\end{aligned}
\tag{14}
$$

The existence of a line of density minima is thus explained by the competition between these two terms (see Fig. 5 (a)): $\alpha_{\rm ph}$ is positive and is the leading term at low temperature; $\alpha_{\rm rot}$ is negative and overcomes the phonon term at high temperature. The respective values at svp of the quantities $(\rho/\mu)(\partial\mu/\partial\rho)_T$, $(\rho/k_0)(\partial k_0/\partial\rho)_T$, and $(\rho/\Delta)(\partial\Delta/\partial\rho)_T$ are -1.6, 0.41, and -0.53, as measured by neutron scattering [30]. These values were used to plot Fig. 5 (a); they differ slightly from the ones used by Atkins and Edwards [37].

We now extend this calculation to the negative pressure region. If the roton minimum still exists near the spinodal (at $\rho \simeq \rho_s = 94.82\,{\rm kg\,m^{-3}}$), Eq. 14 holds. We will use for the roton parameters in Eq. 14 the svp values, except for the gap for which we will take the approximate value $\Delta = 10\,{\rm K}$. As for the phonons, we need to take into account the curvature of their dispersion relation as was pointed out in Sec. 2.2; this leads again to Eq. 11. We keep only the term involving Δ in the curly brackets of Eq. 14 and use the svp value for $(\partial\Delta/\partial\rho)_T$; this favours a larger negative $\alpha_{\rm rot}$. The result near $P_{\rm s}$ is shown in Fig. 5 (b); we have plotted α/χ_T to remove the divergence due to the spinodal. The total expansion coefficient is now positive up to temperatures where the Landau model ceases to be correct. We think this would also be the case if the roton minimum and the maxon peak were merging together.

We need to point out that all these calculations used a temperature-independent value of ρ_s; a self-consistent calculation, allowing a variation of $P_{\rm s}$ corresponding to the calculated α would be of interest, but we have not attempted it yet.

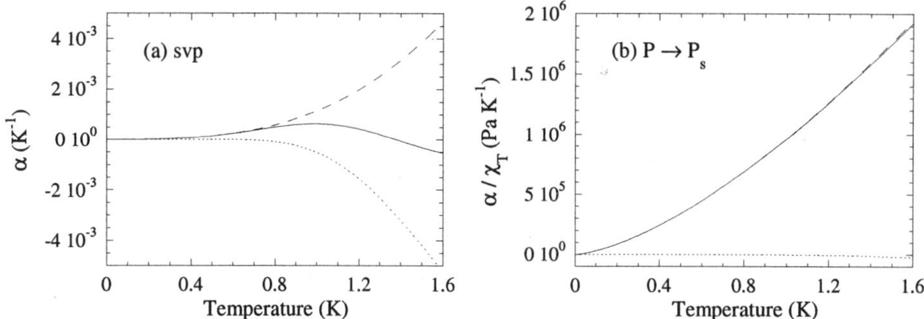

Figure 5. Expansion coefficient of superfluid helium 4 calculated with the Landau model (solid line) at saturated vapour pressure (a) and near P_s (b). The graphs also show the phonon contribution (dashed line) and the roton contribution (dotted line). In graph (b) each quantity is divided by χ_T which diverges at P_s.

Our results support a monotonic spinodal in helium 4, the pressure of which increases with increasing temperature. They also show that the roton minimum does not need to disappear for the spinodal to be monotonic.

3.3. SUPERFLUID TRANSITION

Large attention has been given to the extension of the lambda line in the stretched liquid state. First, measurements of T_λ have been performed at positive pressure *below* the saturated vapour pressure [38, 39]; they show a regular continuation of the lambda curve with the same negative slope. Skripov was the first to propose an extension at negative pressure [40]: he used a shifted Bose gas condensation line, so that the lambda line changes slope at negative pressure and T_λ vanishes before the spinodal is reached. Campbell *et al.* [41] expected a multicritical point at $T = 0\,\text{K}$ between the spinodal and the lambda line because of phonon avalanche near the spinodal (due to the vanishing sound velocity). Hall and Maris [36] claimed that disappearance of the roton minimum would lead to a change in slope of the lambda line. Apenko proposed a Lindemann-like criterion for superfluidity of Bose liquids [42], leading to a transition line with a change in slope between high and low densities (liquid helium and Bose gas region respectively). Bauer *et al.* performed Quantum Monte-Carlo calculations [34]: they obtained a nearly constant T_λ down to the spinodal. Finally, Skripov recently added to his picture the cases of the multicritical point and of a non-zero T_λ at P_s [43].

We have used the Landau model to try to clarify this problem. Of course, as pointed out in Section 3.1, it does not give the exact T_λ; however, this model is valid at low temperature, and with the input of the roton parameters, it predicts correctly the slope of the lambda line, so that it will provide some understanding of what is happening at negative pressure. Using Eq. 13, one can calculate the phonon and roton contributions ($\rho_{\text{n-ph}}$ and $\rho_{\text{n-rot}}$ respectively) to the normal fraction ρ_n/ρ. The result at svp is shown in Fig. 6 (a): $\rho_{\text{n-ph}}$ dominates at low temperature, and it is overcome by $\rho_{\text{n-rot}}$ around $0.6\,\text{K}$; the fluid becomes normal at $T_\lambda = 2.76\,\text{K}$. If one used a linear phonon branch, the vanishing sound velocity at the

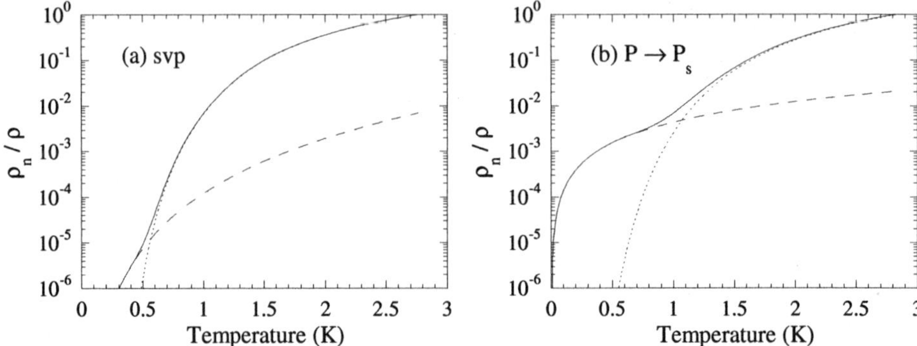

Figure 6. Normal fluid fraction in helium 4 calculated with the Landau model (solid line) at saturated vapour pressure (a) and near P_s (b). The graphs also show the phonon contribution (dashed line) and the roton contribution (dotted line).

spinodal would lead to a phonon avalanche and a multicritical point as proposed by Campbell *et al.* [41]; however, the curvature of the dispersion relation (Eq. 10) prevents this phenomenon to occur and our result is shown in Fig. 6 (b). One can see that $\rho_{\text{n-ph}}$ is much higher than at svp but remains less than 2 % up to T_λ. For $\Delta = 10\,\text{K}$ and $\rho = \rho_s = 94.82\,\text{kg m}^{-3}$, we obtain $T_\lambda = 2.81\,\text{K}$. Concerning the slope of the lambda line, we find here that it remains negative, but choosing $\Delta = 9.5\,\text{K}$ instead of $10\,\text{K}$ would lead to $T_\lambda = 2.65\,\text{K}$ near P_s, and thus to a change in slope. This comes from a competition when one approaches P_s between the increase of Δ which tends to increase T_λ, and a decrease of ρ and increase in $\rho_{\text{n-ph}}$ which tend to lower T_λ.

Our results show that T_λ remains finite down to the spinodal; we would like to emphasize that this comes from the curvature of the phonon branch, and that it is obtained at low temperature where Landau's theory is valid with no approximation. Our calculations are in qualitative agreement with Quantum Monte-Carlo simulations [34]. They also indicate that the lambda line may change slope even if a slight roton minimum remains near P_s.

4. Overpressurized superfluid helium 4

Finally, we turn to the other metastable region for the liquid, at pressures above the melting pressure P_m. Because the liquid-solid transition is first order as the liquid-gas transition, it is possible to observe a metastable overpressurized liquid. Previous experiments only reached an overpressure of a few millibars, but recently we were able to achieve an overpressure of 4.3 bar with a modified version of our cavitation experiment [3, 44]. A theoretical description of the thermodynamics in this region is thus needed.

4.1. INSTABILITY OF THE SUPERFLUID AGAINST THE SOLID

We reconsider a theory proposed by Schneider and Enz [45]. They have suggested that the limit of metastability of the superfluid against the solid is associated with the softening of the roton mode. The roton minimum, which is known to decrease with increasing pressure (see Section 3.1, Fig. 4 (b)), would vanish at some pressure $P_{\Delta=0} > P_{\mathrm{m}}$; the liquid would then become unstable against a density fluctuation of finite wave number k_0, which is close to the reciprocal-lattice vector of solid helium. A linear extrapolation of neutron scattering data for the roton gap (Fig. 4 (b)) leads to an estimate of 150 bar for the stability limit pressure. Edwards and Maris have recently obtained $P_{\Delta=0} \simeq 200$ bar within a density functional approach [35].

4.2. SUPERFLUID TRANSITION

As in the case of the stretched liquid, we may consider the continuation of the lambda line in the overpressurized region. This was mentioned by Skripov [40], but no attempt was made to determine the behaviour of this metastable lambda line.

To do this, we use the Landau model up to the limit where the roton gap vanishes. As the pressure increases above P_{m}, the roton gap decreases and the lambda line should continue with a negative slope as below P_{m}. An interesting issue is to know whether T_λ reaches 0 before the roton instability occurs. Of course, we need to use the simple Landau model with caution, because its range of validity is reduced to lower temperature as $P_{\Delta=0}$ is approached; indeed, to neglect interactions requires a small number of elementary excitations. For any $P < P_{\Delta=0}$, this condition will be fulfilled at sufficiently low temperature; we can use without any approximation the usual formulas for $\rho_{\mathrm{n-ph}}$ and $\rho_{\mathrm{n-rot}}$. These show that close enough to the absolute zero, phonons will dominate the thermodynamics: $\rho_{\mathrm{n}} \simeq \rho_{\mathrm{n-ph}} \ll \rho$; as the sound velocity increases with pressure, $\rho_{\mathrm{n-ph}}$ will even decrease. Therefore metastable liquid helium close to the line $T = 0$ remains superfluid up to $P_{\Delta=0}$; this is a first important result.

We now use the Landau model to predict T_λ, as we did in Section 3.3, keeping in mind that this will not give quantitatively correct results, but will provide a qualitative picture. The main contribution to the normal density near T_λ is due to rotons. Assuming that the energy in the roton region remains quadratic (Eq. 12) [46] and that the effective mass does not go to zero, this gives a vanishing T_λ at $P_{\Delta=0}$. We think that this result will not be qualitatively affected by interactions, because the Landau model gives an overestimate of T_λ. We conclude that the lambda line meets the $P_{\Delta=0}$ line at $T = 0\,\mathrm{K}$.

5. Conclusion

We have shown that in addition to its purity, the quantum nature of helium makes it an ideal system in which to study metastability. Indeed, quantum statistics give a powerful theoretical tool to investigate the metastable regions. In the case of helium 3, they allowed us to explain the origin of the minimum observed in the spinodal line. We are presently adapting our experimental setup to check the existence of such a minimum in liquid water [47]. The use of an analytically tractable model to describe superfluid helium 4 allowed us to predict a monotonous spinodal in this other case. It also gives a qualitative picture for the extension of the lambda line in the metastable regions; further work is in progress to take into account interactions between elementary excitations and make the picture quantitative [35, 48]. In view of our recent experiments on overpressurized liquid helium 4, we have also considered an instability related to the softening of the roton mode.

Acknowledgements

We would like to thank H.J. Maris for stimulating discussions.

References

1. Lambaré, H., Roche, P., Balibar, S., Maris, H.J., Andreeva, O.A., Guthmann, C., Keshishev, K.O., and Rolley, E. (1998) Cavitation in superfluid helium-4 at low temperature, *Eur. Phys. J.* **2**, 381-391.
2. Caupin, F. and Balibar, S. (2001) Cavitation pressure in liquid helium, *Phys. Rev. B* **64**, 064507 (1-10).
3. Balibar, S. and Caupin, F. (2002) The limits of stability of liquid helium, *this conference*, and references therein.
4. Maris, H.J. (1995) Theory of quantum nucleation of bubbles in liquid helium, *J. Low Temp. Phys.* **98**, 403-424.
5. Guilleumas, M., Barranco, M., Jezek, D.M., Lombard, R.J., and Pi, M. (1996) Quantum cavitation in liquid helium, *Phys. Rev. B* **54**, 16135-16138.
6. Caupin, F. and Balibar, S. (2000) Search for quantum cavitation in liquid helium 3, *Physica B* **284-288**, 212-213.
7. Caupin, F., Balibar, S., and Maris, H.J. (2002) Nucleation in a Fermi liquid at negative pressure, *J. Low Temp. Phys.* **126**, 91-96.
8. Speedy R.J. (1982) Stability-limit conjecture. An interpretation of the properties of water. *J. Phys. Chem.* **86**, 982-991.
9. Caupin, F., Balibar, S., and Maris, H.J. (2001) Anomaly in the stability limit of liquid helium 3, *Phys. Rev. Lett.* **87**, 145302 (1-4).
10. Xiong, Q. and Maris, H.J. (1989) Liquid helium at negative pressure : nucleation of bubbles and anomalous phonon dispersion, *J. Low Temp. Phys.* **77**, 347-369.
11. Guilleumas, M., Pi, M., Barranco, M., Navarro, J., and Solís, M.A. (1993) Thermal nucleation of cavities in liquid helium at negative pressures, *Phys. Rev. B* **47**, 9116-9119.
12. Pettersen, M.S., Balibar, S., and Maris, H.J. (1994) Experimental investigation of cavitation in superfluid ^{4}He, *Phys. Rev. B* **49**, 12062-12070.
13. The expression of $V\tau$ obtained in Ref. [12] actually depends on the minimum pressure reached, but we have checked that this does not affect our conclusions.
14. Boghosian, C., Meyer, H., and Rives, J.E. (1966) Density, coefficient of thermal expansion, and entropy of compression of liquid helium-3 under pressure below $1.2\,\mathrm{K}$, *Phys. Rev.* **146**, 110-119.
15. Maris, H.J. (1994) Nucleation of bubbles on quantized vortices in helium-4, *J. Low Temp. Phys.* **94**, 125-144.
16. Abraham, B.M., Chung, D., Eckstein, Y., Ketterson, J.B., and Roach, P.R. (1972) Sound propagation, density and viscosity in liquid ^{3}He, *J. Low Temp. Phys.* **6**, 521-528.

17. Roach P.R., Eckstein, Y., Meisel, M.W., and Aniola-Jedrzejek, L. (1983) Thermal expansion, velocity of sound, and compressibility in liquid ^{3}He under pressure, *J. Low Temp. Phys.* **52**, 433-447.

18. Thermodynamically, the spinodal line is derived from the isothermal sound velocity rather than from the adiabatic one. Both velocities lead to spinodal lines showing a minimum and their difference is less than 30 mbar. We have actually plotted the spinodal obtained with the isothermal data in Fig. 2, and we used it in our discussion.

19. Debenedetti, P.G. (1996) *Metastable liquids*, Princeton University Press, Princeton, and references therein.

20. Ahlers, G. (1976) Experiments near the superfluid transition in ^{4}He and ^{3}He-^{4}He mixtures, in K.H. Bennemann and J.B. Ketterson (eds.), *The physics of liquid and solid helium*, John Wiley and Sons, New York, Part I, Chap. 2, 85-206.

21. Lee, D.M., Reppy, J.D., and Fairbank, H.A. (1958) Evidence for a density maximum in liquid He^3 near 0.5 K, *Bull. Am. Phys. Soc. Ser. II.* **3**, 339.

22. Brueckner, K.A. and Atkins, K.R. (1958) Coefficient of thermal expansion of liquid He^3 near 0.5 K, *Phys. Rev. Lett.* **1**, 315-318 .

23. Landau, L.D. (1956) The theory of a Fermi liquid, *Zh. Eksper. Teor. Fiz.*, **30**, 1058-1064 [(1957) *Sov. Phys. JETP* **3**, 920-925].

24. Greywall, D.S. (1986) ^{3}He specific heat and thermometry at millikelvin temperatures, *Phys. Rev. B* **33**, 7520-7538.

25. Balibar, S., Caupin, F., Roche, P., and Maris, H.J. (1998) Quantum cavitation: a comparison between superfluid helium-4 and normal liquid helium-3, *J. Low Temp. Phys.* **113**, 459-471.

26. Greywall, D.S. (1983) Specific heat of normal liquid ^{3}He, *Phys. Rev. B* **27**, 2747-2766.

27. Lifshitz, I.M. and Kagan, Y. (1972) Quantum kinetics of phase transitions at temperatures close to absolute zero, *Zh. Eksp. Teor. Fiz.* **62**, 385-402 [(1972) *Sov. Phys. JETP* **35**, 206-214]. See also Ref. [10].

28. Landau, L.D. (1947) On the theory of superfluidity of helium II, *J. Phys. (Mosc.)* **11**, 91.

29. Wilks, J. (1967) *The properties of liquid and solid helium*, Clarendon Press, Oxford.

30. Gibbs, M.R., Andersen, K.H., Stirling, W.G., and Schober, H. (1999) The collective excitations of normal and superfluid ^{4}He: the dependence on pressure and temperature, *J. Phys.: Condens. Matter* **11**, 603-628.

31. Abraham, B.M., Eckstein, Y., Ketterson, J.B., Kuchnir, M., and Roach, P.R. (1970) Velocity of sound, density and Grüneisen constant in liquid ^{4}He, *Phys. Rev. A* **1**, 250-257.

32. Atkins, K.R. (1955) Slope of the λ curve of liquid helium, *Phys. Rev.* **98**, 319-320.

33. Dalfovo, F., Lastri, A., Pricaupenko, L., Stringari, S., and Treiner, J. (1995) Structural and dynamical properties of superfluid helium: a density-functional approach, *Phys. Rev. B* **52**, 1193-1209.

34. Bauer, G.H., Ceperley, D.M., and Goldenfeld, N. (2000) Path-integral Monte Carlo simulation of helium at negative pressures, *Phys. Rev. B* **61**, 9055-9060.

35. Edwards, D.O. and Maris, H.J. (unpublished).

36. Hall, S.C. and Maris, H.J. (1997) Thermodynamics and nucleation of bubbles in normal and superfluid liquid helium-4 at negative pressures, *J. Low Temp. Phys.* **107**, 263-282.

37. Atkins, K.R. and Edwards, M.H. (1955) Coefficient of expansion of liquid helium II, *Phys. Rev.* **97**, 1429-1434.

38. Rybarcyk, L.J. and Tough, J.T. (1981) Superheating in He II and the extension of the lambda line, *J. Low Temp. Phys.* **43**, 197-202.

39. Nishigaki, K. and Saji, Y. (1986) Superheating in He II and successive phase transitions in metastable states, *Phys. Rev. B* **33**, 1657-1662.

40. Skripov, V.P. (1994) The metastability boundary in the ^{4}He diagram of state, *Zh. Fiz. Kh.* **68**, 1382-1385 [(1994) *Russ. J. Phys. Chem* **68**, 1252-1255].

41. Campbell, C.E., Folk, R., and Krotscheck, E. (1996) Critical behavior of liquid ^{4}He at negative pressures, *J. Low Temp. Phys.* **105**, 13-36.

42. Apenko, S.M. (1999) Critical temperature of the superfluid transition in Bose liquids, *Phys. Rev. B* **60**, 3052-3055.

43. Skripov, V.P. (2000) Extension of the λ curve of ^{4}He into the region of the metastable state of liquid helium, *Usp. Fiz. Nauk* **170**, 559-563 [*Phys. Usp.* **43**, 515-519].

44. Chavanne, X., Balibar, S., and Caupin, F. (2001) Acoustic nucleation of solid helium 4 on a clean glass plate, *J. Low Temp. Phys.* **125**, 155-164.

45. Schneider, T. and Enz, C.P. (1971) Theory of the superfluid-solid transition of ^{4}He, *Phys. Rev. Lett.* **27**, 1186-1188.

46. Notice that, unlike the case of phonons, the curvature of the dispersion curve is accounted for in the usual calculation of $\rho_{\text{n-rot}}$.

47. Caupin, F. and Fourmond, V. (2002) Ultrasonic cavitation in freon at room temperature, *this conference.*

48. Caupin, F. and Maris, H.J. (unpublished).

ACOUSTIC CAVITATION THRESHOLDS OF OCEAN WATER

V.A. AKULICHEV
*Pacific Oceanological Institute, Russian Academy of Sciences,
43, Baltiyskaya Street, Vladivostok 690041, Russia*

Abstract. The results of experimental measurements of acoustic cavitation threshold are presented for the water of the Atlantic Ocean, the Pacific Ocean, the Indian Ocean, the Arctic Ocean and some other regions of the World Ocean, including the Arabian Sea, the Baltic Sea, the East Siberian Sea, the North Sea, the Philippine Sea, the Sea of Japan, the Sea of Okhotsk and the South China Sea. These measurements were carried out at many ocean expeditions during long time period from 1968 through 1987. It should be noted that acoustic cavitation thresholds of ocean water have different values in different regions of the World Ocean.

1. Introduction

The tensile strength of ocean water determines the limiting values of acoustic power that can be radiated by powerful acoustic sources. Therefore the knowledge of acoustic cavitation thresholds of ocean water at different depths in different regions of the World Ocean is of practical interest. Furthermore, the tensile strength of water determines the values of critical velocities of the motion of different floating bodies in the ocean, which excess leads to the beginning of hydrodynamics cavitation.

Acoustic cavitation arises in sea water or any other liquid under the action of a powerful acoustic field when the acoustic pressure amplitude P_m exceeds some threshold value P_m^* commonly called the tensile strength of acoustic cavitation threshold of water. The P_m^* value in sea water depends on many hydrophysical and hydrochemical parameters. The influence of hydrostatic pressure P_o, linearly growing with depth, is an important factor. The values P_m^* and P_o are bound up by the relationship $P_m^* = P_o - P_c$, where P_c is equal to the hydrodynamic cavitation threshold pressure value in liquid at which cavitation arises. The P_c value is equal to the saturated vapor pressure P_v for the liquids with big sizes of cavitation nuclei. The P_c value can be negative for very pure liquids with small cavitation nuclei. The acoustic threshold pressure P_m^* values are always positive.

Hydrodynamic cavitation is induced by the influence of pressure pulsations arising in water due to rotation of a screw propeller or to flowing along the different bodies in the ocean water. In this case it is customary to call the pressure change value $(P_o - P_c)$ as tensile strength or cavitation thresholds of water.

The acoustic cavitation threshold value P_m^* for ocean water is connected with many hydrophysical parameters of water medium and the acoustic field frequencies f which can change in a wide range depending on the acoustic sources used.

A.R. Imre et al. (eds.), Liquids Under Negative Pressure, 215–230.
© 2002 *Kluwer Academic Publishers. Printed in the Netherlands.*

In the general case the value $(P_o - P_c)$ at hydrodynamic cavitation can differ from the value P_m^* at acoustic cavitation for the same conditions of sea water medium. However, on lowering of acoustic field frequency f to the value of a characteristic average frequency of hydrodynamic pulsations $< f >$ it is to be expected that the value P_m^* at acoustic cavitation will asymptotically tend to the value $(P_o - P_c)$ at hydrodynamic cavitation. This allows us to use the results of acoustic measurements to determine the thresholds of water at hydrodynamic cavitation.

2. Cavitation Nuclei

Usually, the tensile strength of sea water P_m^* grows with sea water depth. However, this growth can obey different laws that are determined, first of all, by the sizes and concentration of cavitation nuclei.

In the upper sea water layer the most characteristic cavitation nuclei are gas bubbles whose sizes and concentration are determined by surface roughness and by hydrophysical parameters that determine gas solubility in water. Usually in each specified water volume the gas bubbles of different sizes R are presented, i.e. there is some statistical bubbles sizes distribution density $g(R)$. The concentration of bubbles of certain size R is defined as $N(R) = g(R) \cdot \Delta R$ where ΔR is the size range. For gas bubbles in sea water a substantial increase in concentration $N(R)$ with the increase of sea roughness and wind speed is characteristic. The problems concerned with determining of gas bubbles sizes and bubbles concentration in ocean water can be solved by means of acoustic methods [1].

Cavitation in sea water can also arise on phase inclusions in the form of zooplankton or phytoplankton. The sizes of zooplankton range from 5 μm for nanoplankton to 1 mm for microplankton and about 5 cm for macroplankton. The sizes of phytoplankton are confined even in a wider range. As the depth of sea water medium increases, the density of plankton population decreases, although at some depth the local increase of density is observed in the form of extended layers known as sound scattering layers. Cavitation can also originate on solid nuclei which get into sea water from the atmosphere, from rivers and others sources. These solid nuclei can have different sizes, form and degree of wetting with water.

In sea water, cavitation nuclei in the form of small vapor bubbles can originate from high energy particles caused by cosmic rays or radioactivity. Primary cosmic rays composed mainly of protons and α-particles get transformed into secondary particles constituted by electrons and μ-mesons at the sea level. Electrons are intensively absorbed in water while μ-mesons possess great penetration power and are poorly absorbed in water. Electrons and μ-mesons interact with electrons of sea water atoms engendering δ-electrons. The local heat release by δ-electrons at energy losses leads to the formation of vapor bubbles the sizes of which are smaller than 10^{-6} cm.

In contrast with charged particles causing ionization, a neutron interacts only with atom nuclei on passing through the sea water medium. In this case, the concentration of free radicals and atoms of oxygen and hydrogen can arise in water. Due to their structure, they pronounce themselves like molecules of dissolved oxygen and hydrogen which can form bubbles owing to coagulation of gas molecules. Great number of both experimental and theoretical works [2] have been devoted to the influence of neutrons and ionizing particles upon the tensile strength of water. Similar effects may be caused by such primary cosmic particle as neutrino.

All the enumerated cavitation nuclei appear in the sea water medium owing to the action of outer forces and disturbances. However, even in case of complete isolation from the external medium the formation of vapor bubbles is possible due to the manifestation of thermodynamic heterophase fluctuations. The size of such cavitation nuclei do not exceed 10^{-7} cm. Under usual conditions in sea water these cavitation nuclei are negligibly small as compared to gas bubbles, plankton and solid particles.

3. Measurements Technique

The determination of tensile strength of sea water by the acoustic method amounts to the measurement of acoustic field threshold amplitude P_m^* exceeding which causes developed cavitation. Characteristic changes of acoustic signals allowing the determination of the acoustic cavitation threshold are presented in Figure 1.

Cavitation was induced by the initial tonal acoustic signal with a basic frequency f and the amplitude P_m. The frequency of exciting initial signal was 10 kHz. The onset of cavitation corresponds to rise of acoustic cavitation noise signal with the summary pressure P_n that contains the discrete harmonic spectral components with the frequencies of nf, where $n=2,3,\ldots$, and also the continuous spectrum component of the received signal. Figure 1 and Figure 2 show an example of experimental measurements of the initial acoustic signal P_m and the cavitation noise signal P_n in the sea water when the electrical voltage V of some high power acoustical source increased. These experimental results were carried out at September 1982 in the Sea of Japan at the point with the latitude about $42°N$ and the longitude about $132°E$.

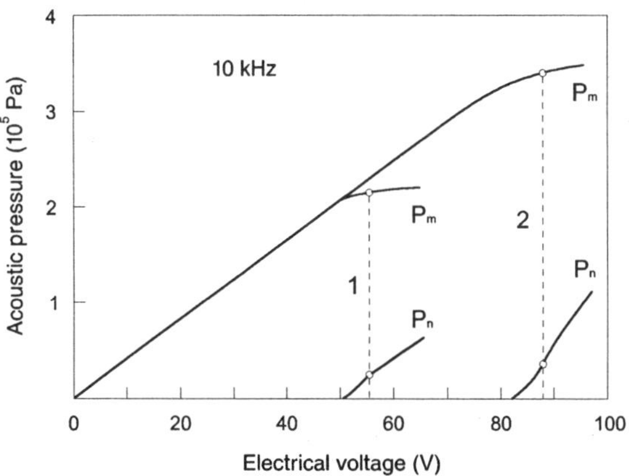

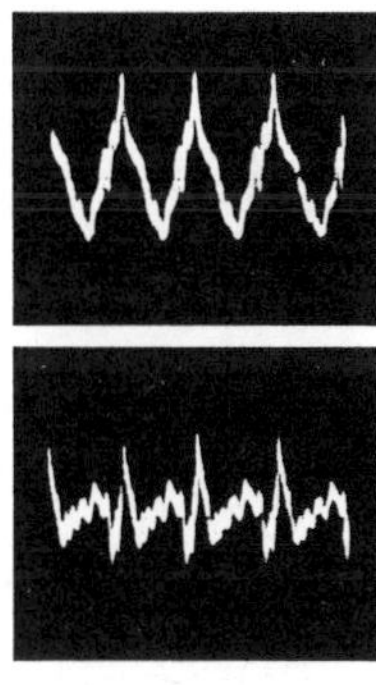

Figure 1. Initial acoustic signal and cavitation noise signal at different depths. 1: depth of 5 m, 2: 10 m.

Figure 2. Initial signal (on top) and cavitation noise signal (below).

The relation of the acoustic cavitation noise signal P_n to the amplitude of the main basic tonal signal P_m determines the coefficient K of nonlinear distortions of the acoustic signal at the beginning of acoustic cavitation, where $K = P_n / P_m$. Experimental investigations into the onset of acoustic cavitation in sea water with different physical and chemical characteristics (temperature, salinity, gas content and others) at different depths showed that the value $K = 0.1$ corresponds to the onset of cavitation at the some threshold amplitude $P_m = P_m^*$. In Figure 1 the cavitation threshold value P_m^* and appropriate noise value P_n are marked with light circles. The idea about a spectral criterion of the onset of acoustic cavitation was used in our work [3]. Henceforth, a similar technique of measurements was used also in the work [4].

In our work [3] it was suggested to use water-filled cylindrical acoustic sources to stimulate the cavitation into the water but not on the radiating surface of the source. Figure 3 and Figure 4 show the two various schemes of the installations for the measuring the acoustic cavitation thresholds P_m^* in the sea water. The techniques of measuring the tensile strength of sea water, presented in Figure 3, are based on the application of cylindrical acoustic sources which are usually manufactured by piezoceramic active materials.

Resonance frequencies f of such acoustic sources are connected with the diameter of the sources d by the formula $f \approx c_p / \pi d$, where c_p is the sound speed in the piezoceramic material. Such sources are suited to application at the frequencies f of more than 1 kHz. With the goal of measuring at lower frequencies f, an excessive increase of value d is required which leads to constructive difficulties. To excite the cavitation in sea water at low frequencies f of the order of hundreds of Hz, it is more convenient to use as sound sources resonance metal tubes open at one end and excited at the other using acoustic vibrator. Such techniques for measuring of the acoustic cavitation thresholds in sea water were suggested in our work [5]. Figure 4 shows a scheme of the installation with the resonance tube for measuring acoustic thresholds P_m^*

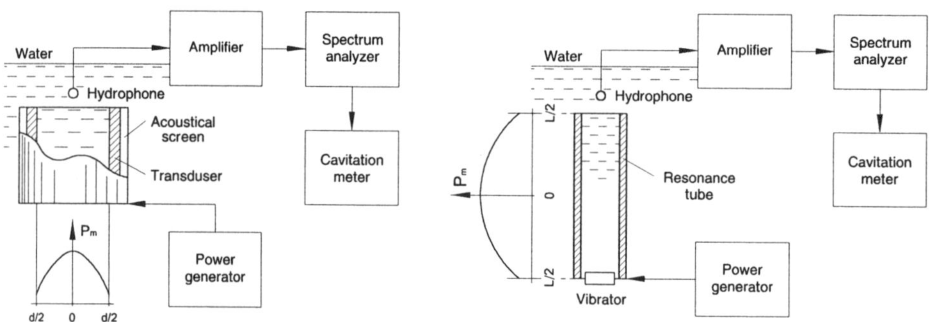

<table>
<tr><td>

Figure 3. Scheme of the installation for the measuring of the acoustic cavitation thresholds using water-filled cylindrical piezoceramic acoustic source.

</td><td>

Figure 4. Scheme of the installation for the measuring of the acoustic cavitation thresholds using the resonance metal tube excited at one end by acoustic vibrator.

</td></tr>
</table>

at which the cavitation occurs. Sea water fills such tube and the onset of cavitation may be observed approximately in the middle section where the amplitude of the acoustic standing wave has maximum. The resonance frequency of such acoustic sources f is connected with the tube length L by the formula $f = c_p / 2L$, where c_o is the sound speed in sea water.

4. Acoustic Cavitation Thresholds in Different Regions of the World Ocean

Below the results of experimental measurements of acoustic cavitation thresholds in different regions of the World Ocean are presented.

4.1. THE SEA OF JAPAN

Figure 5 presents the results of measurements of the cavitation thresholds of sea water P_m^* depending on the depth h obtained at September 1982 in the Sea of Japan at the point with the latitude $42°N$ and longitude $132°E$. Acoustic sources with different frequencies f were used. The measurements at a frequency of 230 Hz were conducted using the resonance acoustic steel tube in accordance with Figure 4. The measurements at other higher frequencies were conducted using water-filled cylindrical piezoceramic acoustic sources in accordance with Figure 3.

As it follows from Figure 5, the values of acoustic cavitation thresholds P_m^* increase on the average with the increases of the depths h. However, one can see the divergence of the value P_m^* below the level of the hydrostatic pressure P_0 which is shown by dashed line. As cited in [6], the acoustic cavitation thresholds P_m^* can be less than the value P_m^* only in case of the action of rectified gas diffusion when the gas cavitation nuclei in the water have such sizes R which resonance frequency f_o is close to the frequency of the exciting acoustic field f. For such gas bubbles the radius R and resonance frequency f_o are connected by next formula:

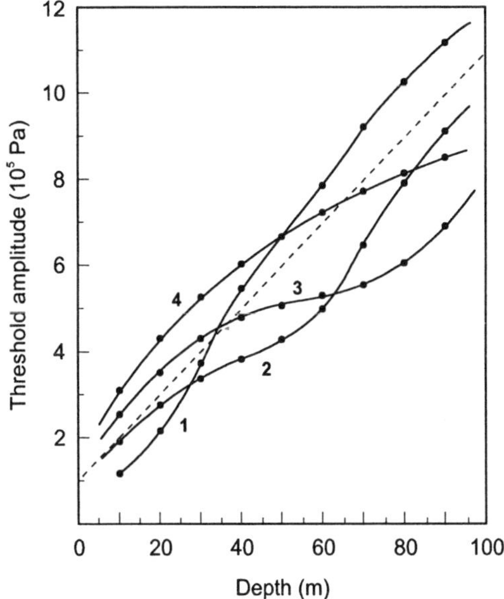

Figure 5. Acoustic cavitation thresholds depending on the depth at different frequencies f.
1: 230 Hz, 2: 750 Hz, 3: 4.5 kHz and 4: 10 kHz.

$$R = \frac{1}{2\pi f_o}\left[\frac{\gamma P_a}{\rho}\left(1+Ah\right)\right]^{1/2}$$

where ρ is the water density, A is a constant value equal to $0.1\,\text{m}^{-1}$, value γ is an adiabatic constant equal to 3/4 for the gas medium, and hydrostatic pressure P_o is shown by the expression $P_o = P_a\left(1+Ah\right)$. Using this formula, one can determine the most probable sizes of cavitation nuclei at selected frequencies f for the depths h which correspond to the most significant divergences of the value P_m^* from the P_o.

220

4.2. THE ARCTIC OCEAN

Particular results were received in the Arctic Ocean. The Figure 6 shows the point in the East Siberian Sea with coordinates 76°N and 164°E where the measurements of acoustic cavitation thresholds were carried out at May 1969. The measurements were conducted from a drifting block of ice with the thickness about 3 m. The air temperature was minus 12°C. The water temperature varied from minus 1.68°C near the water surface to minus 1.55°C at the depth of 50 m.

Table 1 presents the data of measurements of the acoustic cavitation thresholds at different frequencies at the depth of 10 m. Table 1 shows that in arctic water the cavitation thresholds were too big. This result was due to the fact that under conditions of ice coating, the surface roughness of the sea water was practically absent which was reason for small sizes and concentration of gas cavitation nuclei. It is seen that the acoustic cavitation thresholds of ocean water can differ by several times with frequency f. When the frequency f increases, the value of acoustic cavitation threshold P_m^* increases too.

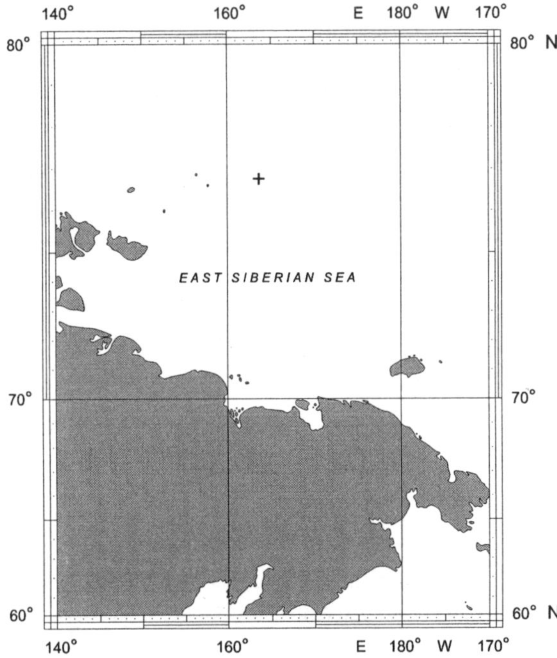

Figure 6. The site of measurements in the Arctic Ocean. The mark + corresponds to May 1969 measurements.

TABLE 1. Acoustic cavitation thresholds at different frequencies in the Arctic Ocean

Frequency (kHz)	2.0	4.0	6.0	8.0	10.0	15.0
Threshold (10^5 Pa)	1.9	2.9	3.8	4.7	5.6	6.9

4.3. THE ATLANTIC OCEAN

The measurements of the acoustic cavitation thresholds were carried out in the northern part of the Atlantic Ocean, northern of equator, including the North Sea and the Baltic Sea. Figure 7 shows the sites of the measurements during two different seasons. The crosses marks correspond to the spring season, during March-April 1968, and the crosses into circles correspond to the summer season, during June-July 1968 [7].

Figure 8 presents the experimental results of acoustic cavitation thresholds as a function of the latitude in the Atlantic Ocean from the equator to the North Sea and the Baltic Sea. The cavitation was excited at the depths of 10 m and 20 m by continuous tonal signal at a frequency of 10 kHz. In Figure 8 the solid circles correspond to the measurements in the spring season, during March-April 1968, and the open circles correspond to the measurements in the summer season, during June-July 1968.

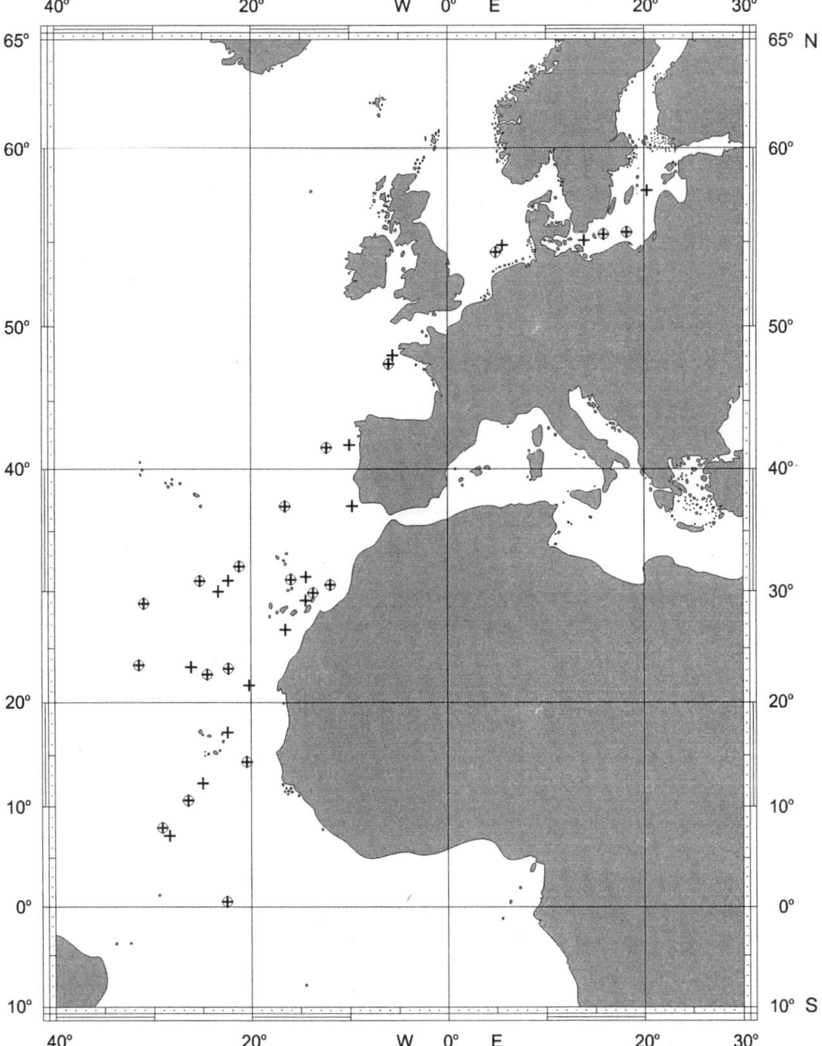

Figure 7. The sites of measurements in the Atlantic Ocean. The marks + correspond to spring season at March-April 1968, the marks ⊕ correspond to summer season at June-July 1968.

It is seen from Figure 8 that in the North Atlantic Ocean some cavitation thresholds variability was observed depending on the latitude. The acoustic cavitation thresholds

222

had higher values near the equator and decreased with growth of latitude. This effect was more manifested at the depth of 20 m than at the depth of 10 m. The points with latitude 54°30'N (and longitude 4°52'E) indicate the results of measurements in the North Sea. Two points with latitude 55°33'N (longitude 15°55'E) and latitude 57°46'N (longitude 20°09'E) indicate the results of measurements in the Baltic Sea. They are shown in Figure 8 as separate points with different values of cavitation thresholds that correspond to the different times of measurements, March and July 1968.

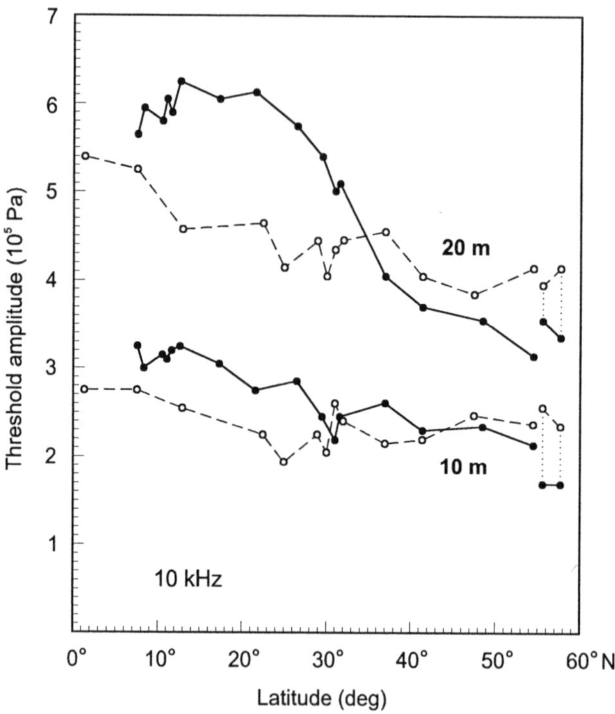

Figure 8. Acoustic cavitation thresholds in the Atlantic Ocean as a function of the latitude at different seasons of 1968. The marks ● correspond to spring, the marks ○ correspond to summer.

Figure 9 shows the results of the measurements of hydrological and hydrochemical parameters that were carried out at one time at the same points of the Atlantic Ocean.

On the Figure 9 the solid marks show the measurement sites at spring season, during March-April 1968, the open marks show the measurement sites at summer season, during June-July 1968. The comparison of Figure 8 and Figure 9 shows that the tendency towards the decrease of acoustic cavitation thresholds in the ocean water on moving away from equatorial region to the high latitude regions connects with the tendency towards the increase of dissolved gas and oxygen into the water on moving to the high latitude regions in the North Atlantic Ocean. The same tendency was observed in the Indian Ocean and in the Pacific Ocean.

4.4. THE INDIAN OCEAN

The measurement of acoustic cavitation thresholds in the Indian Ocean were carried out at March-April 1987, the autumn season of the Southern Hemisphere.

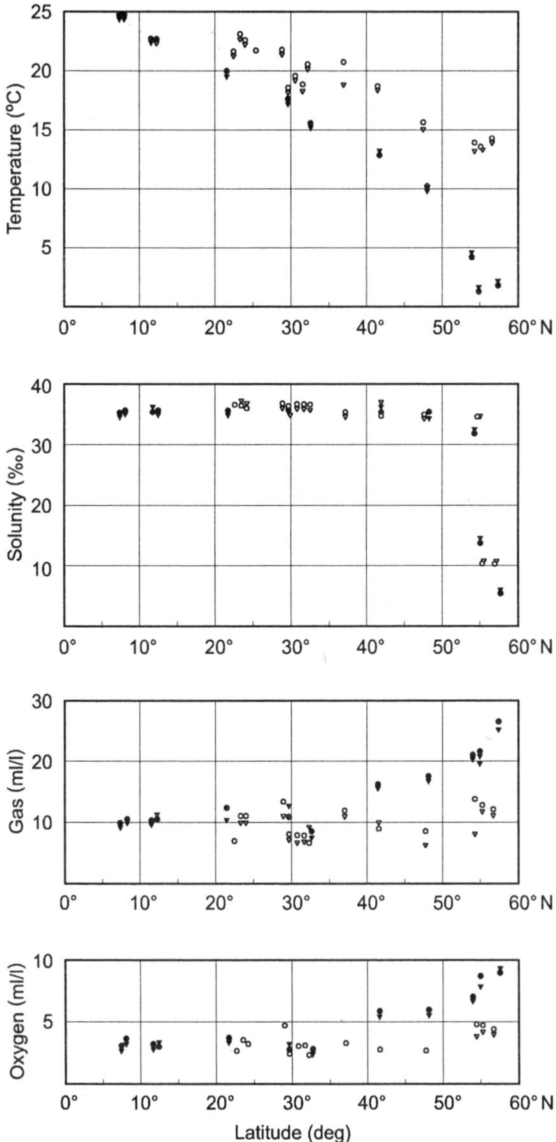

Figure 9. Temperature, salinity, dissolved gas and oxygen at different points in the Atlantic Ocean at different depth. The marks • o corresponds to 10 m, the marks ▼ ▽ - to 20 m.

224

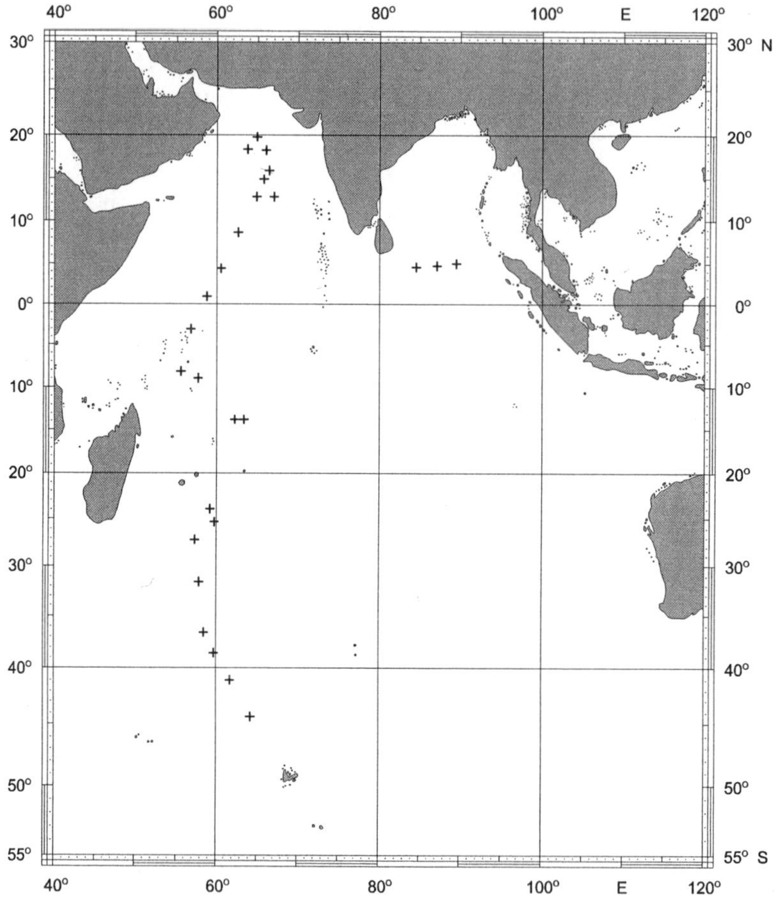

Figure 10. The sites of cavitation measurements in the Indian Ocean. The marks + correspond to March-April 1987 measurements.

Figure 10 shows that measurements were made from the site with latitude about 20°N at the Arabian Sea to the site with latitude about 45°S at the southern part of the Indian Ocean. Figure 11 presents the experimental results of these measurements as a function of the latitude for different frequencies.

The cavitation was excited at the depths of 10 m and 20 m by continuous tonal signal at different frequencies of 750 Hz, 2.3 kHz, 5.4 kHz and 10 kHz. Figure 11 shows that in the Indian Ocean some cavitation threshold variability was observed depending on the latitude too. This variability increased with increasing frequency. The higher variability was at the frequency of 10 kHz. The acoustic cavitation thresholds had higher values near the equator and decreased with growth of latitude to the northern and to the southern directions. This effect was more manifested at the depth of 20 m than at the depth of 10 m.

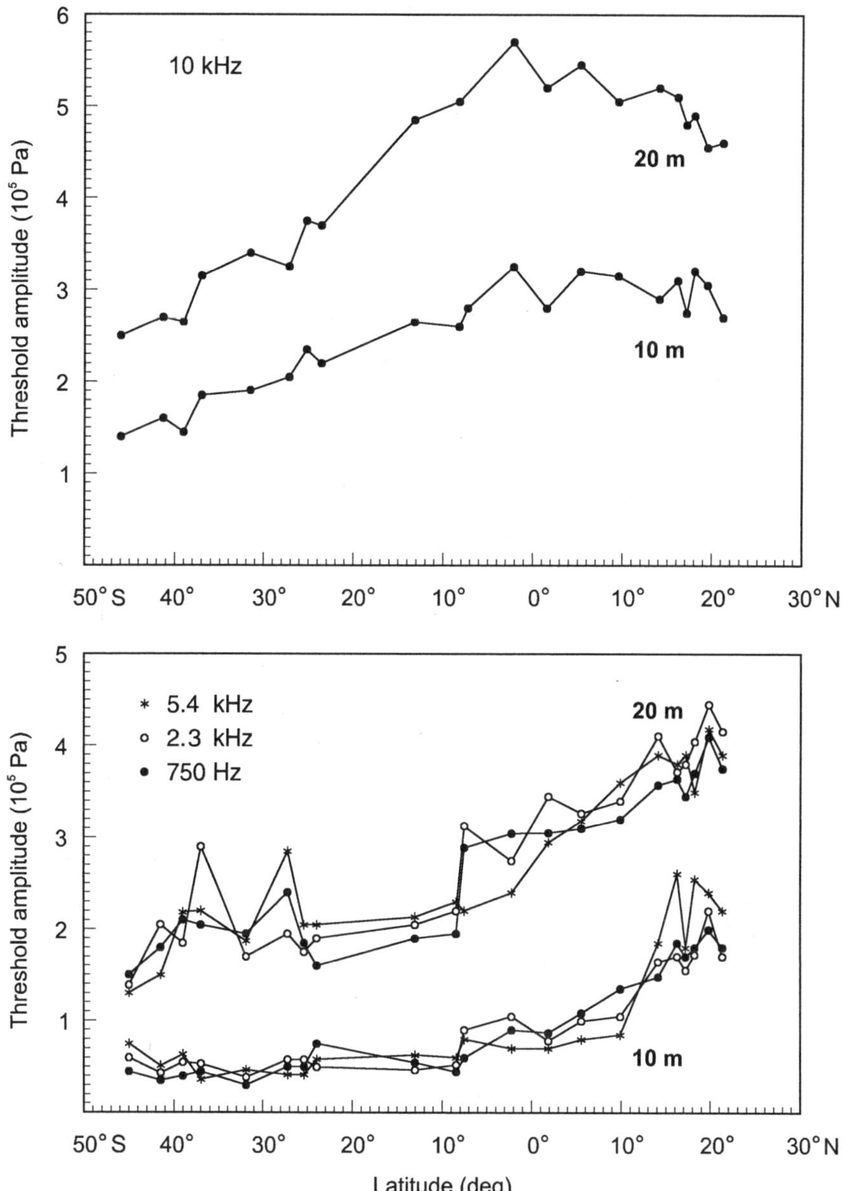

Figure 11. Acoustic cavitation thresholds in the Indian Ocean as a function of the latitude at different frequencies and depths.

226

Figure 12 shows the results of the measurements of the surface temperature and sea roughness parameters that were carried out at one time at the same points of the Indian Ocean. The comparison of the Figure 11 and Figure 12 shows that the tendency towards decrease of acoustic cavitation thresholds in the ocean water on moving away from equatorial region to the high latitude regions connect with the tendency towards increase of dissolved gas that is in inverse proportion with sea temperature and with the tendency towards increase of gas bubbles concentration that is in proportion with sea roughness.

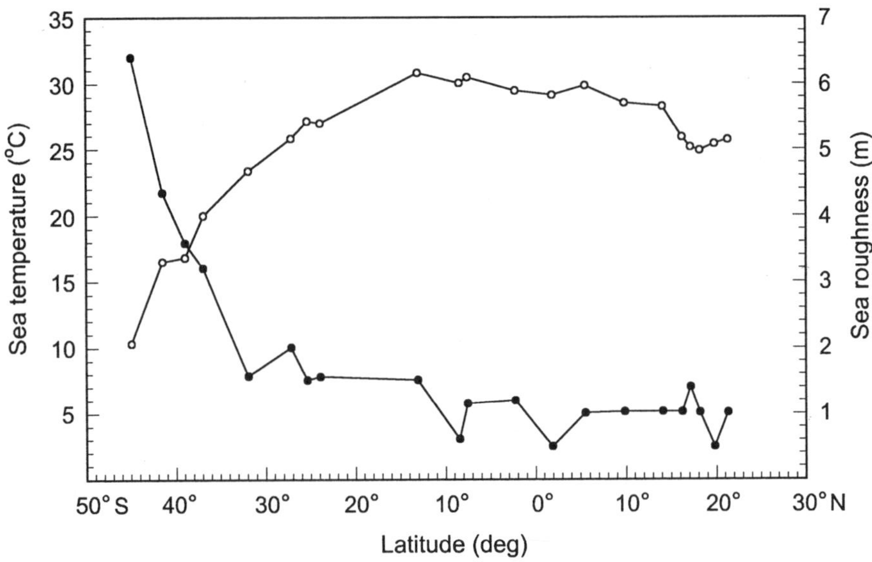

Figure 12. Surface temperature (o) and sea roughness (•) at the time of the measurements in the Indian Ocean at March-April 1987.

4.5 THE PACIFIC OCEAN

The measurements of acoustic cavitation thresholds in the Pacific Ocean were carried out in different years from 1964 through 1985 that are shown on Figure 13. The most extensively measurements were carried out from September to November 1982 and from October to November 1983. These measurements were in the autumn season in the northern part and the spring season in the southern part of the Pacific Ocean. The measurements at different years were carried out at large latitude and longitude ranges.

Figure 14 presents the experimental results of acoustic cavitation thresholds P_m^* as function of latitude obtained at September-November 1982 in the Pacific Ocean from latitude about of 49°N near Kamchatka Peninsula to latitude 15°30′S in the Coral Sea. The cavitation was excited at the depths of 10 m and 20 m by the signals at the frequency of 10 kHz. Figure 14 shows that in the Pacific Ocean the cavitation thresholds variability was observed depending on the latitude too. It turned out, that the acoustic cavitation thresholds at the frequency of 10 kHz was higher near the equator and decreased with growth of latitude in the directions from equator to the northern and to the southern parts of the Pacific Ocean [8].

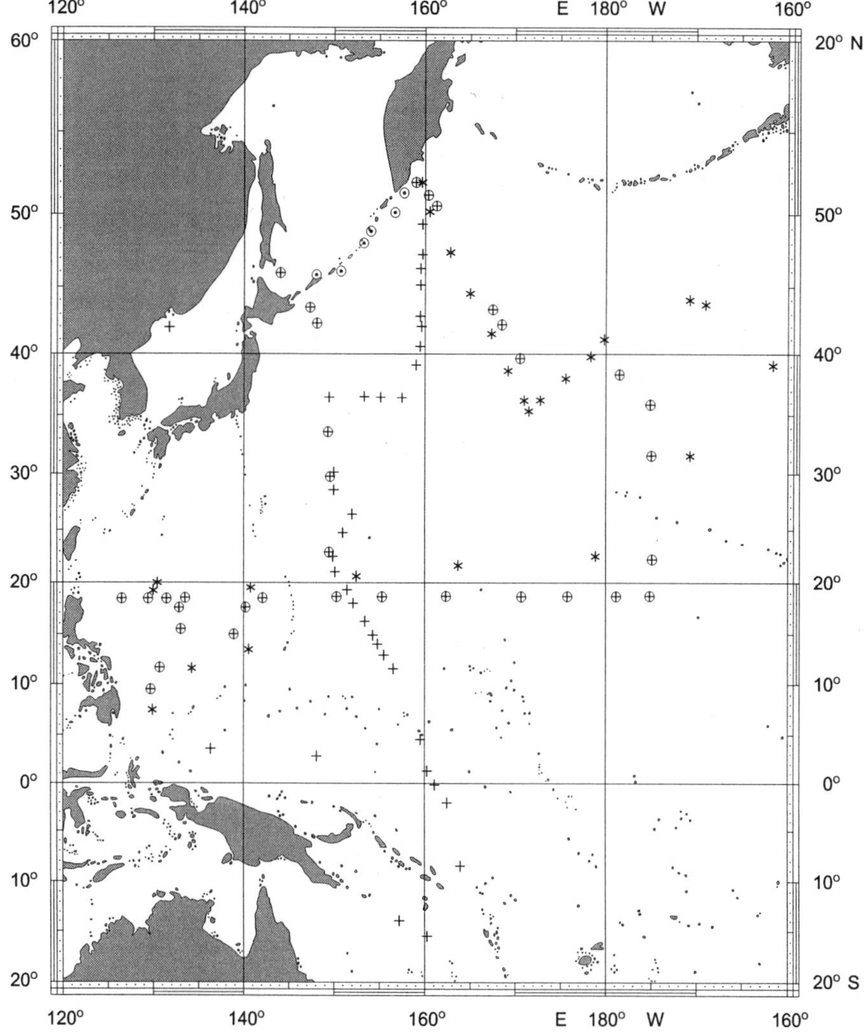

Figure 13. The sites of cavitation measurements in the Pacific Ocean at different years. The marks . correspond to September 1964, the marks + - to September-November 1982, the marks ⊕ - October-November 1983, the marks * - to August-October 1985.

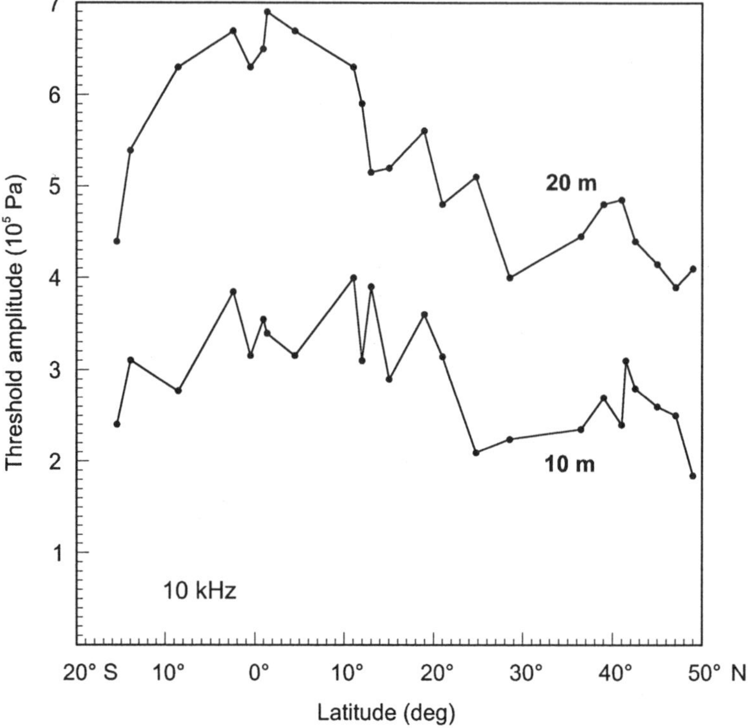

Figure 14. Acoustic cavitation thresholds in the Pacific Ocean as a function of the latitude during September-November 1982.

5. Conclusions

It should be noted that acoustic cavitation thresholds have different values in the different regions of the World Ocean. Table 2 presents the comparative results of acoustic cavitation thresholds P_m^* obtained in the different Oceans and different Seas of the World Ocean at different time from 1968 to 1987. The results at Table 2 correspond to the measurements at a standard depth of 10 m and at the frequency of 10 kHz.

The results of the measurements that are presented in Table 2 were carried out at the expansive areas of water of the Atlantic Ocean, the Indian Ocean and the Pacific Ocean. The values at the Arctic Ocean correspond to the measurements in some area of water in the East Siberian Sea. It is seen that the acoustic cavitation thresholds were the lowest in the Indian Ocean and the highest in the Arctic Ocean. There was seen a big difference between values of the cavitation thresholds for water of different seas. The lowest values were observed in the Black Sea and then in the Sea of Okhotsk and in the Baltic Sea. The highest value was observed in the East Siberian Sea.

TABLE 2. Acoustic cavitation thresholds at different regions
at the depth of 10 m and at the frequency of 10 kHz

Region	Time	$P_m^* \left(10^5 \cdot \text{Pa}\right)$
The Arctic Ocean	May 1969	5.6
The Atlantic Ocean	April 1968	3.2
The Indian Ocean	March 1087	2.8
The Pacific Ocean	October 1982	3.6
The Arabian Sea	March 1987	2.8
The Baltic Sea	June 1968	2.3
The Black Sea	September 1964	1.8
The East Siberian Sea	May 1969	5.6
The North Sea	June 1968	2.4
The Sea of Japan	September 1982	2.6
The Sea of Okhotsk	September 1964	2.2
The Philippine Sea	September 1982	2.5
The South China Sea	September 1985	2.4

The knowledge of the cavitation thresholds P_m^* of ocean water at different regions of the World Ocean is of practical interest. First of all, the value of cavitation threshold P_m^* determines the limited levels of the acoustic intensity J_c that could be radiated by powerful acoustic sources at cavitation by the formula:

$$J_c = K_a \left(P_m^*\right)^2 \tag{5.1}$$

where K_a is a coefficient that determines the type of the acoustic field, for the plane acoustic wave $K_a = \left(1/2\rho c\right)$, where ρ is a density and c is sound velocity of sea water.

Moreover, the value of the cavitation threshold P_m^* makes it possible to determine the value of the critical speed V_c of the motion of some floating bodies in the ocean water. For the motion of a solid body in sea water we could use the formula:

$$V_c = K_h (P_m^*)^{1/2} \tag{5.2}$$

where we introduce the coefficient $K_h = (2 / \rho K_f K_p)^{1/2}$, where the coefficient K_f is defined by the form of floating body, the coefficient K_p is caused by the regime of hydrodynamic pulsations at the floating body.

6. Acknowledgments

These investigations began at 1961 in the town Sukhumi, Republic of Georgia at the Former Soviet Union, where the author started his science activity at the Marine Station that was situated on the coast of the Black Sea. At a later time, the author continued this work at the Institute of Acoustics, in Moscow, and later on at the Pacific Oceanological

Institute, in Vladivostok. The author is greatly indebted to his excellent teachers Dr. Victor I. Il'ichev and Prof. Lasar D. Rosenberg whom he always remembers.

7. References

1. Akulichev, V.A., and Bulanov, V.A. (1987) The study of sound backscattering from microinhomogeneities in sea water, in H.D.Merclinger (ed.), *Progress in Underwater Acoustics*, Plenum Press, New York and London, pp.85-92.
2. Apfel, R.E. (1984) Acoustic cavitation inception, *Ultrasonics*, **22**, 167-173.
3. Akulichev, V.A., and Il'ichev, V.I. (1963) On the spectral indication of the onset of ultrasonic cavitation, *Soviet Physics-Acoustics*, **9**, 128-130.
4. Rusby, J.S.M., (1970) The onset of sound wave distortion and cavitation in water and sea water, *J.Sound and Vibration*, **13**, 257-267.
5. Akulichev, V.A., Penkin, S.I., and Shekhovtsev, D.N. (1984) Acoustic cavitation in tubes and resonators, in A.Nakamura (ed.), *Proc. of the 10th Intern. Sympos. on Nonlinear Acoustics*, Teikohsha Press, Osaka, pp.185-188.
6. Neppiras, E.A. (1980) Acoustic cavitation thresholds and cyclic processes, *Ultrasonics*, **18**, 201-209.
7. Akulichev, V.A. (1994) Cavitation nuclei and thresholds of acoustic cavitation in ocean water, in J.R.Blake et al.(eds.), *Bubbles Dynamics and Interface Phenomena*, Kluwer Academic Publishers, Netherlands, pp.178-178.
8. Akulichev, V.A. (1994) Cavitation nuclei and thresholds of acoustic cavitation in sea water, in H.Kato (ed.), *Proc. of the 2nd Intern. Sympos. on Cavitation*, University of Tokyo, pp.343-348.

CAVITATION IN LIQUIDS BY CLASSICAL NUCLEATION THEORY AND MOLECULAR DYNAMICS SIMULATIONS

C. XIAO, D.M. HEYES AND M.E. PARKER,
Department of Chemistry, University of Surrey,
Guildford GU2 7XH, United Kingdom

and

J.G. POWLES,
The Physics Laboratory, The University, Canterbury,
Kent CT2 7NZ, United Kingdom.

1. Introduction

Transient negative pressures can be produced in liquids flowing in confined geometries, by the passage of ultrasound [1] and by various pressure pulse experiments [2]. Liquids under negative pressure, being thermodynamically unstable, will eventually phase separate, usually via the process of 'cavitation', *i.e.*, the formation of small bubbles which become filled with liquid vapour and any dissolved gases from the liquid phase. On return to positive pressures each bubble collapses to produce in some cases very high temperatures in the gases and liquid vapour molecules inside ($\approx 10^4$ K). Such temperatures are sufficient to produce light ('sonoluminescence' [3]) and even possibly nuclear fusion, as recently reported [4]. The energy deposited, being focussed essentially at a point at the centre of the original cavity, has a 'rebound' effect that produces potentially harmful pressure waves emanating outwards, of such intensity that they can corrode any containing walls. The energy expended in creating cavitation also leads to power loss if occurring near moving parts (*e.g.,* from ship's propellers) [5]. The phenomenon is also perceived as a potential health hazard in body imaging using ultrasound scanners, as the ultrasound can in principle induce damage to body tissues from the after-effects of the cavitation event. Although generally considered to be a problem, there are some benefits that can be derived from the effect, for example, in ultrasonic cleaning of various kinds (*e.g.,* teeth in dentistry), and in the burgeoning field of sonochemistry where ultrasound is used to speed up and control more selectively chemical reactions [6]. Negative pressure tends to be found in liquids in confined geometries such as pores, so water under negative pressure could be found quite widely in the human body.

Cavitation, being a non-equilibrium event, has no exact theoretical description. Effort in providing a theoretical description of these processes has largely concentrated on the predicting the probability of the nucleation of the cavities, as a function of the value of the negative pressure in the bulk liquid, in terms of the ambient temperature and liquid surface tension using Classical Nucleation Theory, CNT [7]. CNT treats cavity formation and subsequent growth as an 'activated event' with a free energy

231

A.R. Imre et al. (eds.), Liquids Under Negative Pressure, 231–242.
© 2002 *Kluwer Academic Publishers. Printed in the Netherlands.*

barrier and associated critical radius that must be achieved for spontaneous growth of the bubble to macroscopic dimension. One of the topics discussed in this report is, we feel, a more self-consistent application of CNT where we have built into the usual formulation an analytic equation of state, the popular van der Waals equation of state and explored the consequences of making systematic simplifications of this in the model.

While the CNT theoretical treatments are informative and provide a useful framework to interpret experimental data, they have nothing to say about the interesting molecular-level pre-nucleation processes that lead to the formation of cavities in the first place. The collapse of the cavities cannot be described by CNT, either and although there are well-established continuum-level fluid mechanics solutions to this aspect of the bubble's behaviour, they do not describe well the final moments of the cavity when it is only a few nanometres across, where one expects the maximum temperatures and pressures inside the bubble to arise. In fact, this is a time and distance domain more suited to statistical mechanics or molecular simulation ('applied statistical mechanics') such as Molecular Dynamics, MD. To our knowledge MD has not been applied to bubble collapse before. We have carried out Molecular Dynamics simulations both of stretched water and, in a novel way, to the collapse of a cavity in a model liquid using the archetypal Lennard-Jones intermolecular interaction.

2. Classical Nucleation Theory

Cavitation in a liquid occurs when the liquid is at a temperature and pressure (or equivalently density and temperature) where the single phase is not thermodynamically stable. The system is inclined to phase separate into coexisting volumes of liquid and vapour, with relative amounts given by the lever rule [8]. Equations of state can be extended into this regime (with some caution) and they have a characteristic ' van der Waals' loop, as illustrated in Figure 1.

The equilibrium pressure for the vapour ($P_v(\infty)$, where the ∞ refers to a flat surface, which has an infinite radius of curvature) and liquid ($P_l(\infty)$) are the same. This equality is used to locate the liquid-vapour co-existence lines ('binodal') in Maxwell's equal area construction. The minimum pressure on the liquid side and the maximum pressure on the vapour side along a given isotherm are known as the liquid and vapour 'spinodals'. Phase separation takes place by a process of cavitation when the system is in the region of density (volume) bounded by the bimodal and spinodal points at a given temperature. This is called the 'metastable' regime. Inside the spinodal boundaries, the phase separation has no activation energy and occurs by a quite different mechanism called 'spinodal decomposition', SD. The phase separation is more co-operative and does not involve the growth of individual cavitation nuclei. SD is probably not important for most occurrences of liquids under negative pressure as extremely low densities are required to enter this regime (less than *ca.* 0.7 g cm^{-3} for water). Classical Nucleation Theory, CNT, is a relatively efficient way of mapping out the general thermodynamic aspects of cavitation for a wide range of state points, although to date it has been limited by the use of relatively simple equations of state (EoS) in its formulation. Any EoS can be included within Classical Nucleation Theory which can be used to calculate the activation energy barriers inside the metastable region, in terms of the pressure in the vapour phase inside the 'critical' cavity of radius R, $P_v(R)$ and the surrounding liquid, P_l. We have used the van der Waals (vdW) EoS,

$$P = \frac{8T}{3V-1} - \frac{3}{V^2} \tag{1}$$

where the pressure, P, volume, V, and temperature, T are given in units of their critical point values (*i.e.,* P=1 at the critical point, *etc.*). We use the fact that the free energy of the vapour molecules must be the same in the cavity as in the liquid. There is a cavity vapour pressure, $P_v(R)$, for a given P_l where R=2 $\gamma/(P_v\text{-}P_l)$ is the critical radius at the free energy barrier and γ is the surface tension. For most practical applications, far away from the critical point, the ideal gas representation of the vapour in the cavity is sufficiently accurate in practice because the vapour pressure in the cavity is so low. This leads to a relatively simple expression for the cavity vapour pressure which is very similar to, and numerically hardly distinguishable from the classical Kelvin equation when applied to cavities. The classical Kelvin equation,

$$\frac{P_v}{P_\infty} = \exp(-\frac{(P_v - P_l)V_l(\infty)}{V_v(\infty)P_\infty}) \tag{2}$$

is often used for the dependence of the vapour pressure in the cavity on the liquid pressure (where $P_v\text{-}P_l$=2 γ/R, Laplace's equation [8]). We find however that along a given isotherm,

$$\frac{P_v}{P_\infty} = \exp(-\frac{(P_\infty - P_l)V_l(\infty)}{V_v(\infty)P_\infty}) \tag{3}$$

where P_∞ is the equilibrium vapour and liquid pressure on the binodal line in a more rigorous treatment. The corresponding molar volumes are $V_v(\infty)$ and $V_l(\infty)$, respectively. We found also that almost irrespective of the level of the approximation, the ratio P_v/P_∞ becomes a simple function of two newly defined parameters which measure the degree of 'stretching' of the liquid, one based on the liquid, s,

$$s \equiv \frac{P_\infty - P_l}{P_\infty - P_l^{sp}} \tag{4}$$

where P_l^{sp} is the 'liquid's' pressure on the spinodal line, and the other based on the vapour pressure in equilibrium with the liquid, y,

$$y \equiv \frac{P_v - P_\infty}{P_v^{sp} - P_\infty} \tag{5}$$

234

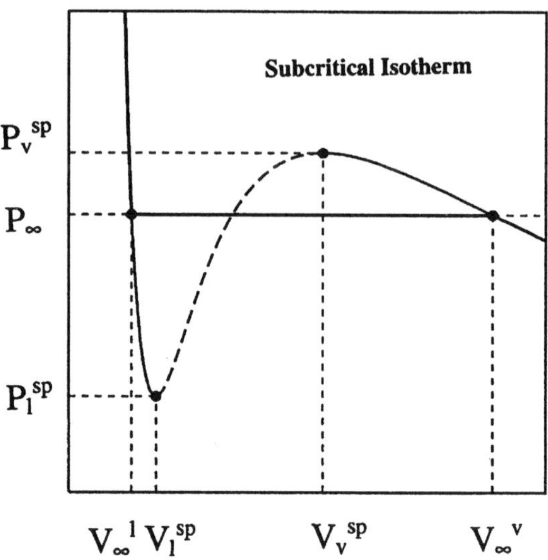

Figure 1. This diagram shows a typical isotherm below the critical temperature. The line joining the two co-existence volume-pressure points is shown (solid horizontal line). The co-existing pressure is P_∞. The minimum and maximum locate the liquid and vapour spinodal points.

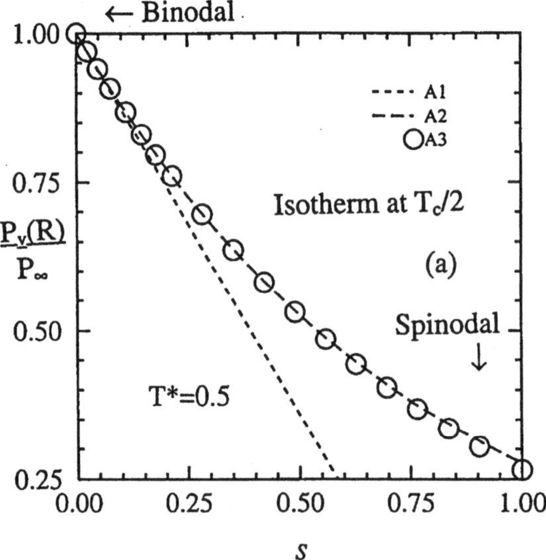

Figure 2. This diagram shows the ratio of the vapour pressure in the 'critical' cavity plotted against the parameter, s, defined in Equation (4). The temperature is equal to half the critical value. The model A1 corresponds to constant vapour and liquid densities, A2 is and ideal gas representation for the vapour and van der Waals equation of state for the liquid and A3 the van der Waals equation of state for both the vapour and the liquid phases. One can see that the difference between models A2 and A3 is negligible at this temperature.

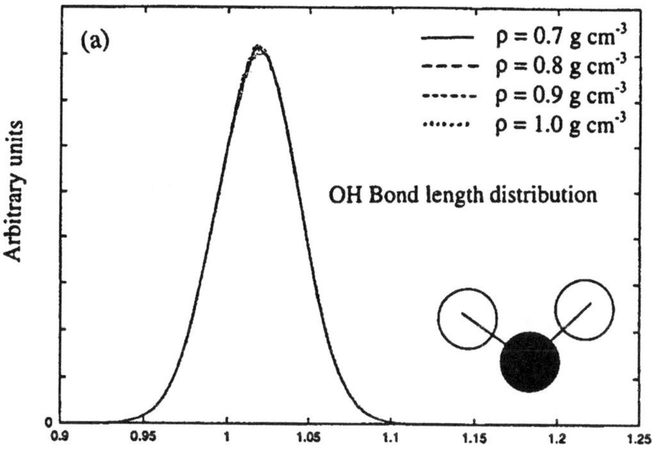

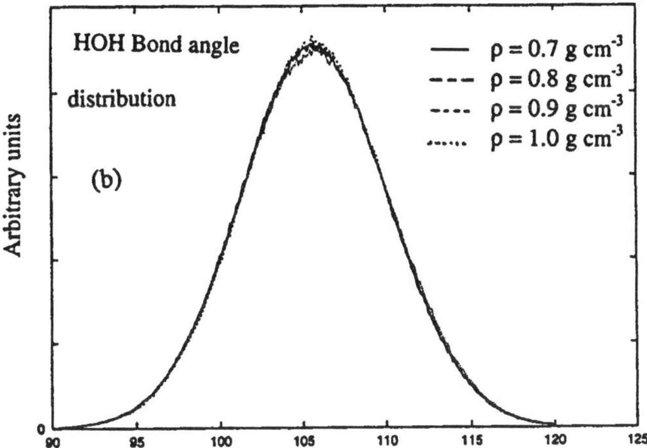

Figure 3. (a) Bond length and (b) bond angle distribution functions for model bulk water, with 256 flexible water molecules in a cubic box replicated with periodic boundary conditions [10]. The distribution functions at the various densities indicated on the figures are shown. As can be seen they show negligible variation with density in the negative pressure region ($\rho < 1$ g/cm^3).

where P_v^{sp} is the vapour pressure on the spinodal line. The dependences of the vapour pressure (see Figure 2) and barrier height on the scaling parameters were found to be quite insensitive to the level of approximation of the equation of state (*e.g.,* simplifications inherent in the vdW EoS), when cast in terms of these dimensionless parameters. The maximum tensile strength, estimated from the spinodal pressure, for liquid water under ambient conditions using the van der Waals EoS close to the triple

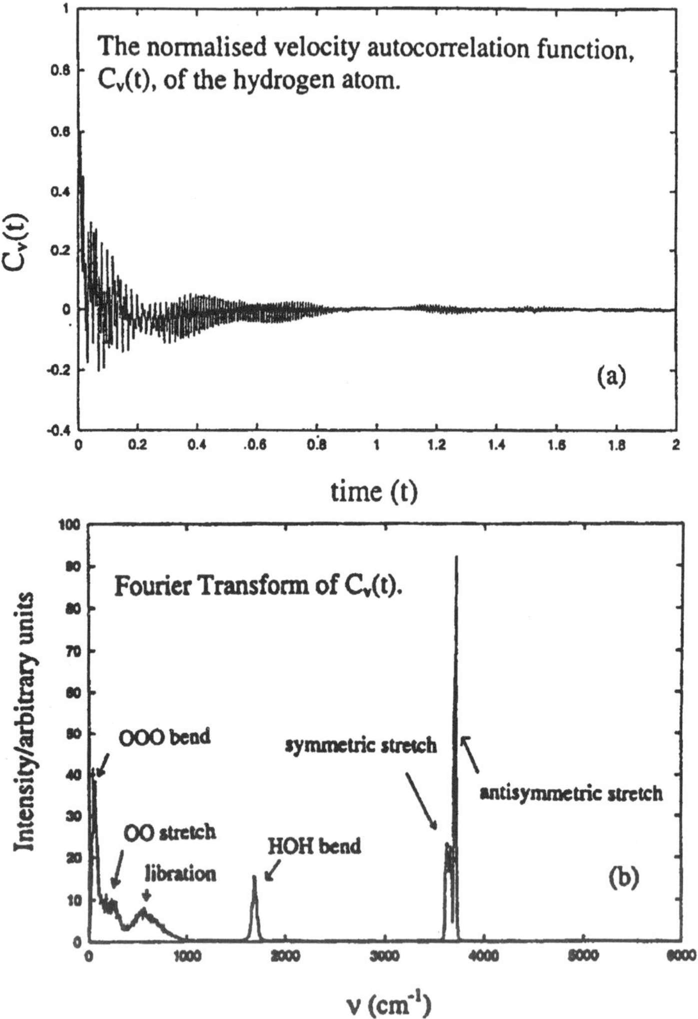

Figure 4. The dynamical behaviour of the hydrogen atoms of the water molecules in the MD simulations of Figure 3, carried out at $\rho = 1.0$ g/cm^3. (a) The normalised velocity autocorrelation function, $C_v(t)$ and (b) the Fourier transform of $C_v(t)$. The peaks in (b) are assigned against the experimental gas phase values, as follows: OOO bend (expt. 60 cm^{-1}; MD 50 cm^{-1}), OO stretch (expt. 170 cm^{-1} ; MD 200 cm^{-1}), libration (expt. 300-1000 cm^{-1}; MD 400-1000), symmetric stretch (expt. 3200 cm^{-1}; MD 3640 cm^{-1}), antisimmetric stretch (expt. 3400 cm^{-1}; MD 3720 cm^{-1}) and HOH bend (expt. 1600 cm^{-1}; MD 1700 cm^{-1})

point was found to be *ca.* 1 kbar. We also considered the Lennard-Jones fluid, as a representative 'realistic' fluid, using a literature-parameterised equation of state. This gives a value of *ca.* 2 kbar for the maximum tensile strength near the triple point. Further details of this work may be found in reference [9].

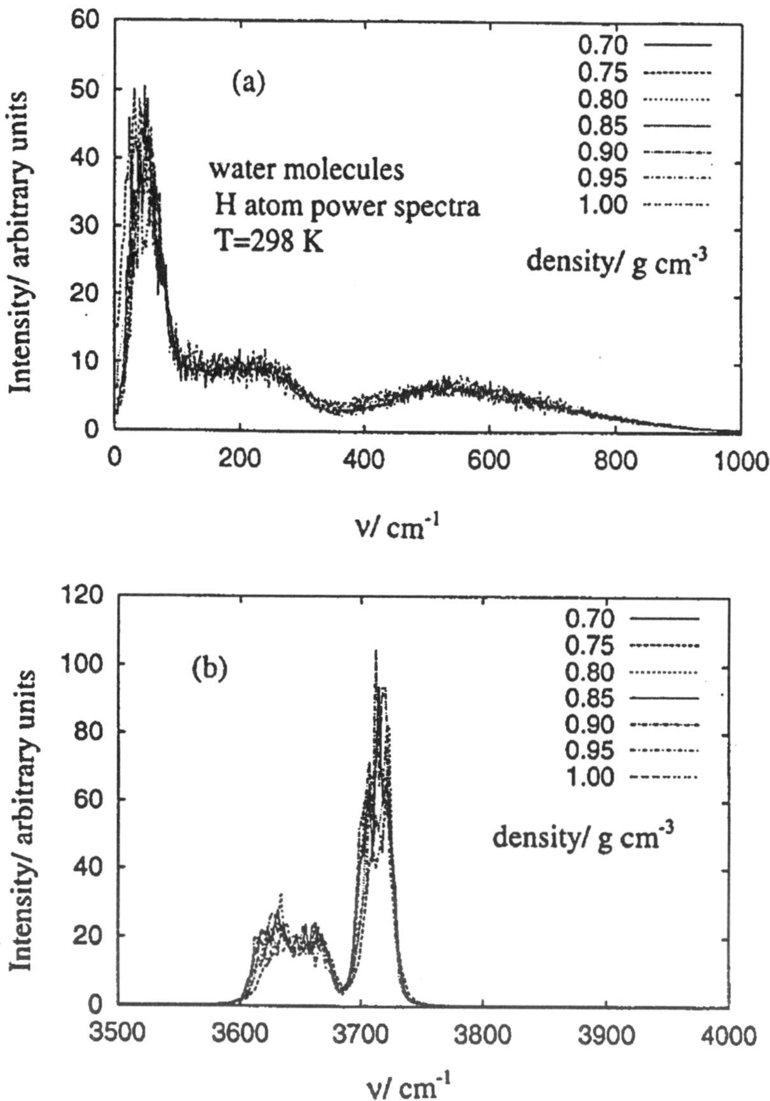

Figure 5. The intensity of the power spectrum of the Fourier Transform of the hydrogen atom velocity autocorrelation function, at temperature of 298 K. The densities of the water are given on the figures. (a) The wavenumber regime between 0-1000 cm⁻¹, and (b) the wavenumber regime between 3500-4000 cm⁻¹.

3. Molecular Dynamics of Uniformly Stretched Model Liquid Water.

One possible procedure to observe cavitation in a Molecular dynamics simulation is to stretch rather quickly the MD cell by a uniform or 'affine' dilation applied to the

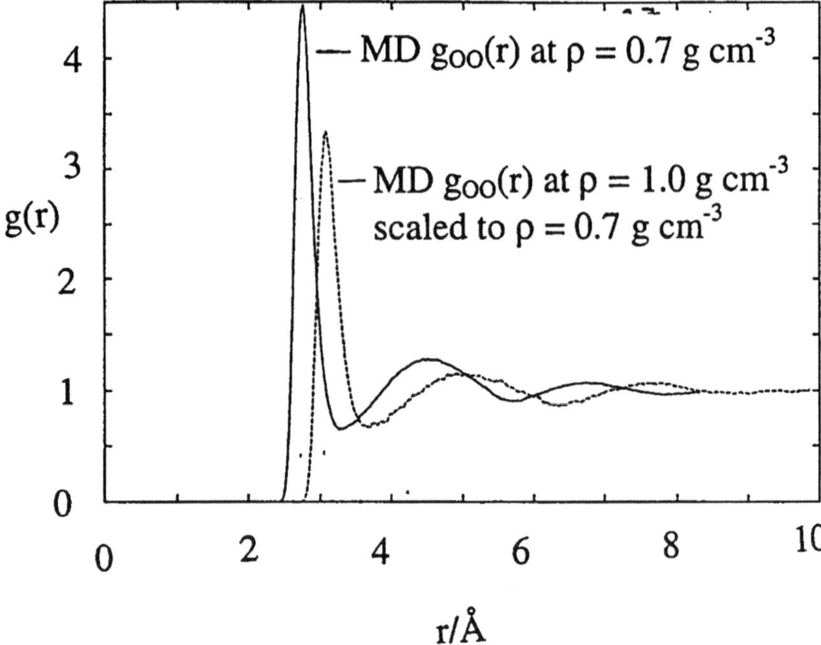

Figure 6. The radial distribution function, $g_{OO}(r)$, for the oxygen-oxygen atoms in model water molecules at temperature of 298 K. Two $g_{OO}(r)$ are shown, (a) one obtained directly my MD carried out at density of 0.7 g/cm^3 and (b) one obtained from a MD simulation at a density of 1.0 g/ cm^3 , which has been affinely scaled to correspond to a mean density of 0.7 g/ cm^3 . This comparison reveals that at lower density the position of the first co-ordination shell is at a smaller radius than would be estimated simply by affine scaling, and is resistant to distortion during the `stretching` process.

centres of mass of the molecules. This was the strategy of one of our earlier studies performed on a system of flexible SPC/E model water molecules under ambient conditions. The dilations used (going from a density of 1.0 to 0.7 g cm^{-3}) carried out over a picosecond timescale were so extreme that the change in the liquid's microstructure reflected more a spinodal mechanism. The model for water adopted was a fully flexible version of the SPC/E potential (both bond stretch and bond angle bend were allowed). 'Snapshots' of the structure showed some evidence of a heterogeneous structure of high and low density regions [10]. The simulation time window and size were probably too small to observe cavitation. We have continued these simulations and found that the water molecule bond length (Figure 3a) and bond angle (Figure 3b) distributions showed no noticeable change during bulk water stretching, which indicates that these density changes are being accommodated by reorganisations in the second and third co-ordination shells, rather than by changes in the geometry of the water molecule.

Probably for the same reason the power spectra of the hydrogen atom velocity autocorrelation function (Figure 4), which has been used as an approximate representation of the IR spectrum, shows no statistically significant density dependence (Figure 5).

There was an increase in peak height and decrease in integrated peak area in the pair radial distribution functions with decreasing density, indicating that there is an increase in local ordering on stretching (which also led to a decrease in the self-diffusion coefficient, D). The relative insensitivity of the first and second peak positions in the $g_{OO}(r)$, $g_{OH}(r)$ and $g_{HH}(r)$ stretching also shows that the dilation of the system is occurring in a non-uniform fashion. This is also shown by a comparison between the actual $g_{OO}(r)$ at 0.7 g cm^{-3} with an 'idealised' radial distribution functions by uniformly scaling that obtained from a 1.0 g cm^{-3} simulation of water to a hypothetical 0.7 g cm^{-3} state (see Figure 6).

4. Molecular Dynamics Simulations of the Collapsing Bubble

We have carried out the first reported Molecular Dynamics simulations of a collapsing bubble or cavity in a simple 'Lennard-Jones' or 'LJ' liquid, which is a popular model liquid in which the model molecules interact with the pair potential,

$$\phi(r) = 4\varepsilon\left(\left[\frac{\sigma}{r}\right]^{12} - \left[\frac{\sigma}{r}\right]^{6}\right) \tag{6}$$

where r is the separation between the particle centres and σ and ε set the distance and energy scales of the particles, respectively. (For a more detailed account of this work, consult ref. [11].) A series of MD simulations was carried out with periodic boundary conditions using a cubic box containing up to 108,000 Lennard-Jones (LJ) particles to create initially an equilibrium bulk liquid. The thermodynamic parameters were, reduced number density, $\rho = 0.8$, the temperature, $k_B T/\varepsilon = 0.97$, and the pressure, $p_l \sigma^3/\varepsilon \approx 1$, where σ is the usual LJ diameter and ε the well-depth. From this bulk liquid a thermodynamically unstable bubble was created at the centre of the box by taking out all the particles within a certain radius, R_b, of a quarter of the box sidelength, ranging typically from 5.6 to 12.8σ. The evolution of the collapsing bubble, which we call a 'Collapse Event' (CE), was followed typically for a 1000 times starting from a statistically independent liquid configuration each time to improve the statistics of the (averaged) radial profiles (which were obtained to a resolution of $ca.$ 0.1σ between neighbouring shells). Radial density, $\rho(r)$, and temperature, $T(r)$, profiles were computed from the centre of the original cavity. The simulations revealed that the filling of the cavity follows a sequence of stages, characterised by (a) a broadening of the initial 'step-function' profile to that of a typical liquid-vapour interface, (b) then the 'hottest' molecules (*i.e.,* from the high energy tail of the Maxwell-Boltzmann velocity distribution) in the boundary region 'evaporate' first into the cavity, (c) Then follows a filling-in stage in which the central region of the original cavity becomes increasingly occupied by molecules, until the density in the centre is near that of the liquid. This becomes an increasingly slower process, as the dense fluid microstructure in the core

240

has to rearrange to accommodate more molecules as the final liquid density is approached.

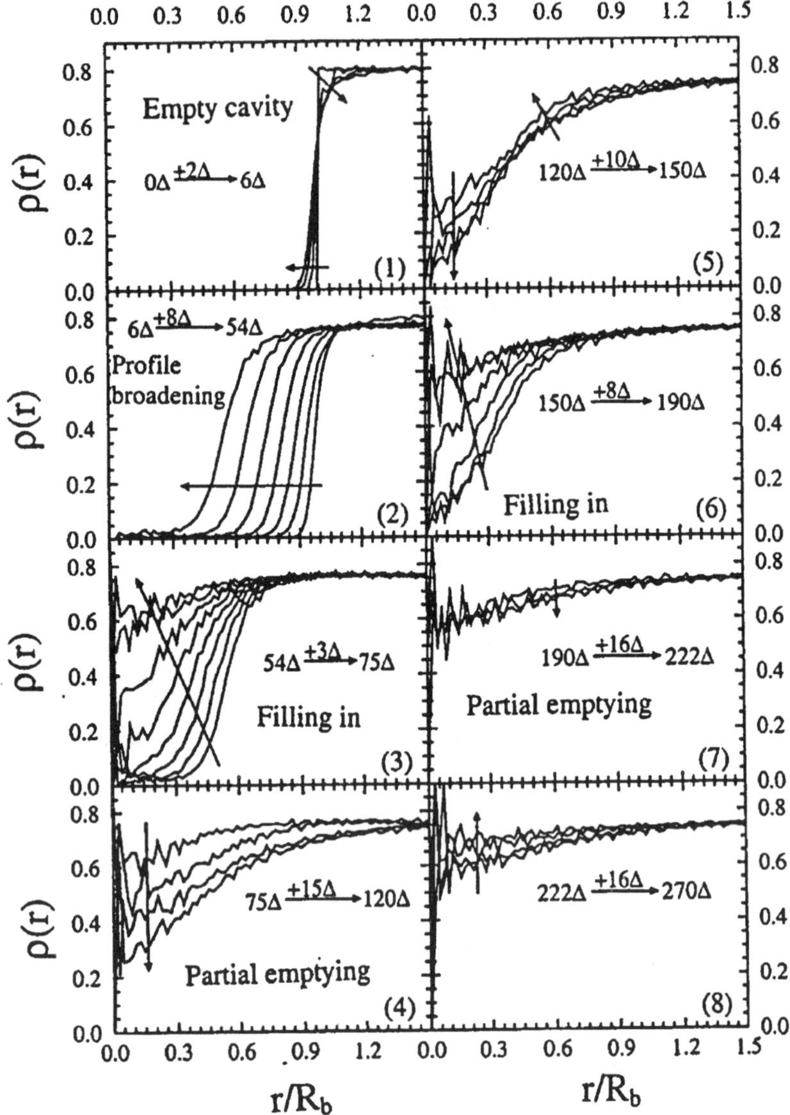

Figure 7. This figure shows the time and radius dependent density, ρ (r), measured from the centre of the cavity, for the filling in of a cavity of initial radius, $R_b=12.8\sigma$. The sequence of frames (1)-(8) shows the initial filling in stage (frames (1)-(3)), the partial evacuation (frame (4)), return to filling in (frame (5)) and so on. The arrow superimposed on each figure gives the general drift direction. The unit of time, Δ, is 0.125 $\sigma(m/\varepsilon)^{1/2}$.

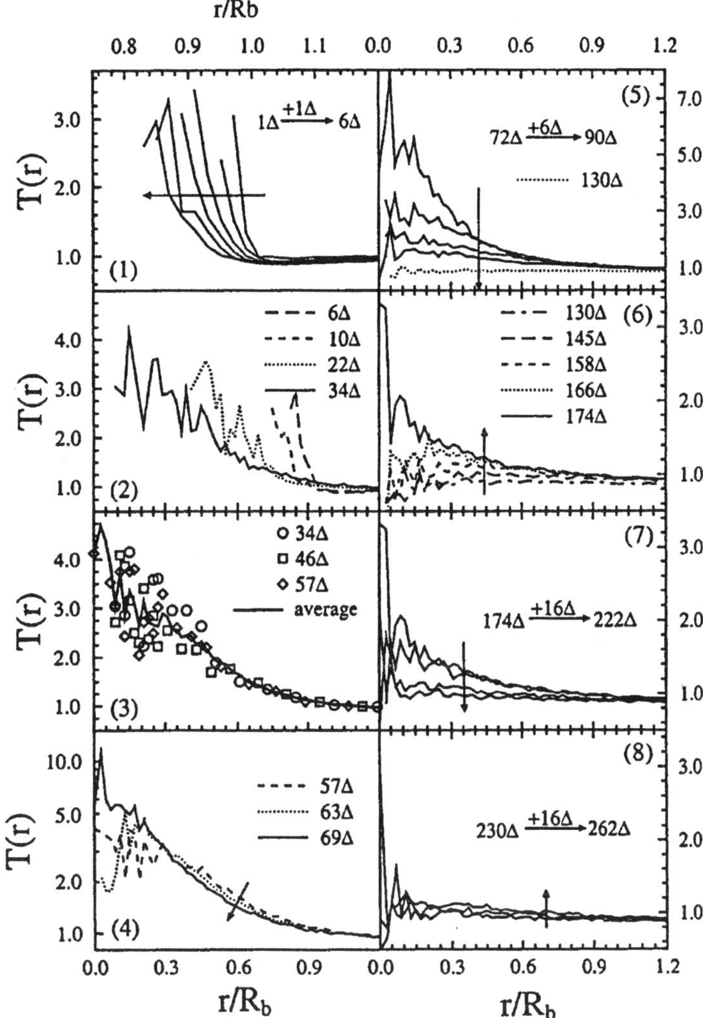

Figure 8. As for Figure, except that the radial dependence of the temperature is shown.

We find for the larger cavities, after almost complete collapse and filling in of the cavity, it proceeds to empty to a certain extent again! The bubble partly fills in, then it partially empties, before filling in some more. The process continues with diminishing amplitude towards the final spatially uniform state. The larger the initial size of the bubble, the higher the final temperature and the more pronounced is the 'oscillatory' nature of the collapse. The density profiles as a function of time are given in Figure 7 and the corresponding temperature profiles in Figure 8.

During the filling in process, the hottest molecules near the centre of the cavity can

242

easily reach a very high temperature, for example in the smallest cavity we found this to be $T^*{\sim}12$, which is $\approx 9T_c$, where T_c is the critical temperature, $T_c{=}1.3$ for the LJ system. For real argon this is about 1,400 K and for water it is 5,800 K. These are the sorts of high temperature values we expect to be relevant to the phenomena of sonoluminescence. They occur quite simply and naturally in our model and, because of the generic nature of the Lennard-Jones interaction, we can only assume are quasi-universal, which suggests a simple mechanism for sonoluminescence.

These results show that the final stages of bubble (cavity) collapse, where one expects the highest pressures and temperatures to occur, and are perhaps the most interesting, to be most realistically described by Molecular Dynamics simulations and not by traditional solutions of the fluid mechanics equations, which is the traditional route [12]. In this region, the usual assumptions of fluid mechanics, such as an incompressible liquid are clearly not realistic.

References

1. Francescutto, A., Ciuti, P., Iernetti, G., and Dezhkunov, N.V., (1999), **47**, 49-55.
2. Brown, S.W.J. and Williams, P.R., (2000) The tensile behaviour of elastic liquids under dynamic stressing. *J. Non-Newtonian Fluid Mech.* **90**, 1-11.
3. Putterman, S. (1998) Sonoluminescence: the star in the jar. *Physics World* **11**, 38-42.
4. Taleyarkhan, R.P., West, C.D., Cho, J.S., Lahey, R.T., Nigmatulin, R.I. and Block, R.C., (2002) Evidence for nuclear emissions during acoustic cavitation, *Science* **295**, 1868-1873.
5. Trevena, D.H. (1987) *Cavitation and tension in liquids*, Adam Hilger, Bristol.
6. Lickiss, P.D. and McGrath, V.E. (1996) Breaking the sound barrier. *Chem. in Britain* March 47-50.
7. Fisher, J. (1948) The fracture of liquids. *J. Appl. Phys.* **19**, 1062-1067.
8. Atkins, P.W. (1987) *Physical Chemistry*, 3rd edition, Oxford University press, Oxford.
9. Xiao, C. and Heyes D.M. (2002) Cavitation in Stretched Liquids, *Proc. Royal Soc. Ser. A,* in press.
10. Parker, M. and Heyes, D.M. (1998) Molecular dynamics simulations of stretched water: Local structure and spectral signatures. *J. Chem. Phys.* **108**, 9039-9049.
11. Xiao, C., Heyes D.M., and Powles J.G. (2002) *The Collapsing Bubble in a Liquid – A Model for Sonoluminescence*, publication in preparation.
12. Leighton, T.G., The Acoustic Bubble (1994) Academic Press: London

A STUDY OF CAVITATION PHENOMENA USING AN ATOMIC FORCE MICROSCOPE

M. S. BARROW, W.R. BOWEN, N. HILAL*, A. AL-HUSSANY, P.R. WILLIAMS, R.L. WILLIAMS AND C. WRIGHT.
Centre for Complex Fluids Processing, School of Engineering, University of Wales Swansea, UK
** School of Chemical, Environmental and Mining Engineering, University of Nottingham, UK*

Abstract. We report work in which an Atomic Force Microscope (AFM) is used to deform a liquid film between a colloid sphere and a plane surface. Under some circumstances, when the sphere and the surface are pulled apart rapidly, an unexpected transient *decrease* of the sphere-surface separation is recorded. Numerical simulations of cavitation bubble dynamics are used to explain how the growth of a cavitation bubble may result in the development of sufficiently large *negative* pressures to account for this phenomenon. The results of this study provide evidence in support of a mechanism mooted by Israelachvilli and co-workers that in the cavitation of liquids within confined spaces, the growth of a cavity may be more damaging than its subsequent collapse.

1. Introduction

The cavitation of liquid films is an important aspect of processes such as lubrication [1] and printing [2]. These, and other industrially important processes often involve meso–scale (0.1–10 μm) thickness films undergoing rapid deformation between separating surfaces. In coating processes involving ink films, cavitational film-splitting may result in the formation of rapidly-stretching filaments, whose breakup leads to unwanted droplet deposition. Filament formation is also a feature of coating flows involving adhesive films but descriptions of the process are still largely qualitative, invoking terms such as 'tack' [2]. A crucial factor is the initial film thickness as even ostensibly low rates of surface separation may provoke the high rates of fluid deformation necessary to generate sufficient tension to overcome the effective tensile strength (or 'cavitation threshold'), σ_c, of a liquid film and result in cavitation [3].

Using a surface forces apparatus (SFA), Israelachvilli and co-workers observed the growth and disappearance of vapour cavities in liquid films between separating mica surfaces [4,5,6]. The growth of a cavity was claimed to represent a 'new' cavitation damage mechanism, insofar as the occurrence of surface damage was thought to occur during cavity inception. This is an interesting finding given that the greatest effort in cavitation damage research has involved the study of bubble dynamics in terms of cavity collapse.

A.R. Imre et al. (eds.), Liquids Under Negative Pressure, 243–254.
© 2002 *Kluwer Academic Publishers. Printed in the Netherlands.*

The asymmetry of cavity collapse leads to potentially damaging phenomena, such as liquid jets [7] but the issue of cavitation damage due to cavity *growth*, as mooted by Israelachvilli and co-workers, is relatively unexplored. In this paper we report experimental and theoretical work involving an atomic force microscope (AFM) which sheds further light on such a mechanism.

2. Experimental: AFM Apparatus and Method

The AFM has been used to study the dynamic properties of systems such as membranes [8], the adhesion of a single biological cell [9] and the properties of liquids [10]. In the present work an 'Explorer' AFM (Vecco Instruments, USA) was used to measure the force between a cantilever and a sample as a function of the scanner displacement (the latter being varied using a piezoelectric crystal. A laser beam reflected from the back of the cantilever falls onto a photosensitive position detector (PSPD) which detects changes in cantilever deflection (figure 1). To generate a force-distance curve, the deflection of the cantilever tip is recorded as a function of the tip-sample separation as the piezo-scanner raises the sample towards the tip. To convert the data into force as a function of true sample-tip separation, it is necessary to know the spring constant of the cantilever and to define zero values of both force and separation distance.

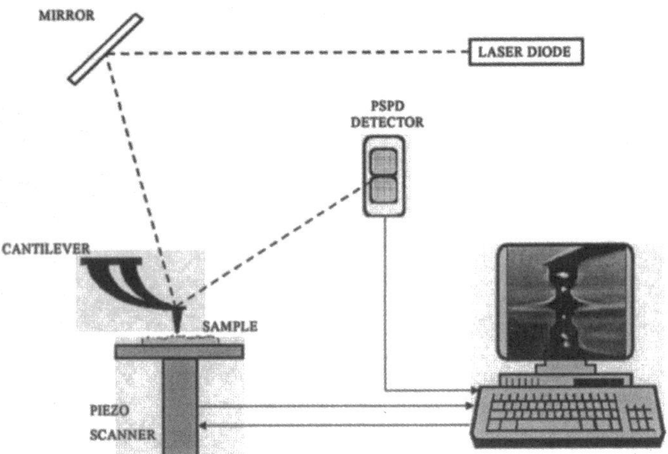

Figure 1. Schematic of apparatus

In the present work the tip was replaced by a 'colloid probe' of known geometry by attaching a 15 μm diameter silica sphere (Polysciences Inc.) to a tipless 'ultralever'-type cantilever (figure 2) whose spring constant K was 2.9 N/m and whose resonant frequency was 160 kHz (Park Scientific Instruments, USA). Additional probes were made by attaching a 15 μm diameter silica sphere to tipless ultralevers 85 μm in length whose spring constant was 2.1 N/m and whose resonant frequency was 320 kHz. Colloid spheres of this type have been shown to have a maximum peak-to-valley roughness of 3 nm over an area of 0.45 μm^2 [10]. The planar surface used in this study was prepared by oxidation of polished silicon wafers to give a 500 nm thick coating (Virginia Semiconductor Inc., USA).

Figure 2. SEM image of colloid probe

AFM force-distance experiments consist of an 'approach' and a 'retraction' stage, in which the maximum distance between the probe and the surface over which force measurements are taken is *ca.* 3 μm. The probe is initially located far from the surface with the cantilever in its equilibrium (undeflected) position for which the corresponding 'zero-force' PSPD output is constant. On approaching the surface, at a constant velocity, the PSPD output *decreases*. This reduction corresponds to a *downward* deflection of the cantilever (an event referred to as 'snap in') which is attributed to short range attractive forces acting upon the probe, prior to the establishment of a physical surface contact [8].

Eventually the probe is arrested by the surface but the other end of the cantilever is attached to the scanner column which *continues to descend*: consequently the cantilever deflection *reverses*, corresponding to a compressive force. The scanner continues to descend until a predefined limit of compressive force is recorded following which the scanner retracts over a specified distance (ca. 3 μm) relative to the final position of approach until the probe-surface contact is broken.

We now consider the general features of the corresponding curve when a thin film of liquid is present on the surface (figure 3). In the presence of a liquid film the retraction of the scanner is not accompanied by a 'snap back' to zero force. Rather, subsequent to the generation of maximum tensile force, a gradual reduction in force is observed, over distances greater than those previously noted for work involving a 'dry' surface. This feature of the retraction curve is attributed to the formation of a liquid film which spans the gap between the probe and the surface. Thus, both the magnitude and the range of the tensile force increase significantly in the presence of a liquid film.

In figure 3, the PSPD output is presented as a function of piezo position relative to the piezo's starting height, not to the planar surface. A change in PSPD output (relative to the zero-force) is produced by a deflection of the cantilever and this causes an additional vertical displacement of the probe which is *independent of scanner position.*

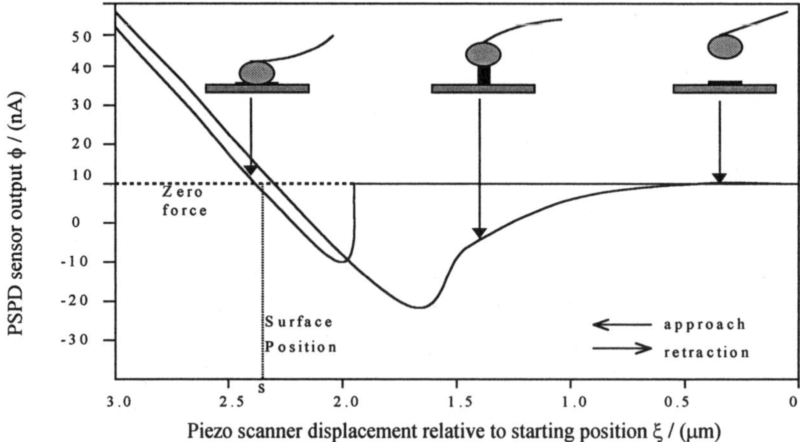

Figure 3. Typical force curves obtained for approach and retraction to and from a surface covered by a fluid film.

The magnitude of the piezo displacement is equal to the magnitude of displacement of the probe *relative* to the piezo (corresponding to a compressive deflection of the cantilever). Under such circumstances, the relationship between PSPD output and cantilever deflection is calculated from the gradient of the constant compliance region. This relationship is assumed to be valid for deflections produced by the action of tensile forces on the cantilever. The cantilever deflection, α, is determined from the PSPD output, ϕ, according to equation (1), for which a *negative* value of α indicates a *downward* displacement of the probe.

$$\alpha = \frac{\phi - \phi_0}{|q|} \tag{1}$$

In equation (1), q is the gradient of the region of constant compliance (nA/nm) and ϕ_0 is the sensor output (nA) in the 'zero-force' region.

The position of the surface indicated in figure 3 is inferred from the force profile and the value of the 'zero-force' as follows. The cantilever deflection is measured relative to its equilibrium position (being also relative to the position of the piezo column). The position of the surface may then also be related to the position of the piezo column: this is deemed to be the point at which the cantilever is in contact with the surface *and* in its equilibrium (undeflected) state. This 'zero distance' corresponds to position 's' in figure 3, from which the separation distance of the probe from the surface, β , at any given scanner position, ξ , is given by

$$\beta = (s - \xi) + \alpha \tag{2}$$

Prior to experiments involving liquid films, the scanner surface and colloid probe were cleaned by rinsing with ethanol and flushing copiously with deionised water. The liquid sample was subsequently deposited by placing a drop of liquid on the silicon wafer which was then wiped to produce a liquid film. The colloid probe was brought into contact with the liquid and removed, prior to an approach being made on a clean area of the silica surface. Once the colloid probe and surface were in contact, the probe was retracted at a specified velocity, v, in the range 5 μm/sec to 5000 μm/sec.

3. Results

3.1 AFM FORCE-DISTANCE MEASUREMENTS.

The results obtained in experiments on a Newtonian 12 Pa.s silicon oil (Dow Corning 200, BDH Chemicals UK) are shown in figures 4 $(a) - (d)$. The following regions of the force *vs.* distance curve may be identified (see figure 4(a)). In region '1' of this approach curve the separation distance x between the scanner and the wetted sphere is gradually reduced by lowering the probe. The cantilever registers no force as x decreases in the range 350 nm $< s <$ 1800 nm, the latter distance being the maximum separation. As x decreases further, the cantilever at first registers an attractive (tensile) force, F, as the scanner comes into contact with the liquid (region '2'). At position '3' the sphere is at a separation of 50 nm, F at this position being 1000 nN. As the probe is pushed further downwards, F rapidly diminishes (region '4') due to the lubrication forces generated by the viscous liquid film between the scanner surface and the sphere.

The retraction curve of figure 4(b) was generated by reversing the direction of motion of the probe (the latter being pulled upwards). The retraction curve begins at the equilibrium separation identified in the corresponding approach curve (figure 4(a)) and which is denoted feature '1' in figure 4(b). As the probe moves upwards at the lowest attainable speed of 5 μm/sec, F increases with increasing separation distance, reaching a peak value F_{max} of 2500 nN ('2' in figure 4(b)) at a separation of approximately 350 nm. It is important to note that at this separation no force was registered during the corresponding approach curve. Thereafter, as x continues to increase, the cantilever records a gradually decreasing tensile force. At a separation of 900 nm, F is 2000 nN, double the corresponding value recorded in the approach curve (see '3' in figure 4(b)).

As the probe continues to retract, the cantilever registers a sudden increase in F which is accompanied by a rapid *reduction* of the distance between the scanner and the sphere, the latter moving downwards over a distance of *ca.* 100 nm (region '3–4'). From the peak tensile force (at '4'), which is approximately equal in magnitude to that registered at position '2', F decreases rapidly thereafter following a reversal in the direction of motion of the sphere, x increasing rapidly in this region to a maximum of 1400 nm, at which distance no force is registered by the cantilever.

The approach curve (figure 4(c)) and corresponding retraction curve (figure 4(d)) show the same sequence of events, these being the results of further experiments involving a clean sphere and a fresh liquid sample, in which v was increased to 500 μm/sec. The peak tensile force of 1300 nN registered in this approach curve is slightly

248

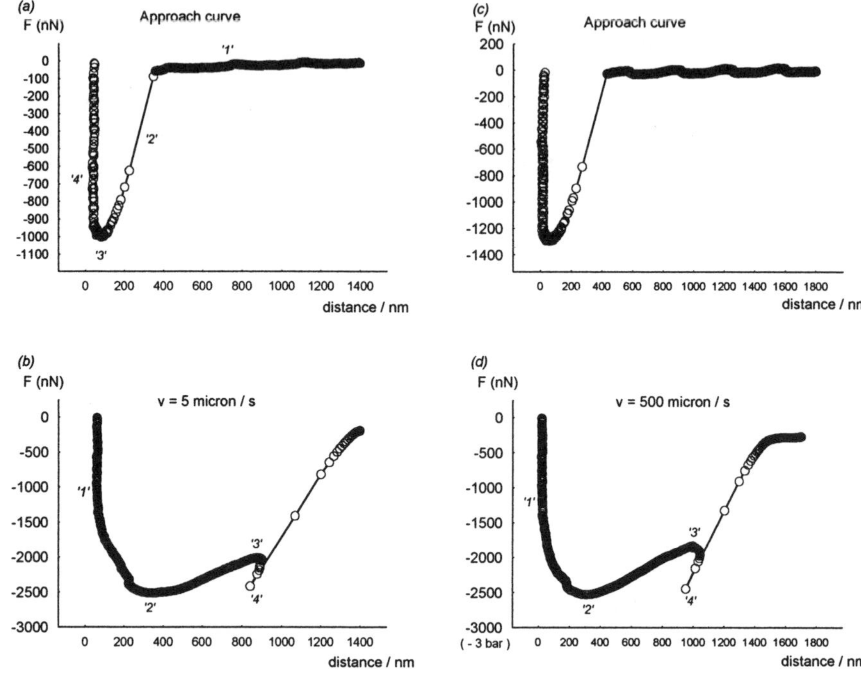

Figure 4. Force-distance curves obtained for a 12 Pa.s silicone oil, (a) and (b) show results for consecutive scanner approach and separation at 5 micron / s respectively, (c) and (d) show results for consecutive scanner approach and separation at 500 μm s^{-1} respectively.

higher than in that described previously although the zero force separation distance isessentially the same as in figure 4(*a*). At the higher speed, the event labelled '3–4' occurred at a slightly greater separation distance (*ca.* 1000 nm) than in the slower experiment. Further work was conducted over a wide range of *v*, using fresh liquid and clean spheres. In all cases, the separation of the surfaces led to the sphere suddenly moving downwards in response to the rapid development of tensile force as identified in feature '3–4' in figures 4(*b*) and 4(*d*).

We next consider the results of similar experiments on a lower viscosity, non-Newtonian liquid – in this case a 50 p.p.m. aqueous solution of 'Magnafloc 10' polyacrylamide, PAA (Ciba, UK). The results are summarised in figures 5(*a*)–(*d*), for four different speeds. In figures 5(*a*) and 5(*b*), *x* increases *continuously* during probe retraction, at speeds of 5 μm/sec and 200 μm/sec, respectively; and F_{max} increases from 60 mN/m at 5 μm/sec, to 95 mN/m at 200 μm/sec (the peak force being divided by the diameter of the sphere). At the two higher speeds the unusual retraction event previously encountered in work involving the higher viscosity liquid was evident. At 300 μm/sec, *x* reached 1500 nm prior to a rapid decrease of *x* over 100 nm: at the highest attainable speed of 5000 μm/sec, *x* reached approximately 1800 nm prior to the sphere's sudden downward movement. The value of F_{max} registered at 300 μm/sec and 5000 μm/sec was 95 mN/m in both cases – the same as that recorded at 200 μm/sec.

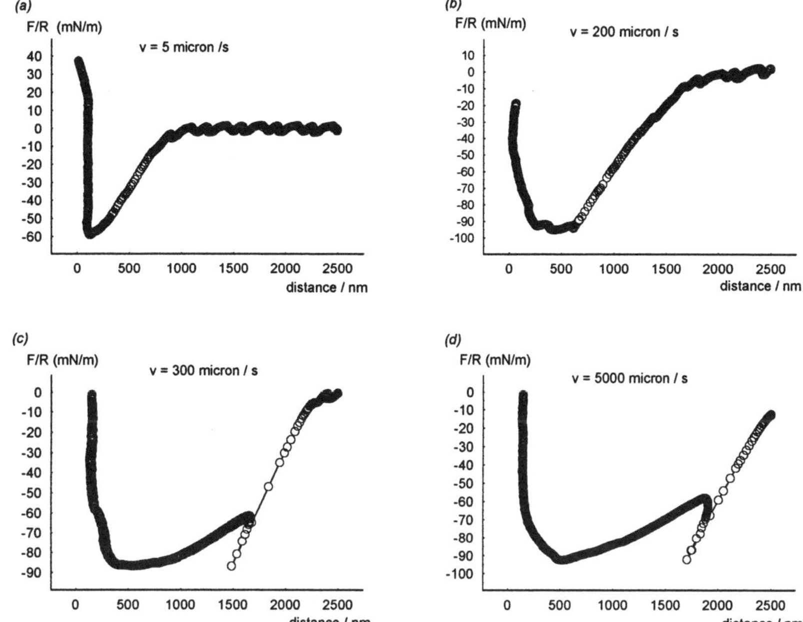

Figure 5. Force-distance curves obtained for 50 ppm aqueous PAA solution, figures (a), (b), (c) and (d) are obtained for the respective separation scanner speeds of 5, 200, 300 and 5000 μm s^{-1}.

As far as we are aware, the unusual feature recorded in the retraction curves (i.e. a transient increase in tensile force which is accompanied by a rapid reduction of separation distance) has not previously been described in AFM work. In what follows, we consider whether cavitation provides a basis for explaining the unusual retraction event in AFM work.

4. Cavitation bubble dynamics

4.1 GROWTH OF A CAVITY DUE TO THE RAPID DEVELOPMENT OF TENSION IN A LIQUID.

We begin by considering a cavity growing spherically in an unbounded liquid [11]. In a spherical coordinate system (r, θ, ϕ), with the origin at the bubble's center, the only velocity component is radial and, in order to satisfy continuity,

$$u = \left(\frac{R(t)}{r} \right)^2 \dot{R}(t) \tag{3}$$

250

where $R(t)$ is the bubble radius at time t, and $\dot{R}(t) = dR/dt$. The radial component of the momentum equation is given by,

$$\rho\left(\frac{\partial u}{\partial t} + u\frac{\partial u}{\partial r}\right) = -\frac{\partial u}{\partial r} + \frac{\partial \tau_{rr}}{\partial r} + \frac{2\tau_{rr} - \tau_{\theta\theta} - \tau_{\phi\phi}}{r} \tag{4}$$

where ρ is the liquid density and τ_{rr}, $\tau_{\theta\theta}$, $\tau_{\phi\phi}$ are the normal components of the deviatoric stress tensor related by,

$$\tau_{rr} + \tau_{\theta\theta} + \tau_{\phi\phi} = \tau_{rr} + 2\tau_{\phi\phi} = 0 \tag{5}$$

giving, after integration between infinity and R,

$$\rho\left(R\ddot{R} + \frac{3}{2}\dot{R}^2\right) = p(R) - p_\infty(t) + \tau_{rr}(\infty) - \tau_{rr}(R) + 3\int_r^\infty \frac{\tau_{rr}}{r}dr \tag{6}$$

where $p(R)$ and $p_\infty(t)$ are the pressure at the interface on the liquid side and the pressure far from the bubble, respectively. The pressure at the interface on the liquid side is given by,

$$p(R) = p_v + p_g - \frac{2\gamma}{R} + \tau_{rr}(R) \tag{7}$$

where p_g is the non condensible gas pressure within the bubble, p_v the vapour pressure and γ the surface tension. By determining p_g from the initial conditions of the nucleus of radius R_0 and gas pressure p_{g0}, expression (7) becomes,

$$p(R) = p_v + \left(p_0 - p_v + \frac{2\gamma}{R_0}\right)\left(\frac{R_0}{R}\right)^{3k} - \frac{2\gamma}{R} + \tau_{rr}(R) \tag{8}$$

where k is the polytropic index and p_0 the pressure at infinity at $t = 0$. Substituting (8) into (6) we obtain the generalized 'Rayleigh-Plesset' equation [12],

$$\rho\left(R\ddot{R} + \frac{3}{2}\dot{R}^2\right) = -(p_\infty(t) - p_v) + (p_0 - p_v)\left(\frac{R_0}{R}\right)^{3k} + \frac{2\gamma}{R_0}\left[\left(\frac{R_0}{R}\right)^{3k} - \frac{R_0}{R}\right] + 3\int_R^\infty \frac{\tau_{rr}}{r}dr \tag{9}$$

from which the time evolution of the cavity radius as a function of $p_\infty(t)$ may be computed under the assumption that the liquid is incompressible and that thermal and mass transfer effects are negligible. For a Newtonian liquid we write,

$$3\int_{R}^{\infty}\frac{\tau_{rr}}{r}\,dr = -4\mu\frac{\dot{R}}{R} \tag{10}$$

where μ is the shear viscosity of the liquid [13]. For a non-Newtonian 'power-law' fluid we write,

$$3\int_{R}^{\infty}\frac{\tau_{rr}}{r}\,dr = -4(2\sqrt{3})^{n-1}\frac{m}{n}\left|\frac{\dot{R}}{R}\right|^{n-1}\frac{\dot{R}}{R} = \eta(R,\dot{R},m,n)\frac{\dot{R}}{R} \tag{11}$$

where m and n are the power law parameters [14] and $\eta(R, \dot{R}, m, n,)$ the apparent viscosity. Fruman [13] has reported results of the integration of the generalized Rayleigh-Plesset equation with the integral term of equation (11) for a Newtonian 'reference' fluid and three power law fluids. (the latter having power-law indices of 0.8, 0.65 and 0.4, respectively). For these fluids the evolution of cavity radius with time was practically indistinguishable from the reference Newtonian fluid ($n = 1$).

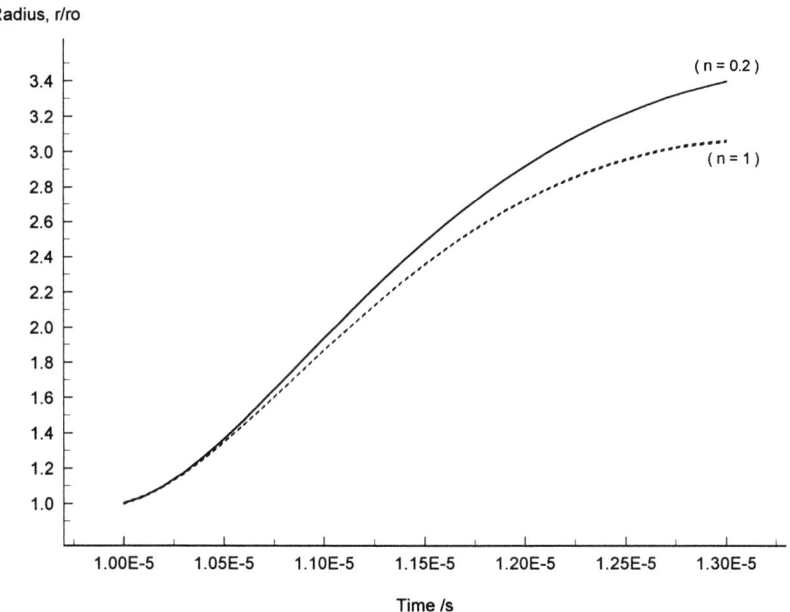

Figure 6. Cavity growth, r/r$_o$, as a function of time for a Newtonian (*n*=1) and a shear thinning (*n*=0.2) fluid

In figure 6 we show the results of integrating the 'Rayleigh-Plessett' equation (using a fourth-order adaptive Runge-Kutta method) for a cavity which is initially at rest within a Newtonian fluid and a shear-thinning fluid which is characterised by a power-law index of 0.2. In response to the sudden development of a pulse of tension within the liquid, the resulting differences in the growth of the cavity in both fluids appears only marginal but in what follows we seek to establish that the relatively small differences in the rate at which the cavities approach their maximum radius may have significant consequences in terms of the development of hydrodynamic pressures (specifically *negative* pressures) about the growing cavity.

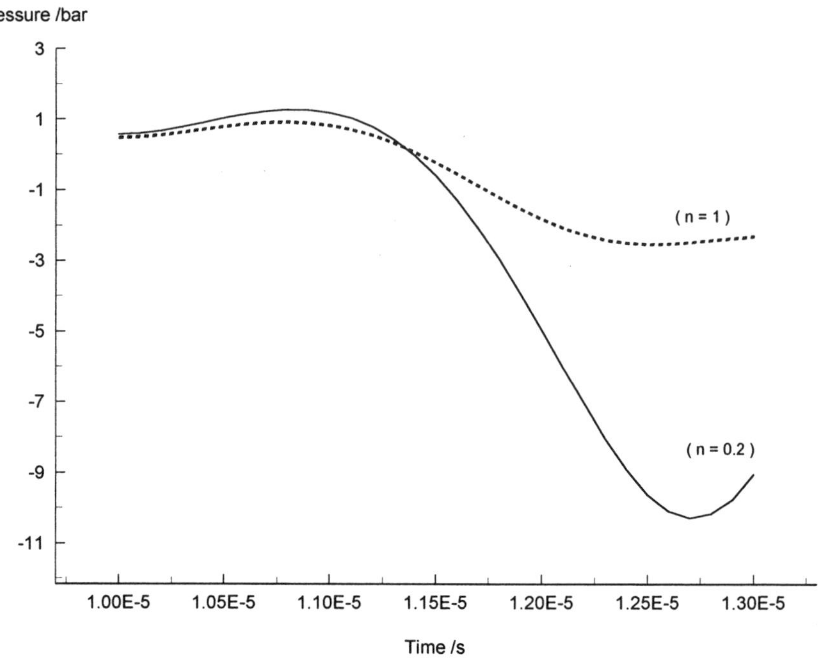

Figure 7. Hydrodynamic pressure at a distance r (1.5 times maximum cavity radius), as a function of time, resulting from cavity growth in a Newtonian (n=1) and shear thinning (n=0.2) fluid.

The hydrodynamic pressure p at a point r in the liquid in the neighbourhood of the bubble is given by the general Bernoulli equation [15]:

$$p = \rho \, d\psi/dt - 1/2 \, (\rho \upsilon^2) + p_\infty \tag{12}$$

where p_∞ is the pressure in the undisturbed water, ρ is the density and $\upsilon = - \, d\psi/dr$ is the radial velocity at that point, with velocity potential ψ. The pressure variation Δp at r due to cavity motion is

$$\Delta p = p - p_\infty = \rho/r \, d/dt \, (R^2 \, dR/dt) - \rho \, R^4/2 \, r^4 (dR/dt)^2 \tag{13}$$

Overton & Trevena [16] used the Bernoulli equation to account for the pressure-tension cycles recorded in their bubble oscillation experiments and in so doing associated the peak values of tension with the attainment of maximum cavity volume, V. When V reaches a maximum, d^2V/dt^2 is negative and this results in the development of negative pressure due to the fact that, for values of r large enough to discount the second term in equation (13), $d^2V/dt^2 \propto \Delta p$. Subsequent work by Williams & Williams [17] established that the maximum negative values of d^2V/dt^2 (and hence the maximum tension) may precede the attainment of maximum cavity volume. From this it follows that a pulse of negative pressure is generated about a cavity during its growth phase.

Figure 7 shows the variation of pressure resulting from the cavity growth which is shown in figure 6. The pressure is calculated at a distance r of 1.5 times the maximum cavity radius. From figure 7 it may be seen that the growing cavity initially produces a pulse of positive pressure but this is immediately followed by the development of tension. As the distance between r and the cavity surface decreases, the pressure in a thin 'shell' of liquid between the growing cavity and r becomes increasingly negative with the result that a growing cavity may develop sufficient negative pressure (at least several bar) in the liquid between its surface and an adjacent surface to cause the latter to be 'sucked' rapidly downwards, over a distance of 100–200 nm. This corresponds closely to the unusual retraction event recorded in the AFM work described in Section 3.1.

5. Discussion

The theory developed in Section 4 provides an explanation for the unusual retraction event in terms of the development of negative pressure in the liquid film about a growing cavity. This cavitation results from the development of tensile stress within the film, due to the relative motion of its bounding surfaces (the colloid sphere and the plane surface of the AFM scanner).

An interesting point which emerges is that in this situation, the magnitude of the effect is significantly larger in the case of cavity expansion in a non-Newtonian (shear-thinning) fluid than in its Newtonian counterpart. This latter finding may have signifance in the context of the lubrication of natural joints, which are sometimes referred to as 'biological bearings' [18]. In joints, which involve meso-scale (0.1-1 μm) fluid films, the lubricating film is synovial fluid, a dialysate of blood plasma with the addition of a long chain polymer (hyaluronic acid) which is responsible for the characteristic high viscosity. Normal healthy synovial fluid is highly shear-thinning but only a small volume of joint fluid is available for study from a healthy human joint (ca 1 ml). As much higher volumes of joint fluid are produced in the presence of pathological conditions, previous studies have been heavily biased towards pathological joint fluid (which is essentially *Newtonian*) due to the ease of collecting such fluid. In terms of the specific cavitation mechanism considered herein, the results of the present work suggest that the consequences of choosing to study pathological fluid rather than healthy fluid may be significant. Further work is now planned in order to investigate this issue in greater detail.

Given the ability of growing cavities to generate substantial tension in liquids in close proximity to their surfaces, the present work lends support to the arguments advanced by Israelachvilli and co-workers as a result of their SFA-based cavitation work. A significant difference which emerges as a result of our work is that we envisage the potential for sufficiently rapidly growing cavities to damage adjacent structures by the development of tension – rather than by the evolution of positive pressure and this mechanism may be relevant to instances of cavitation *in vivo*, in which a liquid film's bounding surfaces may be formed from biological membranes. In such a situation the consequences of the evolution of a large tension in an adjacent liquid may be as damaging as the development of large positive pressures.

References

1. Dowson, D. and Taylor, C.M. (1979) Cavitation in Bearings, *Ann. Rev. Fluid Mech.*, **11**, 35-66.
2. Zang, Y.H., Aspler, J.S., Boluk, M.Y. and De Grace, J.H. (1991) Direct measurement of tensile stress ("tack") in thin ink films, *J. Rheol.*, **35**, 345-361.
3. Joseph, D.D. (1998) Cavitation and the state of stress in a flowing liquid, *J. Fluid Mech.*, **366**, 367-378.
4. Chen, Y.L. and Israelachvili, J. (1991) New mechanism of cavitation damage, *Science*, **252**, 1157-1160.
5. Chen, Y.L., Kuhl, T. and Israelachvili, J. (1992) Mechanism of cavitation damage in thin liquid films: collapse damage *vs.* inception damage, *Wear*, **153**, 31-51.
6. Kuhl, T., Ruths, M., Chen, Y.L. and Israelachvili, J. (1994) Direct visualisation of cavitation and damage in ultrathin liquid films, *J. Heart Valve Disease*, **3**, 117-127.
7. Tomita, Y. and Shima, A. (1986) Mechanisms of impulsive pressure generation and damage pit formation by bubble collapse, *J. Fluid Mech.*, **169**, 535-564.
8. Bowen, W.R., Hilal, N., Lovitt, R.W., Sharif, A.O. and Williams, P.M. (1997) Atomic force microscope studies of membranes: force measurement and imaging in electrolyte solutions, *J. Mem. Sci.*, **126**, 77-89.
9. Bowen, W.R., Hilal, N., Lovitt, R.W. and Wright, C.J. (1998) Direct measurement of the adhesion of a single biological cell using atomic force microscopy, *Colloids and Surf. A: Physicochem. and Eng. Aspects*, **136**, 231-234.
10. Ducker, W.A., Senden, T.J. and Pashley, R.M. (1992) Measurement of forces in liquids using a force microscope, *Langmuir*, **8**, 1831-1836.
11. Rayleigh, Lord. (1917) On the pressure developed in a liquid during the collapse of a spherical cavity, *Phil. Mag,* **34**, 94-98.
12. Plessett, M.S. and Prosperetti, A. (1977) Bubble Dynamics and Cavitation, *Ann. Rev. Fluid Mech.*, **9**, 145-185.
13. Fruman D.H. (1999) Effects of non-Newtonian fluids on cavitation: in *Advances in the Flow and Rheology of non-Newtonian Fluids*, Part A. Eds. Siginer DA, De Kee D and Chhabra RP. Elsevier.
14. Barnes, H.A., Hutton, J.F. and Walters, K. (1989) *An Introduction to Rheology.* Elsevier Science Publishers B.V., Netherlands.
15. Temperley, H. N. V. and Trevena, D. H. (1978) *Liquids and their Properties.* Ellis Horwood, Chichester.
16. Overton, G. D. N. and Trevena, D. H. (1981) Cavitation phenomena and the occurrence of pressure-tension cycles under dynamic stressing, *J.Phys.D: Appl.Phys.*, **14**, 241-250.
17. Williams, P.R. and Williams, P. M. (1996) Pressure-tension cycles induced by dynamic stressing and cavitation in liquids, *J. Phys. D: Appl. Phys.*, **29**, 1904-1909.
18. Trevena, D.H. (1987) *Cavitation and Tension in Liquids*, Adam Hilger, Bristol.

CAVITATION THRESHOLDS, FREE SURFACE AND CAVITY CLUSTER DYNAMICS IN LIQUIDS AT SHOCK WAVE REFLECTION

A. BESOV and V. KEDRINSKII

Lavrentyev Institute of Hydrodynamics, Novosibirsk, Russia

J. DE VRIES and M. KLOOSTERMAN

University of Twente, Enschede, The Netherlands.

Abstract. Cavitation cluster dynamics and cavitation thresholds under plane short shock waves reflected from the free surface of various liquids is investigated experimentally By means of capacity and light absorption methods it is shown that the free-surface velocity is directly connected to the volume concentration of cavitation bubbles in the cavitating area and an expression for it was found. An effect of the time of contact of water and atmosphere on the dynamics of cavitation development is revealed. It was found that one hour is enough to change the free-surface velocity dynamics from monotonous to oscillating, which is characteristic for spalling damage of metals. High-speed filming showed that in the first case, the bubble distribution over the volume is uniform, while in the second case a dense layer of bubbles is formed under the free surface and partly reflects the rarefaction wave which continues moving freely in the formed liquid layer. This effect is probably associated with the changes in the structure of a near-surface layer due to air saturation. It is shown experimentally that for transformer oil, the previous loading history and the time interval between the experiments are essential considerations and the intensity of cavitation is increased under multiple sequential loading.

By means of the capacitance method, which determines the cavitation thresholds and intensity of cavitation processes with high accuracy, the dependence of the intensity of cavitation processes on water temperature was investigated experimentally. It is shown that the intensity of cavitation processes is extreme and increases with temperature from 18,5 ^{0}C to 42 ^{0}C, dropping to an initial value at 52 ^{0}C and again increasing abruptly from 66 ^{0}C. A mechanism which accounts for this phenomenon is proposed.

1. Introduction

Numerous investigations of the cavitation strength of water under the action of negative pressures and reasons for its reduction in comparison with expected one began in the 1950's [1,2,3] and continue to the present day [4,5,6]. The latter studies confirm the complexity and many-faceted aspects of this problem. It is necessary to emphasize two particular aspects that make this problem complicated:

1. The complex initial liquid state that is determined by the physical nature, size distribution and concentration of cavitation nuclei on which actually cavitation development in real liquids begins. This state determines the initial dynamics of cavitation processes there.

255

A.R. Imre et al. (eds.), Liquids Under Negative Pressure, 255–270.

2. The complex non-linear dynamics of both bubbles developing from cavitation nuclei and cavity clusters formed therefrom. The behavior of such systems has nothing in common with that of separate bubbles. They can easily transmit shock waves of large intensity and at the same time they can immediately extinguish or unrecognizably distort negative pressure pulses. Here, the dynamics of such formation greatly depend on the pressure decay speed in a rarefaction wave that results in a wide scatter of cavitation threshold magnitudes as measured experimentally using different methods.

2. Initial state

The investigation of a liquid's initial state is a complex matter requiring detailed investigation in each particular case. Due to the presence of various admixtures, samples of water from various water-supply systems and natural sources are essentially different [4]. A similar problem results from water purification as a consequence of multiple distillation, filtration and deionization stages. It is almost impossible for different investigators to provide the same degree of purification of samples for investigation. Consequently, as a conventional standard which may be easily obtained in laboratory conditions, it is suggested that water resulting from a single distillation be used without any additional or special purification. Such water contains about 10^5-10^6 cm^{-3} of microheterogeneities of submicron size and about 10^3-10^4 cm^{-3} of microbubbles with radius $\approx$ 1.5 μm in different combinations with microparticles [5]. These are just microbubbles that determine cavity strength and the initial dynamics of cavity processes in distilled water. The use of such water allows comparisons of data on cavity water strength that were obtained by different investigators and by different methods. However, preliminary washing and purification of experimental equipment (cuvette, shock tubes, etc) from admixtures should be particularly taken into account because admixtures and chemical agents in water result in changing of the equilibrium radius and microbubble size distribution. The latter may considerably change the cavitation characteristics of water [7].

3. Obtaining negative pressures in liquids

It is possible to induce negative pressures in liquids by thermal means i.e. to heat it in a sealed-off vessel until residual air dissolves and, then, to cool it down until rupture of the liquid occurs. Dynamic loading can be cyclical (in the field of an acoustic transducer or concentrator), monotonic (as in the the breaking of a liquid under the influence of centrifugal forces in a rotating glass tube [4]) or it may involve impulses, as occurs when a pressure pulse, propagating towards free surface and reflecting from it as a rarefaction wave, is introduced into a liquid [8]. In such a rarefaction wave, growing cavity nuclei begin to result in cavity cluster formation and changes of the dynamics of free surface displacement, pulse changes of passing light absorbing and its scattering. The duration of a pulse, its amplitude and the pulse decay slope determine further dynamics of the cavity zone. There are other ways of obtaining negative pressures that are presented in [4], however, from the point of view of handling a certain problem each of the above mentioned techniques has its own particular advantages and disadvantages. The method of shock wave reflection from the free surface of a liquid was chosen as the most suitable one in the present work.

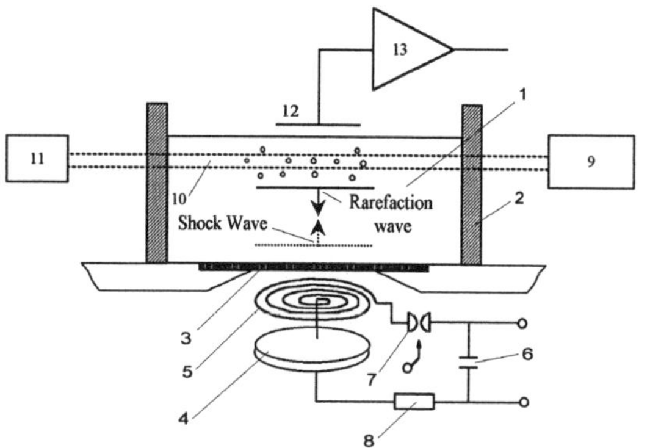

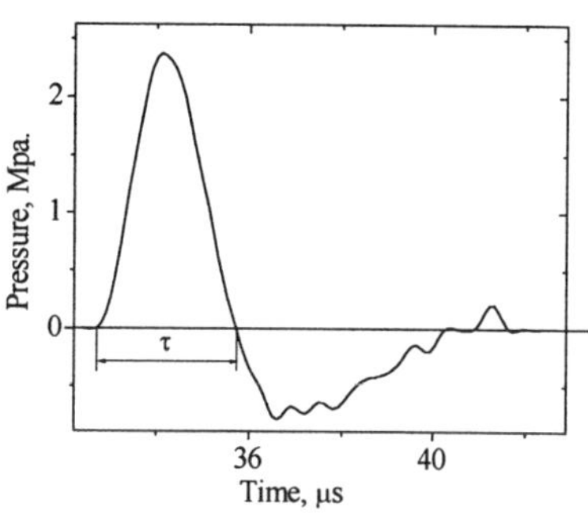

Figure 1. Experimental set up and principal circuit of the electromagnetic shockwave-generator. 1 – Liquid; 2 – Light transparent walls of the cuvette; 3 – The shock membrane; 4– Supporting disk from copper; 5 – Archimedes spiral coil; 6 – The capacitor-bank. 7 – The driving spark gap; 8 – Additional resistance; 9 – He-Ne laser; 10 – Laser beam; 11 – Photomultiplier; 12 – Capacitor microphone, 13 – High impedance amplifier.

Figure 2. Typical shape of the pressure pulse

The experimental set-up and the block-scheme of the electromagnetic shock wave generator are presented in Fig. 1. in which 1 is liquid; 2 is the cuvette with diameter 70mm made of glass and filled with liquid; 3 is the membrane made from duralumin with thickness 0.6 ÷2 mm, squeezed on the contour with diameter 30mm; 4 is the support disk of copper with thickness 2 mm; 5 is the Archimedes spiral coil with diameter 30mm; 6 is the low inductive capacity bank (2μ Fx10 kV, L= 2,5 10^{-8} H); 7 is the controlled spark gap; 8 is the resistance R = 0,25 Ω; 9 is a He-Ne laser; 10 is a laser beam, passing through the cavity cluster; 11 is a photo multiplier; 12 is a capacitive transducer; and 13 is a high input impedance amplifier.

A flat, short shock wave (whose profile is shown in Fig.2) is created in the liquid cuvette 2 (filled to a level of 20-30 mm) by a pulsed magnetic field pressure on a conducting membrane (3) which passes a pressure impulse into the liquid. A magnetic field appears as a result of the discharge of a low-inductive capacity bank (6) to flat coil (5) placed between a shock membrane (3) and a supporting copper disk (4). Controlled spark gap 7 allowed us to discharge the capacitor bank at a range from 4,5 kV up to 10 kV. Resistance 8 provides the non-periodical mode of discharge and the generation of a single pressure pulse. The amplitude and duration of the shock wave is determined by a voltage magnitude on the capacity bank and the thickness and material of membrane 3. The amplitude of the pressure pulse can be easily varied from tenths of an atmosphere to hundreds atmospheres practically without changing impulse form. The pulse duration can be changed in not so wide a range but at least with a factor of two (see later on Fig. 12.). A shock wave propagates as a flat disk with initial diameter approximately 25-30 mm toward the free surface of water (Fig.1, Fig. 3a). The needle with diameter 0.5 mm and wire with diameter 50 μm being pulled down through the free surface for scale. After reflection from the free surface, the shock wave transforms into a rarefaction wave and propagates downwards, initiating the growth of cavitation bubbles from nuclei (Fig. 3b). Bubbles, that can be seen before rarefaction wave front, are out of focus, which means that they are probably growing on the boundary of the disk-shaped shock wave propagation trace and steady state liquid. Some of them, nonetheless, may grow in the negative phase of the shock wave (Fig. 2.) but could not be seen immediately after the shock wave had passed (Fig. 3a.).

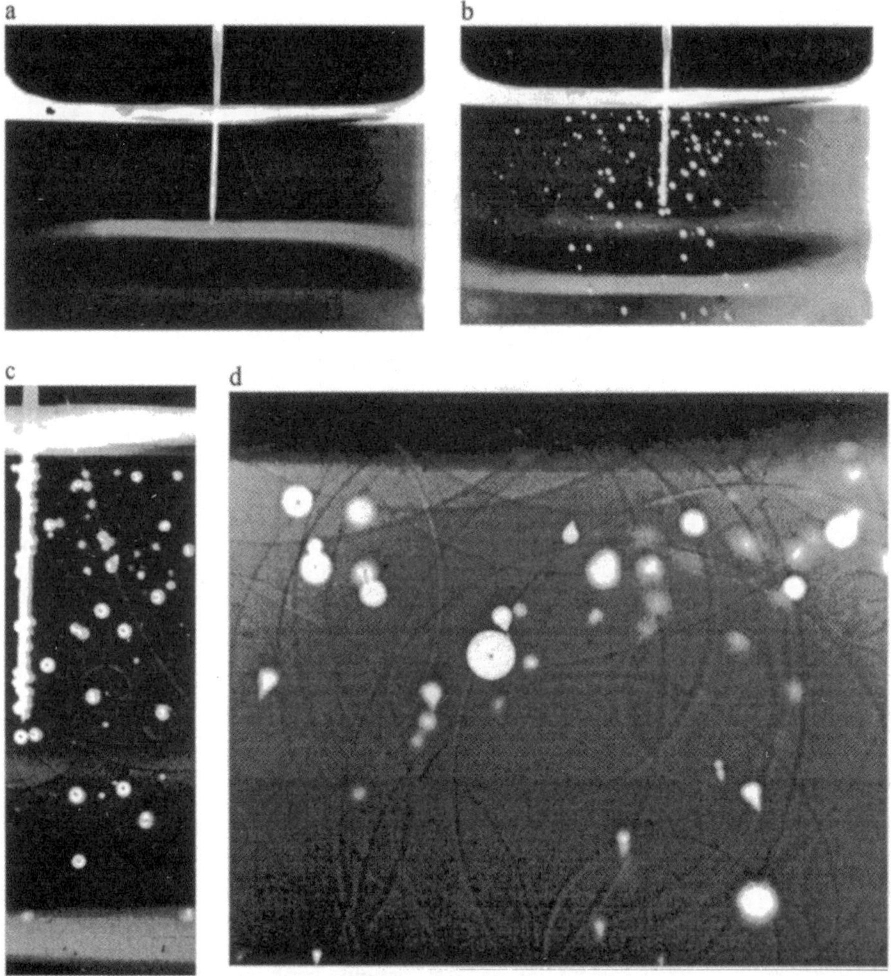

Figure 3. a). Approach of the plane shock wave toward the free surface of water. b). Rarefaction wave, traveling downward from the free surface and cavity cluster after it. c). Scaled-up part of Fig. 3.b. d) Frame of some cavity cluster area with size approximately 4.5mm x 5.5mm. The radius of the biggest bubble is 0.18 mm and jet from the collapsing near it another one has about the same length.

Fig.3b shows the rarefaction wave travelling downward from the free surface rarefaction wave and the cavity cluster formed after it. It may be seen in Fig. 3.b. and the magnified section of it (Fig.3c.) that some of the bubbles are growing but some are collapsing with jets and shock waves generated simultaneously. To reveal the structure of the cavity cluster, a more detailed picture with bubbles, shock waves and cumulative jets is shown in Fig.3d from which it is possible to conclude that the thickness of the shock waves from the collapsing bubbles do not exceed 30-45 μm (duration 20-30 ns) and that the length of the cumulative jets may exceed the maximum radius of the cavitation bubbles.

Free surface displacement is registered by the capacitive transducer 12. Transducer 12 is a cap made from brass with aluminized mylar film with thickness 20 μm on its surface. The film is put into immediate contact with water and moves together with the free surface, resulting in a change of distance between the film and the comparatively immobile central electrode with diameter d =15mm, causing a change of capacity and corresponding to it change of voltage on the transducer. The transducer works in the regime wherein Q=Const (Q is electric charge on it), does not require calibration and allows to the time dependence of the absolute magnitude of the free surface displacement X(t) to be obtained for different intensities of loading (Fig. 4, 5).

The method has several advantages:
1. It allows a plane, short shock wave to be obtained, which makes observation and modeling of cavity phenomena easier.
2. The choice of a cuvette (rather than a shock tube) allows the exclusion of the influence of walls at the initial stage of cavity cluster development.
3. The pressure pulse duration ($\tau \approx 3 - 5\mu$s, Fig. 2) has the same order of magnitude as that of the natural (self frequency) of the microbubbles, which are the basis of cavity nuclei [5]. This feature promotes an efficient (and possibly resonant) interaction with them.
4. Small volume of samples and the possibility of their easy replacement facilitates work with expensive liquids.
5. Small sizes, mobility and almost momentary readiness of shock wave generator for the next experiment.
6. Stability and repeatability of amplitudes and form of generating shock waves, which are determined by stability of capacitor bank voltage by the time of discharging, let us reach fractions of percent in repeatability of loading.

However, there are certain disadvantages to be noted:
1. Intense electrical pulse stray pick-up complicates measurements of electrical signals (several mV amplitude).
2. Preloading of the liquid sample by passing a pulse which may considerably change its initial state.
3. The presence of a rarefaction phase in the propagating pressure pulse (Fig. 2).

4. Cavitation thresholds and methods of their registration

Cavitation in liquids is a phenomenon which is essentially statistical in nature, and is difficult to measure quantitatively. Actually, the inference of a certain cavitation

260

threshold is a phenomenon stipulated by an explosive growing of cavitation bubbles, which is registered by a definite method. The resulting inferred value of cavitation threshold is then determined by the sensitivity of the particular method [9.]. We now briefly consider some of these methods. It is obvious that maximal sensitivity can be obtained by observing the dynamics of growing cavity nuclei with the use of light – scattering methods [10]. The changing of a bubble's radius by a factor n results in a corresponding change of light-scattering intensity of between n^4 and n^6 [11]. Thus, even if the part of light scattered by such bubbles in whole intensity of scattered light does not exceed 10%, an increase of their radius by a factor of two almost doubles the scattered light intensity. Such large changes provide a good basis for the detection of cavitation inception, but analysis of this data involves inverse problems which are too complicated and the level of achievable precision is too low. As a result it can be more effective to use a light absorption method as a less sensitive and more simpler one [12]. With increasing of the method sensitivity, less significant intensity changes can be detected and its sensitivity limit can be defined (operationally) in terms of a cavitation threshold.

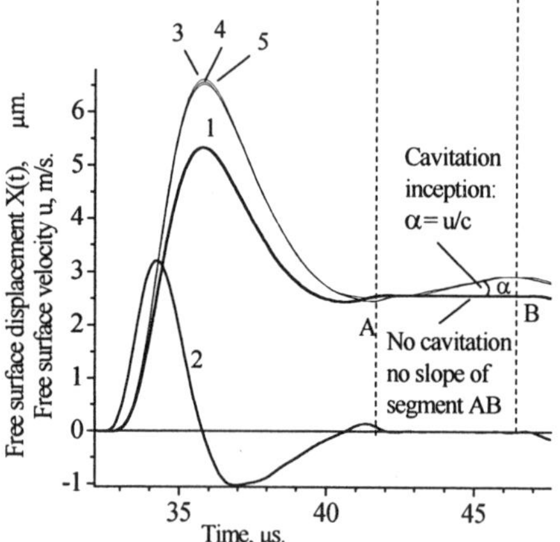

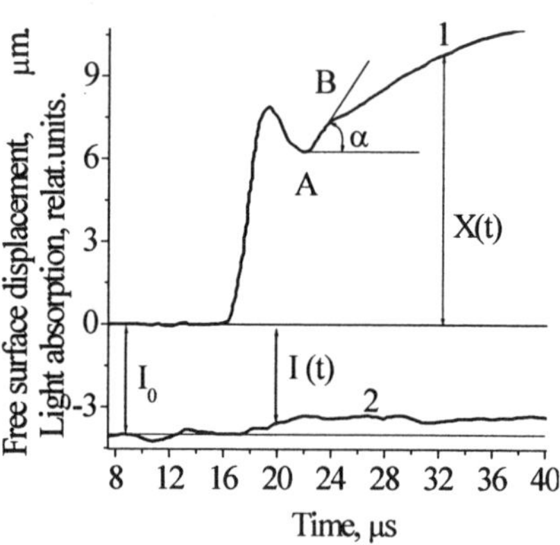

Figure 4. 1 – Representative caver of the free surface displacement when no cavitation happens, 2 – Velocity of the free surface, 3,4,5 – Representative cavers of free surface displacement at threshold of cavitation inception. The three cavers (3,4,5) are demonstrate stability and recurrence of the data.

Figure 5. Simultaneous pictures of the displacement of the free surface X(t) and changing of the intensity of the passing laser beam I(t) (diameter 2.5 mm, at a distance 6mm bellow the free surface).

5. Cavity cluster and free surface dynamics

The results of theoretical work [13] have shown that the behavior of the free surface reflects the main pecularities of the bubbly cluster dynamics developing near it. The capacity method [5] (Fig. 1, 12) registers the displacement of a liquid's free surface, which reflects cavity cluster dynamics developing underneath it [13]. This method is not so sensitive as light–scattering methods and cavitation threshold magnitudes registered by its use are higher [9], however, it gives very stable results that could help to establish the method as a basic one. Fig.4 (curve 1) presents typical dynamics free surface coordinate $X(t)$ of a distilled water sample at reflection of the pressure pulse (Fig. 2) with an amplitude of 2.4 MPa (in the absence of cavitation) and at some threshold pressure value (2.9 MPa) at which cavitation phenomena begin (Fig. 4, curves 3,4,5). The absence of the slope of curve 1 on the segment **AB** corresponds to zero speed of the free surface after completion of the process of shock wave reflection Fig.4(curve 2). Curve 2 is the derivative of line 1, representing the velocity of the free surface. The rarefaction wave travels downward, inducing tension in the stretched water sample, which preconditions its free surface raising. The growth of cavitation bubbles (beginning after the attainment of a definite threshold pressure), results in the free surface rising on the segment **AB**, and the slopes of the curves 3,4,5 (Fig. 4) correspond to some speed with which the free surface has been moving after completion of the pressure pulse reflection. The achievement of good repeatability in shape and the close position of the curves 3,4,5 (Fig. 3), corresponding to consequent loading, confirms the method's stability. The downward displacement of the free surface after point B corresponds to the arrival (into the investigated area) of the unloading wave from the cuvette walls.

In order to obtain information about the current size of, and the cavity bubble concentration within, the cavity cluster both the capacitive method and light – absorbing method should be used (Fig.1.). Simultaneously with the free surface displacement registered by the capacitive transducer 12, the changing intensity of the light beam 10 from the laser 9 passing through the cavitation cluster (aided by the multiplier 11) is registered. Increasing the intensity of the incident shock wave, up to 3.7 MPa, results in significant changes in the free surface dynamics of the water sample (Fig. 5, curve 1). The appearance of curve slope $X(t)$ on segment **AB** is directly connected to the increasing of specific volume α of cavity bubbles and can be presented as following:

$$\alpha = n \; 4\pi r^3/3 \tag{1}$$

where **n** – volume density of the bubbles, **r** – radius of it.

The first relation between **n** and **r** can be taken from the data of absorption of the laser beam [14].

$$I = I_0 \; exp(2\pi r^2 nd) \tag{2}$$

where d – diameter of the cavitation cluster, . I_0 – initial intensity of the passing light, I – initial intensity of the passing light at any stage we are interesting in.

The rarefaction wave propagates in the liquid with the speed of sound c and influences a volume dV in a time dt, which can be written as $dV=S{\cdot}c{\cdot}dt$ where S is the area through which the rarefaction wave propagates. Assuming a characteristic number of initial small nuclei per volume-unit n, the total number of initial small nuclei N in volume dV, which equals the number of bubbles growing due to the passing of the rarefaction wave, will be:

$$N = n \cdot S \cdot c \cdot dt \qquad (3)$$

Multiplying this number to the volume of a single bubble the total volume of these bubbles dV_b can be found:

$$dV_b = n \cdot S \cdot c \cdot V_{bubble} \cdot dt \qquad (4)$$

This volume should be equal to the increase of the volume of the liquid-gas mixture. This increase of volume will be equal to the rise of the free-surface dX multiplied with the same area of shock wave propagation as before S:

$$dV_b = S \, dX \qquad (5)$$

Setting these two equations equal to each other, the final result will be:

$$n \cdot c \cdot V_{bubble} \cdot dt = dX \qquad (6)$$

but $dX/dt = u$ – velocity of the free surface, $V_{bubble} = 4\pi r^3/3$ and finally from (1) and (6) we have

$$\alpha = u/c. \qquad (7)$$

Further from *(1)*, *(2)* and *(7)* we have:

$$r = (3/2) \cdot (u/c) \, d/\ln(I_0/I) \qquad (8)$$

and

$$n = (2/9\pi) \cdot (c/u)^2 \cdot (\ln(I_0/I)/d)^3 \qquad (9)$$

For our particular case: $u \approx 0.7 \ m/s$; $I/I0 \approx 0.85$; $c \approx 1500 \ m/s$, $d \approx 30 \ mm$ thus and so $r \approx 0.130 \ mm$ and $n = 50 \ bubbles/cm^3$.

Thus, simultaneous use of both methods allows us to obtain full information concerning the cavity zone dynamics at every instant. In order to check the resulting data fit to the real characteristics of the cavity cluster (at a time corresponding to the end of segment AB) an optical picture of the cavity zone was taken (Fig. 6.). Despite considerable bubbles radius dispersion from 70 μm up to 170 μm, its average magnitude is about 130–140 μm and the bubble concentration corresponds to that which was estimated above. It must be noted that the appropriate formula has a narrow band of application due to the fact that for light intensity changes $(I-I_0)/I < 5$ % the measurements are complicated and at $(I-I_0)/I > 20$ % the conditions of single scattering are violated (i.e. the formula cannot be applied). Even observing the condition (I-I_0)/I<20 % the presence of coupled bubbles (Fig. 6) influences the correspondence of calculated values and real magnitudes. Therefore the application of the method described above requires an additional optical control of reliability of obtaining data.

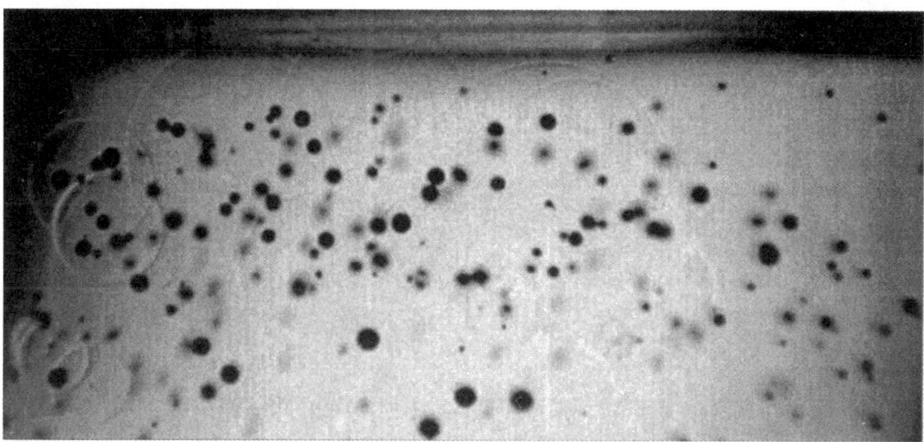

Figure 6. High speed photo of the cavitation cluster bellow the free surface at shock wave reflection. Flash time synchronized with the end of the segment AB (Fig .5.). The average radius of the bubbles 130 –140 μm with the spread from 70 μm up to 170 μm.

Since pictures are taken in experiment it's important to have an estimation of the size of the bubbles to expect. It can be done backward from the Rayleigh equation for the time of a bubble's collapse $\tau \approx 0.915 \cdot R_{max} \sqrt{(\rho/P)}$ [4]. If for approximation of the sine wave (Fig.2.) we will take square-wave with duration $\tau \approx 3.0 \cdot \mu s$ and arithmetic mean pressure $P \approx P_{max}/2$ where P_{max} =3.7 MPa (Fig.5.), $\rho \approx 1000$ kg/m^3 an estimation give us expected value $R_{max} \approx 0{,}14$ mm which is in close correspondence with the magnitude of the experimental results.

6. Effect of the surface layer saturation

An effect of the contact time between the studied samples of water and atmosphere on the dynamics of cavitation development is apparent. Thus the cavity development dynamics in the water, settled before experiments for more than 24 hours, has been found to depend on its time of contact with the atmosphere Fig. 7a (curve 1,2). Curve 1 presents free surface dynamics of a sample of distilled water immediately after placing it into the cuvette and curve 2 corresponds to the same sample 1 hour later. There are considerable differences in the dynamics and intensity of the cavitation processes. On curve 2 there are characteristic oscillations which can be seen more clearly after time differentiation within a narrow band (to avoid distortion of the signal by the flash lamp's noise at 32-33 μs) from 22 μs to 32 μs (Fig.7b.). According to [15,6] this oscillation is typical for spalling damage of solids. In order to check the possibility of spalling destruction of the water sample, photographs were taken of the cavity cluster at the points of time marked on the graphs as noise (Fig.8a,b.).

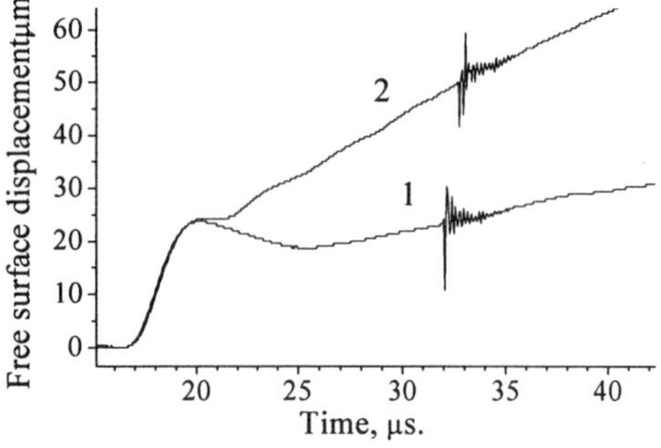

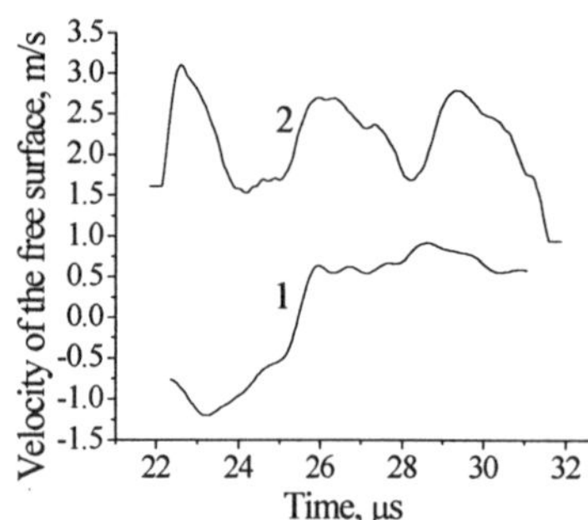

Figure 7a: Displacement of the free surface at 6.0 MPa loading.. 1– Immediately after placing, 2 – One hour later.

Figure 7b. Velocity of the free surface, calculated from Fig. 7a. at short band.

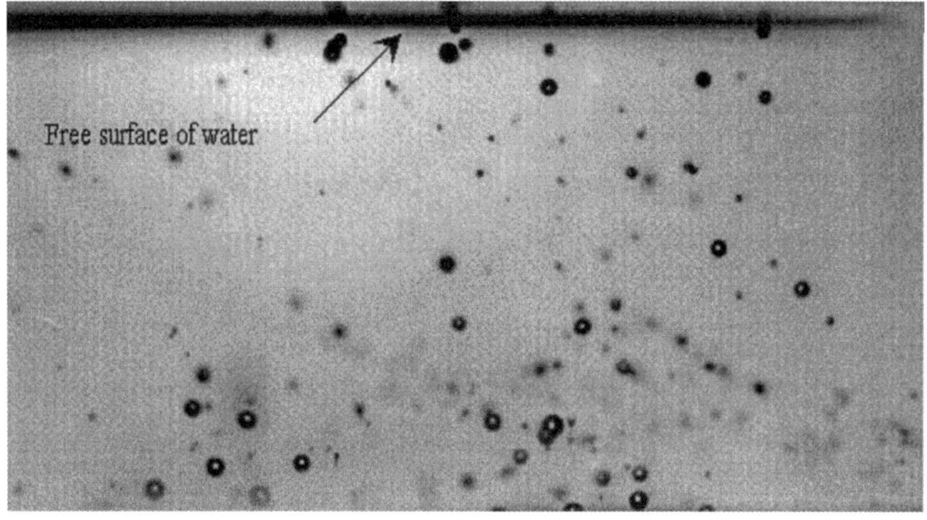

Figure 8a. Photo of cavity cluster in distilled water sample immediately after placing it into cuvette at the point of time marked on the graphs (Fig.7a, curve 1) as noise.

High-speed filming showed that in the first case, the bubble distribution over the volume is uniform, while in the second case a dense layer of bubbles is formed under the free surface (2.5–3 mm), which partly reflects the rarefaction wave. This wave continues moving freely in the formed liquid layer. As can be seen from the photographs, despite the apparent similarity in behavior, there is no evidence of spallation damage to the water. The effect of contact time is probably due to the change in the structure of a near-surface layer because it can be saturated with air, which can result in an increase in both the concentration and the average equilibrium radius of cavitation nuclei.

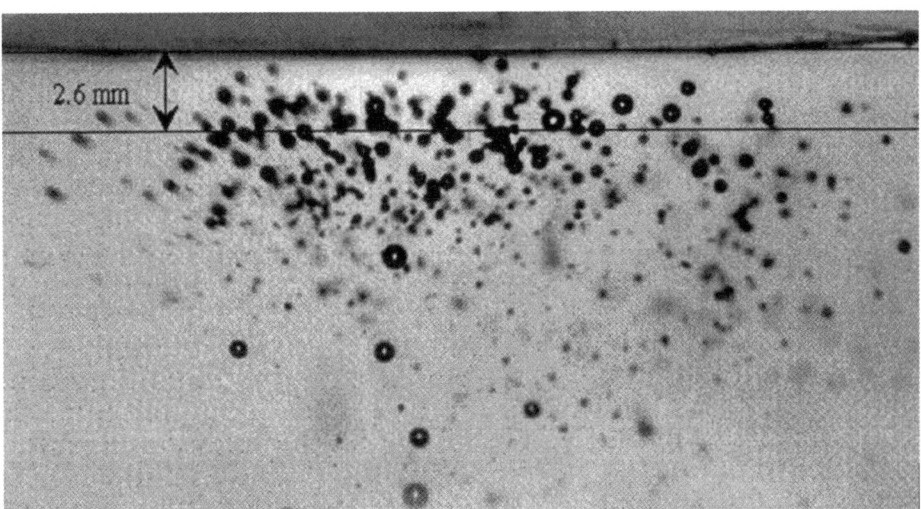

Figure 8b. Photo of cavity cluster in distilled water one hour later after placing it into cuvette at the point of time marked on the graphs (Fig.7a, curve 2) as noise.

7. Development of cavitation in the different liquids

To see the general picture of cavitation development some measurements with different liquids have been done (Fig.9a-e.). The smallest pressure corresponds to a voltage of 4 kV on the capacitor bank of the shockwave generator, the largest pressure corresponds to 8,5 kV. The displacement of the free surface increases with increasing pressure and for cavitation inception, the characteristic changes of the dynamics of the free surface can be seen clearly. The first sample to be investigated was water.

The threshold of cavitation inception for the water sample placed immediately into the cuvette (Fig.9a, curve 2) and 1 hour later (Fig.9b, curve 2) as measured by the capacitance method (details as in Fig.1.) is similar at − 2.7 MPa. At the second level of loading (2.7 MPa, Table 1) the appearance of the slope of the curves after completion of shock wave reflection can be seen. The further increasing of pressure leads to more significant changes and the appearance of oscillations in the curves (Fig.9b., curves 6,7,8,9).

The results of experiments conducted with Hexane provide more evidence of cavitation. According to the data presented in Fig.9c. cavitation inception definitely takes place only at amplitudes of the incident shock wave of between 10.4 MPa-12 MPa. Even at such intensive loading, the rate of development of the cavitation cluster in not so high as for water (Fig. 9a,b.). This may be explained by the low activity and concentration of cavitation nuclei.

The threshold of cavitation for an emulsion of butanol and octanol in water (Fig. 9d.) is similar to that found for pure water (Fig. 9a,b.), but the elevation of the free surface of the emulsion as a result of the growing cavitation cluster growing is less. The reason could be the high concentration of cavitation nuclei, because butanol and octanol arise as small droplets in the water. The mean size of these droplets is

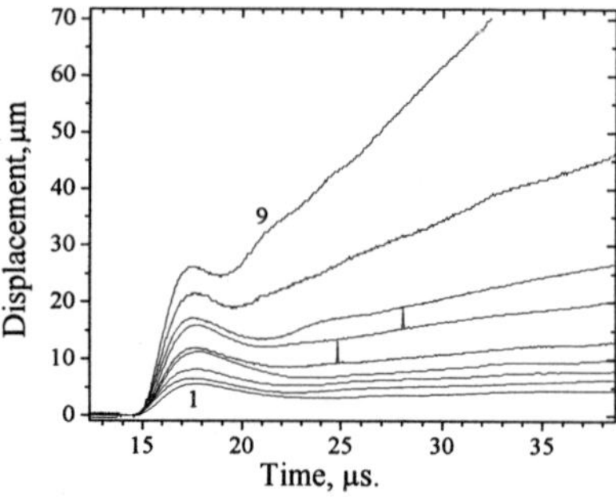

Figure 9a: Water (0h), 2.1 – 12.6 MPa

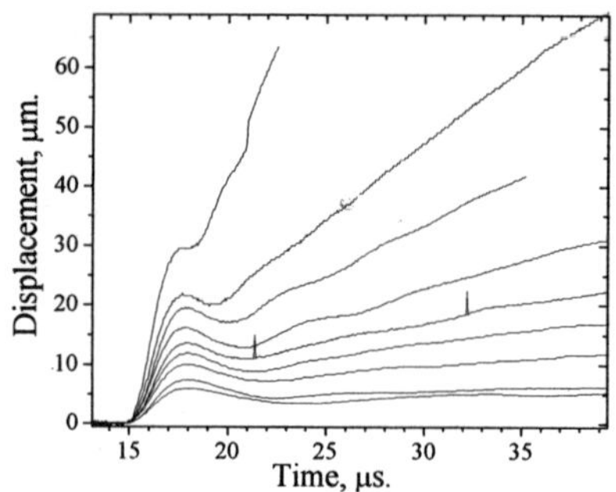

Figure 9b: Water (1h), 2.1 – 12.6 MPa

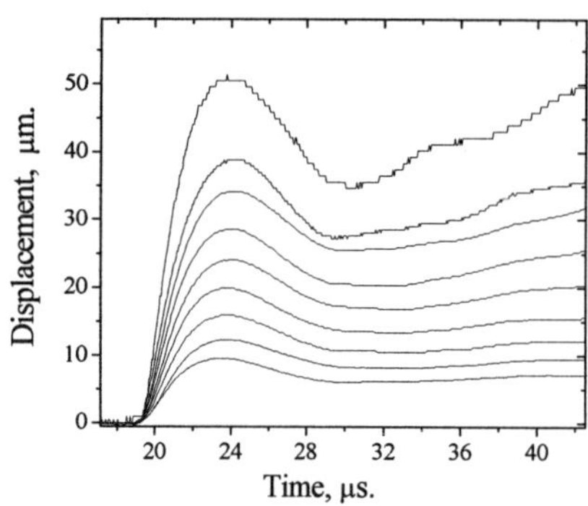

Figure 9c: Hexane, 3.1–16.5 MPa

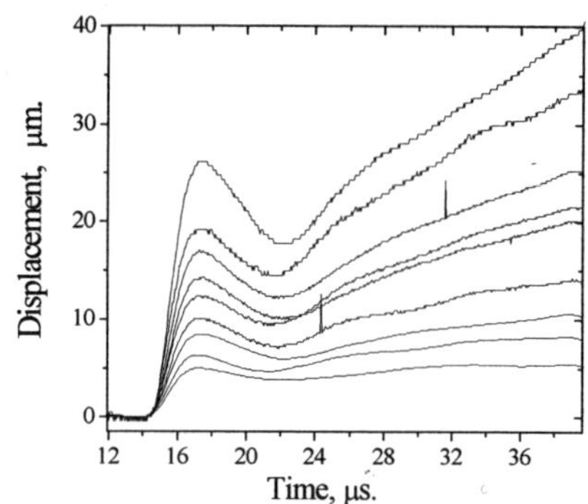

Figure 9d: Emulsion, 2.1–12.6 MPa

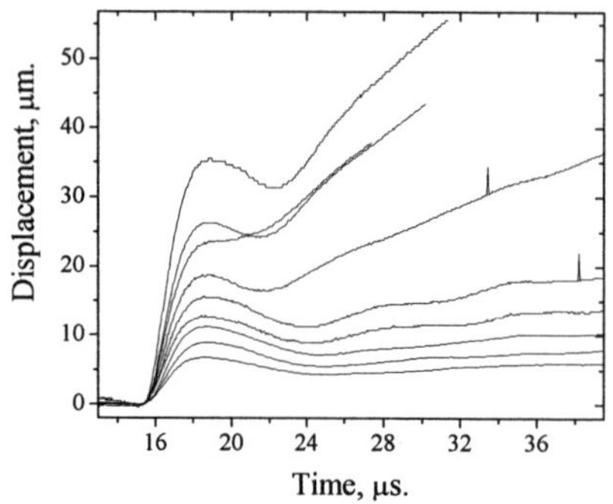

Figure 9e: Transformator oil, 2.3–12.3 MPa

Figure 9.a,b,c,d,e.: Measurement with different liquids of the displacement of the free surface as a function of pressure (a, c, d, e). The smallest displacement corresponds to the minimum pressure and the largest displacement corresponds to the maximum pressure. The name of the liquid, the minimum pressure and the maximum pressure are given for figures a, c, d, and e. For other values of the pressure see Table 1.

approximately 2 μm. The volume-concentration of the drops is about $4,19*10^{-4}$ and the number density is about 10^8 droplets per cm^3. A significant amount of energy associated with the shock wave can be lost in interfacial interactions during the growth of the bubbles. As for the cavitation strength of the emulsion, it seems to be determined by the properties of the cavitation nuclei within the water matrix.

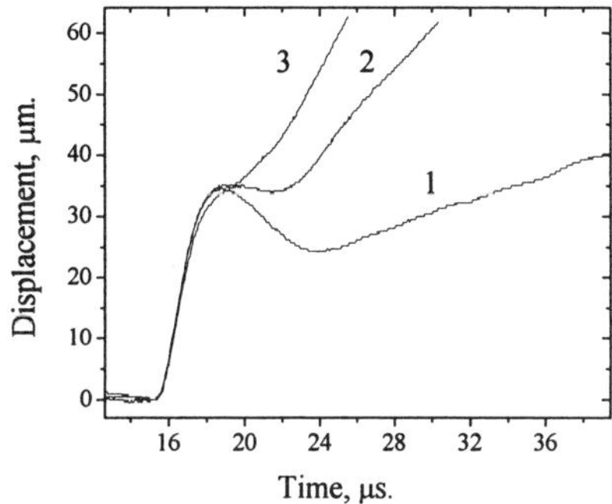

Figure 10: Displacement of the transformator oil free surface as a function of time. Loading with pressure 12.3 Mpa every 1 minute

Table 1: Values of the loading pressure pulses in different liquids.

Water and emulsion (MPa)	Hexane (MPa)	Transformator oil (MPa)
2.1	3.1	2.3
2.7	4.0	3.0
3.5	5.1	4.0
4.3	6.4	4.6
5.3	7.2	5.7
6.0	8.9	6.8
7.5	10.4	8.1
8.9	12.0	9.2
12.6	16.5	12.3

For transformer oil the threshold of cavitation inception lies between 4.0 MPa and 4.6 MPa (Fig.9e.). The most interesting point is the crossing of the curves of displacement of the free surface at the 8.1MPa and 9.2MPa pressure levels. Such behavior indicates a dependence on the previous loading, which changes the initial state of oil. The interval between loading was 3-5 minutes, which appears to be insufficient for relaxation of the structure of the oil sample to the initial state. The results of additional experiments confirm this point. It was shown experimentally that for motor oils, compressor oil and especially for transformer oil, the previous loading history and the time interval between the experiments are essential considerations. Because of the high viscosity of the medium, the formation of nuclei is cumulative, which leads to an increase in intensity of cavitation under multiple sequential loading. The result of one of these experiments is presented in Fig.10. The measurements were done with a pressure of 12.3 MPa. Measurements 2 and 3 were interspersed with one minute delays. (It is interesting to note that the capacitive method can be used even in the case of liquids which are not transparent).

8. Dependence of cavitation processes in water on temperature

Using the capacitive method, which determines the cavitation thresholds and intensity of cavitation processes with high accuracy, we are able to study the dependence of cavitation processes in water on temperature. It was shown that the intensity of cavitation has a marked dependence on temperature and the data obtained are presented in Fig. 11. Here curve 1 refers to the displacement of the free surface of water at 18.5^{0}C in the absence of cavitation; curve 2 is the derivative of it (zero slope indicates no displacement of the free surface). With increase in temperature from 18,5 ^{0}C (Fig. 11, curve 3) to 33^{0}C (curve 4) up to 42 ^{0}C (curve 5), the intensity of the cavitation processes (in terms of the slope of curves, which is equal to U = dX(t)/dt) increases. When the temperature reaches 52 ^{0}C (curve 6), the intensity drops down to the initial level, corresponding to 18,5 ^{0}C, while at 66 ^{0}C (curve 7), it increases abruptly again (curve 8 at 74^{0}C and 9 at 84^{0}C) even if initial intensity of applied shock wave is going down at fixed voltage at capacity bank. Equivalent increasing of intensity of applied shock wave does not change the general picture of this phenomenon and leads to more intensive cavitation processes above 52 ^{0}C.

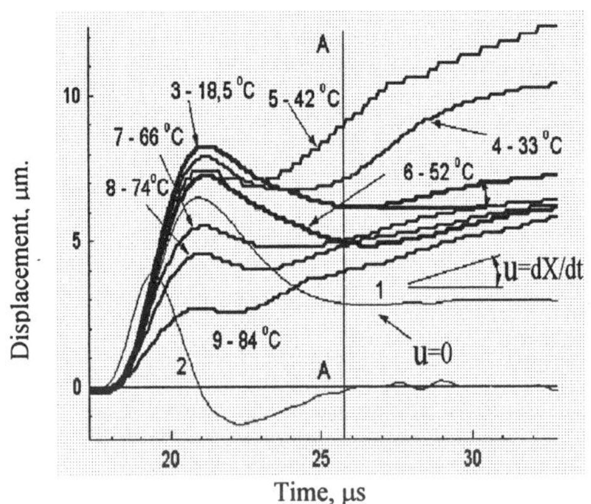

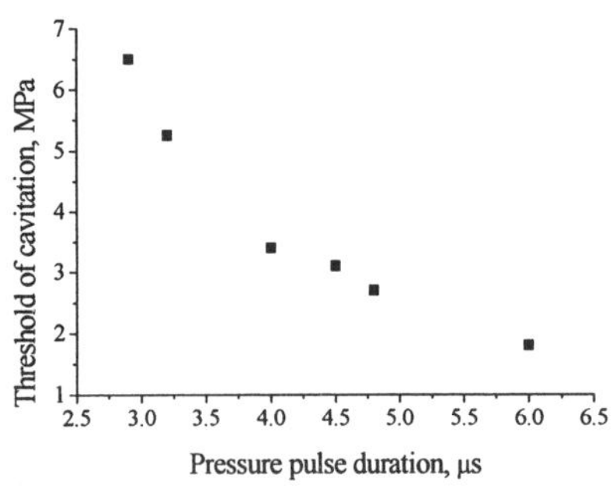

Figure 11. Temperature dependence of displacements of free surface of water.

Figure 12. Dependence of cavitation threshold in water on pressure pulse duration.

The mechanism of this phenomenon implies the occurrence of two competing processes: the first is the temperature-induced growth in the equilibrium radius of cavitation nuclei [16], which causes them to float and, hence, an increase in the cavitation strength of water; and the second is an increase in the convective flux rate, which hampers the floating of cavitation nuclei at a temperature of 66 ^{0}C and higher.

From consideration of Archimedes and Stokes force we can find the floating velocity $U=2\rho g R^2/9\eta$, where ρ –density, η – viscosity of liquid, R – radius of bubble, g – acceleration of gravity. Assuming that the gas inside the bubble is ideal, the number of molecules N is constant and do not change with the temperature it is possible to write down: $(P_0+2\cdot\sigma/R)\ 4\cdot\pi\cdot R^3/3 = N\cdot k\cdot T$, wherefrom, under condition $2\cdot\sigma/R >> P_0$, the bubble radius changes with increasing temperature as $R\sim T^{1/2}$, but if

$2 \cdot \sigma/R \ll P_0$ then $R \sim T^{1/3}$. Here k – Boltzmann's constant, P_0 –atmospheric pressure. For the stable nuclei $R_0 \approx 1.5$ µm, $2\sigma/R_0 \approx P_0$ and during forward growing of it, conditions $2 \cdot \sigma/R \ll P_0$ and $U \sim T^{2/3}$ will take place. The result means that the rate of growing of convection flow velocity, which is proportional to temperature [16], is greater than the rate of the immersion speed of the bubbles and finally one will exceed another. According to [16], for similar conditions, the convection flow speed at 40-60 ^{0}C exceeds a few mm/s. It is sufficient to involve the bubbles with size 10-30 µm in circular motion and keep them inside the cuvette. This could be the reason for the abrupt increasing of the absorption of the shock wave at a temperature of 66 ^{0}C and higher. It is interesting to point out that the maximum of the cavitation erosion also takes place in vicinity of 52 ^{0}C [17]. An additional consideration is the link to the compressibility of water which decreases with increasing temperature up to 50 ^{0}C [18].

9. Dependence of cavitation thresholds in water on pulse duration

Additional experiments involving the dependence of cavitation thresholds on the duration of pressure pulse were conducted. These experiments established that there is a strong dependence of cavitation threshold on the duration of the pressure pulse (Fig. 12.). Increasing the length of the pressure pulse by as much as a factor of two leads to a decrease of the cavitation threshold by more than a factor of four. These findings are in agreement with the conclusions of other work [19] i.e. that the development of cavitation depends strongly on the rarefaction wave front.

10. Conclusions

1.   It is shown that the repeatability and precision of the capacitive method makes it a valuable tool in investigation of thresholds and cavitation dynamics in various liquids when short plane shock waves are reflected from their free surfaces. It is shown that the free-surface dynamics reflects the cavity cluster dynamics not only qualitatively but also quantitatively.

2.   A quantitative ratio of the free surface velocity to the volume concentration α of the bubbles formed in the rarefaction wave is found

3.   An effect of the contact time between the studied samples of water and atmosphere on the dynamics of cavitation development is reported. This effect is probably due to the change in the structure of a near-surface layer because the latter is saturated with air, which can result in an increase in both the concentration and the average equilibrium radius of cavitation nuclei.

4.   The cavitation thresholds of different liquids under the same circumstances were measured. It was determined experimentally that cavitation strength of hexane against pulse loading almost three times excel it for water.

5.   It is shown experimentally that for transformer and motor oil, the previous loading history and the time interval between the experiments are essential. Because of the high viscosity of the medium, the formation of nuclei is accumulative, which leads to an increase in the intensity of cavitation under multiple sequential loading.

270

6.   By means of capacitance method the temperature dependence of intensity of cavitation processes in water is investigated. It is shown that it is an extreme dependence. With increase in temperature from 18,5 ^{0}C to 42 ^{0}C, the intensity of cavitation processes increases. When the temperature reaches 52 ^{0}C, the intensity drops to the initial level, corresponding to 18,5 ^{0}C, while at 66 ^{0}C, it increases abruptly again. A mechanism which accounts for this phenomenon is proposed.

7.   It is shown experimentally that there is a strong dependence of the cavitation threshold on the duration of the pressure pulse. Doubling the length of the pressure pulse results in a fourfolddecrease in the cavitation threshold of water.

References

1.   Briggs, L.J., (1950) Limiting negative pressure of water, *Journal of Applied Physics* **21**, 721-722.
2.   Briggs, L.J., (1955) Maximum Superheating of Water as a Measure of Negative Pressure, *Journal of Applied Physics* **26**, p.1001–1003.
3.   Bull, T.H., (1956) The tensile strengths of viscous liquids under dynamic loading, *British Journal of Applied Physics* **7**, p.416-418.
4.   Knapp, R.T., Daily, J.W., Hammit F.G. (1970) *Cavitation*, McGraw-Hill Book Company, New York, London, etc.
5.   Besov, A., Kedrinskii, V., (1994) Dynamics of bubbly clusters and free surface at shock wave reflection *Proc. of Int. Symp. on Bubble Dynamics and Interface Phenomena*, Birmingham, UK, , p.93-103.
6.   Bogach, A.A. Utkin, A.V. (2000) Strength of Water under Pulsed loading, *Journal of Applied Mechanics and Technical Physics*, Number 4, p. 752-758.
7.   Fox, F.E., Herzfeld, K.F. (1954) Gas bubbles with organic skin as cavitation nuclei, *Journal of Acoustic Society of America*, 26, No.6, November.
8.   Besov, A.S., Bichenkov, E.I., Kedrinskii, V.K., Palchikov, E.I. (1984) Investigation of Initial Stage of Cavitation by Diffraction Optic Method. *IUTAM Symposium on Optical Methods in Dynamics of Fluids and Solids (Czechoslovakia, Liblice Castle, September 17- 21)*, Ed. Pichal M. -Berlin, Heidelberg, New York, Tokio: Springer-Verlag.P.129-135.
9.   Besov, A., Kedrinskii, V. Palchikov, E. Palchikov, E. *(1989)* On Threshold Cavitation Effects in Pulse Rarefaction Waves. *Proc. of 13-th Int. Congress on Acoustics.- Yugoslavia.- Belgrad. P. 355- 358.*
10.   Besov, A., Berngardt, A. Kedrinskii, V., Pal'chikov, E. (1985) Diffraction optic and x-ray techniques of cavitation research, *J.Acoust. Soc. Am. Suppl.1*,Vol.77, spring.
11.   Boren, C., Huffman, D. (1983) *Absorption and scattering of light by small particles*, John Wiley & Sons, Inc., New York.
12.   Besov, A.S., Zaitcev, B.B. (1997) Investigation of the initial stage of cavitation fracture of suspensions in pulse rarefaction waves. *Proc. of 4-th Seminar of SNG on Acoustic of heterogeneous medium*, Novosibirsk, 28-31 of may 1996. Dynamics of continuous medium **112**, p. 43-54, (in Russian).
13.   Kedrinskii, V., Plaksin, S., (1984) Interaction between nonstationary shock wave and free surface in real liquid, Proceedings of 10th Intern. Sympos. on Nonlinear Acoustics, 24-28 July, Kobe, Japan, pp.181-184.
14.   Hulst V. (1957) Light scattering by small particles. John Wiley and Sons, Inc. New York, Chap 9,10.
15.   Kanel, G., Razarenov, S., Utkin, A., Fortov, B. (1996) *Shock-wave phenomenon in condensed mediums.* Yanus-K, Moscow, 154-155.
16.   Besov A.S., Kedrinskii V.K., Palchikov E.I., Matsumoto Y., Ohashi H. (1992) Microinhomogeneity Structures and Histeresis Effects in Cavitating Liquid, *Internat. Aqoustic. Congr.* China.
17.   Leontiev A.I., Kirdyashkin, A.G. (1968) Experimental investigation of free convection in horizontal and vertical layers of liquids, *Proc. of Third USSR Symp. on heat and mass transfer.* V. 1, Moscow, p. 661-664.
18.   Flynn, H. G. (1964) Physics of acoustic cavitation in liquids, in W.P. Mason (ed.), *Physical Acoustics Principles and Methods,* Vol.1, Part B, Academic Press New York and London.
19.   Zatsepina, G.N. (1987) Physical properties and structure of water, Moscow State University.
20.   Kedrinskii, V.K. (1975) *Journal of Applied Mchanics and Technical Physics*, Number 5, p. 68-78.

RELAXATION EFFECTS AND DISINTEGRATION PROBLEMS OF CAVITATING LIQUIDS AT PULSE LOADING

V. K. KEDRINSKII
Lavrentyev Institute of Hydrodynamics, Siberian Branch of the RAS
Novosibirsk, 630090, Russia

1. Introduction.

Paper presents the results of experimental and numerical studies of bubbly cavitation development as well as of tensile stress structure arising in a liquid under dynamic loading.

Experimental studies of liquid macrostructure have shown that a real liquid contains a lot of microinhomogeneities as microbubbles of free gas, solid microparticles and their combination which can play the role of cavitation nuclei [1]. So the notion of a two-phase system can be applied to real liquids and hence a dynamics of tensile stresses and a cavitation development can be described by corresponding two-phase mathematical models [2].

A two-phase approach to a description of liquid state dynamics under negative pressure has allowed a number of principal questions to be resolved. In particular, these questions are relevant to magnitude of maximum negative pressure which can be admitted by a real liquid and to the estimation of relaxation times of tensile stresses in cavitating liquids [3,4].

If pulse tensile stresses applied are rather intense a cavitation zone development is characterized by practically unlimited inertial expansion of bubbles to the stages when a cavitating liquid is transformed into a structure of foam type and then into a gas-droplet system [5]. Special experimental methods and set-ups such as the shock tube and pulse x-ray technique were applied to study these effects [6].

The model of instantaneous relaxation was suggested to calculate the later stages of cavitation development [7]. The approach made it possible to study the dynamics of mean density of cavitation zone and distribution of mass velocities in it. Numerical studies have shown that mass velocity profiles are preserved during an evolution of cavitation zone. This effect is in a good agreement with the experimental data obtained by the method of lead foil traces and pulse X-ray photography [8].

The process of pulsed fracture of a liquid under negative pressure pulse as an inversion of its two-phase state, with transition from a "liquid-bubbles"- to a "gas-droplet"-structure, was studied using the effect of disintegration of liquid drop as a result of its interaction with a ultrashort shock wave [9].

All these problems involve hydrodynamic aspects of this workshop topic "Liquids under negative pressure."

271

A.R. Imre et al. (eds.), Liquids Under Negative Pressure, 271–283.
© 2002 *Kluwer Academic Publishers. Printed in the Netherlands.*

272

2. State of real liquids.

We study an origin and evolution of so called bubbly cavitation in samples of distilled or settled water as an example of real liquid with controlled properties. A negative pressure is created in a liquid in a result of underwater shock wave reflection from a free surface of liquid sample (fig.1). Underwater shock waves are generated in a result of underwater explosions of High Explosive charges (HE) or in hydrodynamic shock tubes. It must be noted that the rarefaction wave parameters are determined by the corresponding parameters of incident shock waves, for example, by a steepness of a shock wave front, its maximum amplitude and profile. It's well known that rarefaction wave is a reflection of the shock wave profile until a cavitation threshold arises.

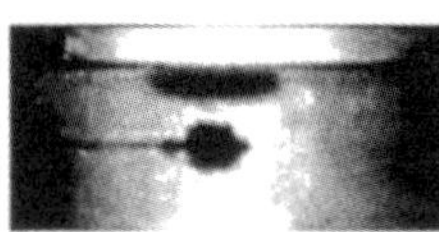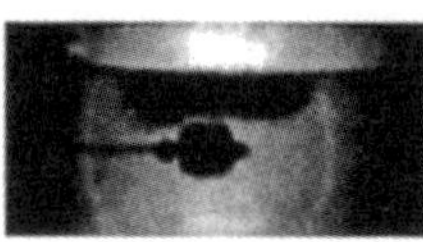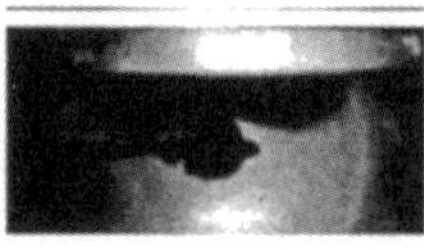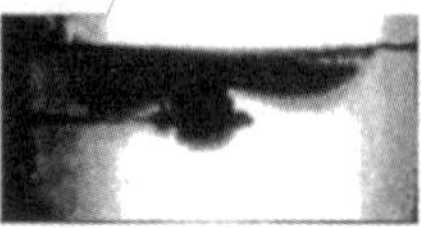

Figure1. Reflection of shock wave (white spherical front) from plane free surface and bubbly cavitation zone (black one) evolution for time instants 16, 32, 48 and 64 μs after shock wave reaches free surface.

Fig.1 shows a few sequential frames of high speed registration of cavitation development near free surface at underwater explosion of 1.2 gr charge of HE (fig.1, black cavity with detonation products) on depth of 5 cm under free surface. Studying later stages of this process we have discovered unexpected effect : cavitating layers (see black hands in fig.2) has detached from main cavitation zone (fig.2, black zones in lower part of frames) [10,11]. This effect is similar to well known effects of brittle solid fragmentation under strong shock wave loading. In the case of liquid sample these layers have a foam-like structure which is later transformed in a liquid drop system and in result we have so-called "gas-droplet" two-phase system. We determine this process as an inversion of two-phase state of liquids under negative pressures. To understand a dynamics of liquid state during an inversion process we need to solve some problems.

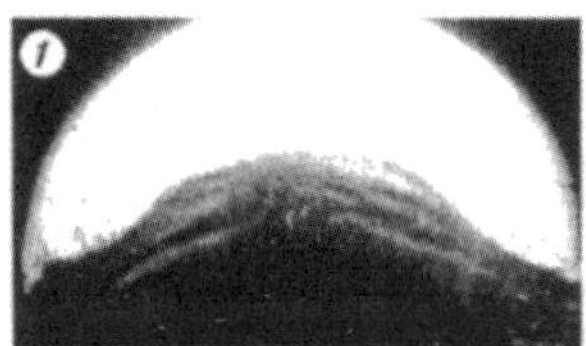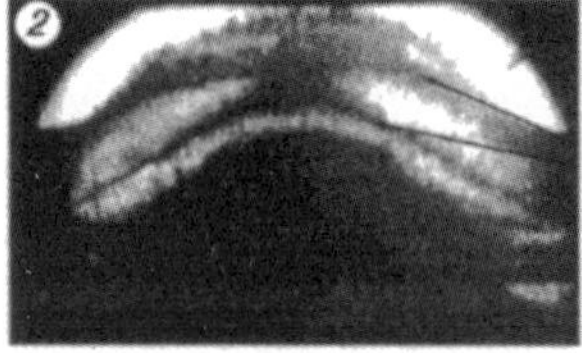

Figure 2.. Two sequential frames from high-speed registration of cavitation zone fragmentation on separate layers (shown in frame 2 by arrows) at underwater explosion near free surface (2D-statement, plane case).

What is the initial state of a liquid ? What is the mechanism of bubbly cluster development ? What kind of mathematical models can be suggested to describe both cavitation development and wave processes in cavitating liquid ? And finally, what is the mechanism of inversion of two-phase liquid state ?

It is well known that magnitude of negative pressure required for an isothermal fracture of liquid is equal approximately to 10^4 atm. But experience shows that such a value for normal liquids is unreal. The reason is very simple: liquids always contain microinhomogeneities which play a role of cavitation nuclei. Until recently the acoustic and other methods of diagnostics predicted rather low density of microinhomogeneities, in particular, such as microbubbles of free gas in water at normal conditions. According to Gavrilov's data, [12], the bubble radii can occupy a value interval $R \approx 0,5 \div 50\,\mu m$ at their density about 1 bubble per cm^{-3} and volumetric concentration $k_0 \approx 10^{-8} \div 10^{-12}$. The experimental data of Hammit et al, [13], gives $3 \div 6\,\mu m$ and $1 \div 100\,cm^{-3}$ correspondingly. Practically all results on density concern only gas nuclei.

But the experimental studies carried out by Sirotyuk [14] have shown that a bubble density in a focus of ultrasonic concentrator during a cluster development can be increased by 5-6 orders. That was a base to suggest a so-called "ultra-sonic pumping" mechanism of the avalanch-like nuclei multiplication as a result of bubble shape instability at collapse, and their fracture into separate fragments, each of which was considered as a new nucleus. The experimental data mentioned seemed to confirm this model: a number of bubbles in a focus zone increases from period to period of ultrasonic field oscillation. However such an explanation of high density of bubbles contradicts the experimental data presented in fig.1 where we can observe the development of cavitation zone (cluster) as a result of action of a solitary rarefaction wave which represents a single negative pressure pulse.

To resolve this contradiction detailed experimental studies were carried out using the methods of light scattering on inhomogeneities and of shock tubes [1]. The light scattering of He-Ne laser with $\lambda = 0,63\ \mu m$ was studied in a cylindrical plexiglass tube filled with distilled water. An angle distribution of a scattering light intensity was measured by the photomultipliers in the static and dynamic regimes. In the first case for the angle resolution a telescope with a diaphragm was used and the special cover plate of plexiglass was mounted on the outside of a tube. To reduce an undesirable scattering the probing beam was put in the liquid sample through special diaphragms. The polarization plane of a probing beam has been taken to be perpendicular to the observation plane of a scattered light.

The stationary scattering indicatrix was characterized by zero maximum, one explicit lateral maximum at $17,5^0$ and by two smooth maxims which were registered in the vicinity of 28^0 and 37^0. These results correspond to the distribution of intensity induced by a spherical particle about $1,5\,\mu m$ in radius. The indicatrix smoothness was explained by a presence of particles distributed by sizes in an interval about $1,3 \div 1,7\,\mu m$. The number of scattering centers was estimated according to the relation of measured intensities of scattered and incident light. It turned out to have order $10^5\,cm^{-3}$. This result ($10^5 \div 10^6\,cm^{-3}$) for distilled water was confirmed by measurement of a number of microinhomogeneities of any origin as the density of diffraction spots in the liquid volume illuminated by a laser beam.

To prove a presence of gas microbubbles (compressible particles) among the scattering centers in a liquid the shock wave generator was used and two directions for photomultipliers were chosen so that, during propagation of shock wave above a sample studied, the intensity of scattered light has to be increased in one direction but to be decreased in another when gas microbubbles collapsed under the shock wave. A calculation has shown that angles of 10^0 and 15^0 correspond to the conditions mentioned. Indeed, the experimental oscillograms show the antiphase changes of two photomultiplier signals. Detailed experimental studies have shown that real water (settled or distilled one) contains microinhomogeneities as microbubbles of free gas, solid microparticles and their combination (fig.3) [1].

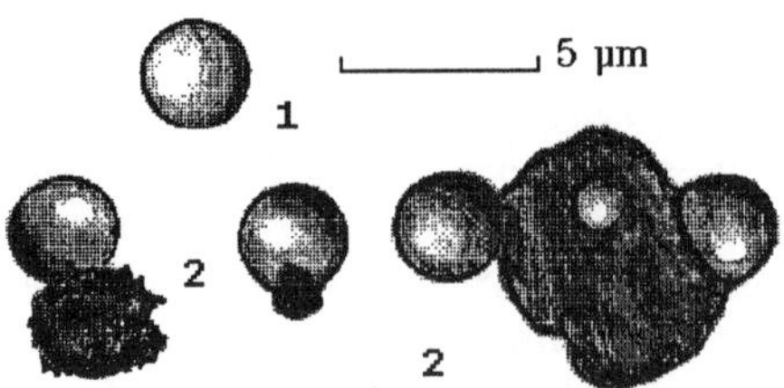

Figure 3. Structure of microinhomogeneities: (1) is the free gas bubble, (2) is the combined structures.

The experimental data received in refs. [15], [13] and [1] were generalized as the theoretical spectrum of cavitation nuclei $N_i(V_i)$ [4]:

$$N_i \approx N_0 \cdot (\frac{V_i}{V_*})^2 \cdot \frac{1}{[1+(V_i/V_*)^4]}$$

where index "star" corresponds to N_i^{max}. Taking into account that a vapor cover can be formed on a solid microparticle surface under negative pressure, we can consider that a distilled water contains vapor-gas nuclei about $N_0 \simeq 10^5 \div 10^6$ cm^{-3} having the initial radius $R_0 \simeq 1,5$ μm and volumetric concentration $k_0 \simeq 10^{-8} \div 10^{-12}$ [16].

So the notion of two-phase system and a corresponding two-phase mathematical model can be applied to real liquids. Two-phase approach allows a number of principal problems such as describing a dynamics of tensile stresses, wave processes in cavitating liquids and bubbly cavitation to be resolved.

3. Tensile stress relaxation in cavitating liquids.

Two-phase mathematical system of equations for average pressure p in cavitating liquid and relative volumetric concentration of gas phase k has the following form:

$$\Delta p - c_l^{-2} \frac{\partial^2 p}{\partial t^2} = -\rho_l \cdot k_0 \cdot \frac{\partial^2 k}{\partial t^2} \, , \tag{1}$$

$$\frac{\partial^2 k}{\partial t^2} = \frac{3\,k^{1/3}}{\rho_l\,R_0^2}(p_0 k^{-\gamma} - p) + \frac{1}{6k}(\frac{\partial k}{\partial t})^2 \qquad (2)$$

where $k = k_b / k_0 = (R/R_0)^3$, k_b is the current vapor-gas phase concentration, p_0 is the initial pressure inside cavitation bubbles, γ is the adiabatic index.

Numerical simulation of cavitation zone development near free surface at underwater explosions (in an axisymmetrical statement) shown in fig.1 was carried out within the framework of two-phase system of equations (1)-(2) for the following initial data: $k_0 = 10^{-11}$, $R_0 = 0,5\ \mu m$, $Q = 1,2\ g$ is the charge weight, $h = 5,3\ cm$ is the depth of charge location, $p(0) = 40\ kbar$ is the initial pressure in a cavity with detonation products (fig.4, extending spherical cavity below cavitation zone). A charge explosion was simulated by an instantaneous detonation at constant volume of charge. The notion of visible size of cavitation bubbles was introduced, i.e. such size which can be detected by the experimental methods used. In our case $R_{vis} = 0,1\ mm$. The initial parameters of the incident rarefaction wave were determined by the superposition method from an imaginary source. The model began to work in any point investigated only when a rarefaction wave arose there.

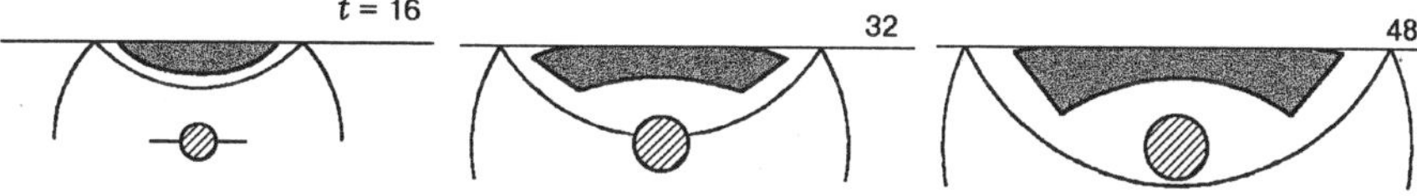

Figure 4. Numerical simulation of experimental data presented in fig.1 (time in μs)

Calculations show that in the zone of developing cavitation, the k value increases by 7-8 orders under negative pressure. It is clear that rarefaction wave parameters will be changed (fig.5). The numerical results presented in Fig.5 confirm this assumption [3]. Three profiles of shock waves "registered" in points $r = 2,5$, $3,5$ and $5R_{ch}$ on the axis including a rarefaction phase (dotted lines for one-phase model) are presented in Fig.5. According to two-phase model, rarefaction waves are very fast absorbed by developing cavitation zone (fig.5, solid lines for negative phases in points mentioned above).

Within the framework of system (1)-(2) we can estimate a relaxation time of a tensile stresses (in the vicinity of $t \to 0$). We assume that terms $(\partial k / \partial t)^2 / 6k$ as well as $k^{-\gamma}$ can be neglected, liquid layer is sufficiently thin to consider that $k \neq k(z)$ and liquid component of two-phase medium may be considered incompressible.

In this case the system mentioned is reduced to the system (3)-(4) where the signs are changed in both equations as the dynamics of negative pressure in cavitation zone is analyzed [4]:

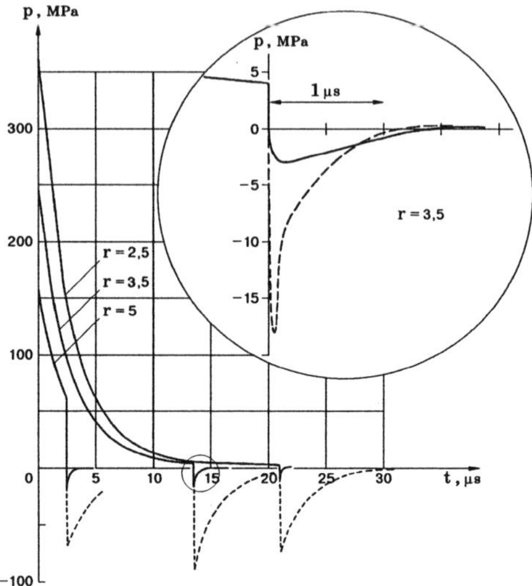

Figure 5. Profiles of shock waves including the rarefaction phases for the different points on a symmetry axis for different steepness of front τ ; ring frame: $\tau = 0,1\,\mu s$ - dotted line, $\tau = 1\,\mu s$ - solid one.

$$\frac{\partial^2 p}{\partial z^2} = \rho_l \cdot k_0 \cdot \frac{\partial^2 k}{\partial t^2} \quad , \tag{3}$$

$$\frac{\partial^2 k}{\partial t^2} = \frac{3\,k^{1/3}}{\rho_l\,R_0^2}\,p \quad . \tag{4}$$

Substitution (4) in (3) gives

$$\frac{d^2 p}{d\varsigma^2} = p \quad , \tag{5}$$

where $\varsigma = z \cdot \alpha \cdot k^{1/6}$ and $\alpha = \sqrt{3k_0 / R_0^2}$. Solution of eq.(5) is $p = b \cdot \exp(-\varsigma)$.

Let's consider the process of rarefaction wave formation when a piston is pulled out with acceleration $a(t)$ from a tube filled with water. In the vicinity of a piston the tensile stresses arise. Then the boundary conditions will be determined in the following form

$$\frac{\partial p}{\partial z}\Big|_{z=0} = -\rho_0 \cdot a(t) \tag{6}$$

Hence,

$$p = \frac{\rho_0 \cdot a(t)}{\alpha \cdot k^{1/6}} \cdot e^{-zak^{1/6}} \tag{7}$$

After substitution (7) in (2) the later is converted to simple and integrated form. For case $z = 0$ it takes the form

$$\frac{d}{dt}(k^{-1/6} \cdot \frac{dk}{dt}) = \frac{\sqrt{3} \cdot a(t)}{R_0 \cdot \sqrt{k_0}} \tag{8}$$

The eq.(8) solution is

$$k^{1/6} = (1 + \frac{5}{6} \cdot \frac{\sqrt{3}}{R_0 \cdot \sqrt{k_0}} \cdot \int_0^t \int_0^t a(t)dt)^{1/5} \tag{9}$$

Two cases are considered: a) acceleration $a = a_* = const$, b) $a(t) = a_* \cdot (t/t_{fr})$, where t_{fr} is the steepness of rarefaction wave front [4].

a) If the acceleration is constant the solution for $k^{1/6}$ and correspondingly for relaxation time of negative pressure in cavitating liquid (for point $z = 0$) can be determined by the following formula :

$$k^{1/6} = (1 + 0.72 \cdot \frac{a_*}{R_0 \sqrt{k_0}} \cdot t^2)^{1/5} \tag{10}$$

$$p = \rho_0 \cdot a_* \cdot \alpha^{-1} \cdot (1 + 0.72 \cdot \frac{a_*}{R_0 \sqrt{k_0}} \cdot t^2)^{-1/5} \tag{11}$$

Initial amplitude of negative pressure is $p_* = \rho_0 \cdot a_* \cdot \alpha^{-1}$. Then a relaxation time τ_* (when $p = p_* / e$) is determined as

$$\tau_* \approx (200 \cdot \frac{R_0 \cdot \sqrt{k_0}}{a_*})^{1/2} \tag{12}$$

For example for $R_0 = 10^{-4}\, cm$, $k_0 = 10^{-10}$ and $a_* = 5 \cdot 10^7\, cm/s^2$ (the data correspond approximately to $-300\, bar$ in a front of rarefaction wave) the relaxation time will be very small: $\tau_* \approx 6 \cdot 10^{-8}\, s$.

b) The steepness (t_{fr}) effect of rarefaction wave front can be simulated by the linear function acceleration $a(t) = a_* \cdot (t/t_{fr})$ for time interval $t = 0 \div t_{fr}$. In this case, eq.(13) allows the dynamics of negative pressure amplitude to be analyzed

$$p = \rho_0 \cdot a_* \cdot \frac{t}{t_{fr}} \alpha^{-1} \cdot (1 + 0.24 \cdot \frac{a_*}{R_0 \cdot t_{fr} \cdot \sqrt{k_0}} \cdot t^3)^{-1/5} \tag{13}$$

The case $t = t_{fr}$ corresponds a limit value of tensile stress in cavitating liquid. The calculations carried out have shown that steepness of a rarefaction wave front restricts essentially a magnitude of maximum negative pressure which can be admitted by a real liquid. For example, according to eq. (13) for $t_{fr} = 1\,\mu s$, $0,1\,\mu s$ and $0,01\,\mu s$ the limit negative pressure can reach only -44, -111 and $-280\,bar$, instead of $-300\,bar$ applied to liquid. We can conclude that the relaxation time is very short, characteristic times of cavitation development turn out to be much more than relaxation time and hence cavitation process is developing mainly *inertially*.

4. Relaxation of average density of cavitation zone at pulse loading.

On the basis of the results obtained, the model of instantaneous relaxation (MIR) of tensile stresses in a cavitation zone was developed [7]. According to this model, when the stresses decrease to a saturated vapor pressure the latter becomes constant and uniform in all cavitation zone behind a rarefaction wave front. Bubbles in such zone are growing inertially up to a formation of a foam type structure. This problem was considered within the framework of MIR. Let's consider a charge of HE (initial radius is r_0) surrounded by a liquid shell (radius r_1). It assumed that $R_o = 1,5\,\mu m$, a density of microbubbles is equal to $N_0 = 10^4\,cm^{-3}$ (corresponds to a value of $k_0 = 10^{-7}$). A detonation wave is initiated by explosion of a charge and reaches the interface "charge-shell" at the instant $t = 0$.

The calculations were performed for a charge of TNT/RDX (cylindrical symmetry) with a specific volume $v_* = 6,25 \cdot 10^{-4}\,m^3/kg$, specific explosion energy $Q = 4,88 \cdot 10^3\,kJ/kg$ for detonation parameters: pressure $p_j = 2,19 \cdot 10^4\,MPa$, specific volume $v_j = 4,7 \cdot 10^{-4}\,m^3/kg$ and mass velocity $u_j = 1,85 \cdot 10^{-3}\,m/s$. Three main characteristics were studied : a distribution dynamics of average mass velocity and density as well as characteristic times τ^* required for bubbles to reach the "bulk density" $\left(k^* \approx 0,5 \div 0,7\right)$. The numerical analysis shows that: 1) the liquid shell structure is characterized by two zones - external cavitation zone and internal zone which bounds with detonation products ; 2) the mass velocity profile in cavitation zone is "frozen" ($u = u(\xi)$=const, where ξ is the Lagrangian coordinate); 3) the $k \approx k^*$ value is achieved when the outer radius of the shell has a value $r \approx 1,5 r_1$ and the corresponding time is $t \approx 12\,\mu s$ after reflection of the shock wave from the free surface [17].

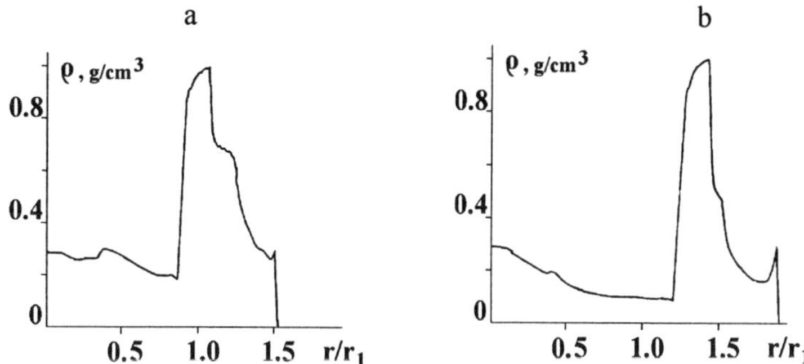

Figure 6. Density distribution in cavitation zone for t =16 (a) and t =24 µs (b).

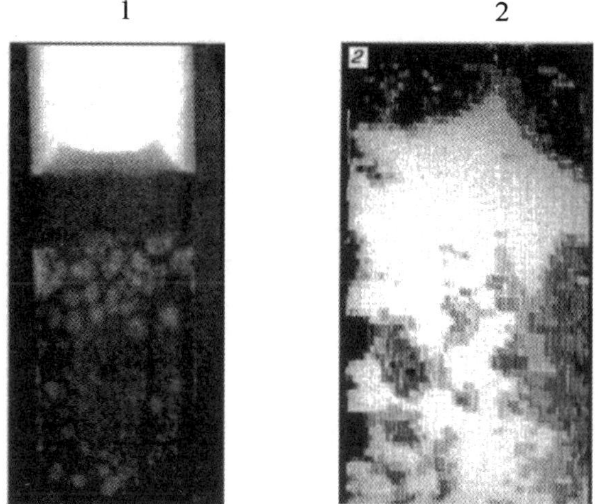

Figure 7. Inner structure of cavitation zone: X-ray instant picture (1) and its computer version (2).

Fig.6 shows the distribution of density in detonation products and in both zones of liquid shells where density between them is changed by leap. It is important to note that the cavitation zone consists of two parts which differ by density gradients (fig.6a, for t =16 μs) which are smoothed when the maximum value of ρ in a cavitation zone tends to 0,4 (fig.6b).

As mentioned, the later stages of cavitation processes in intense rarefaction waves were studied by shock tube and pulse X-ray methods described in detail in [6]. The shock wave was initiated in a liquid by impact of a piston. The registration system was synchronized with three X-ray apparatus which were triggered by pressure gauge signals with different assigned lags. These methods allowed ultra-high-speed X-ray pictures of the inner structure of the cavitation zone to be obtained. The mean photon energy was about 70 keV and radiation pulse duration 80 ns. Computer processing of density distribution of X-ray negatives allowed the experimental data on the structure dynamics of the cavitation zone to be reconstructed. We can compare (Fig.7) the X-ray photo (1) and the computer version (2) of the liquid sample structure for instant 800 μs after   shock wave reflection from its free surface. Note that here the computer picture presents   a half of cavitation zone (its left boundary is the axis, right one is the tube wall).

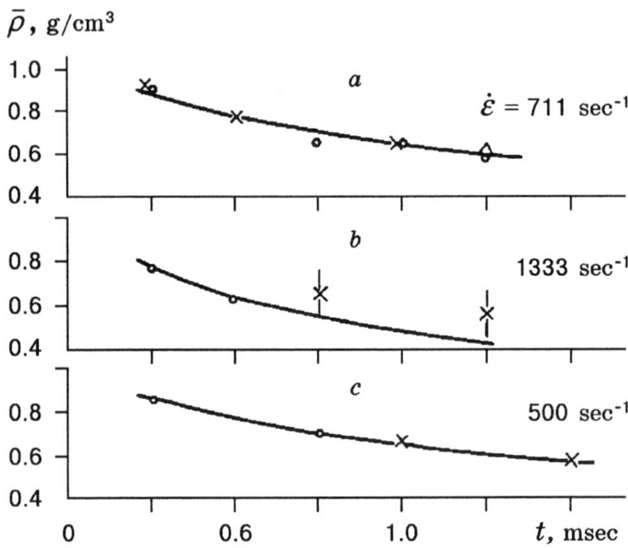

Figure 8. Relaxation of average density in cavitation zone for different strain velocity (solid lines are the theoretical data).

The results of high-speed photography involving the X-ray system, together with computer-processing of the X-ray negatives, allow us to image the internal structure of the opaque cavitation zone, as well as obtain unique data pertaining to the dynamics of its local and average density (fig.8). The analysis of average density dynamics shows (fig.8) that it relaxes to the state of bulk density for the times τ_{rel} which depend on a strain velocity $\dot{\varepsilon}$: $\tau_{rel} \simeq 700\,\mu s$ for $\dot{\varepsilon} = 1330\,s^{-1}$ and $\tau_{rel} \simeq 2000\,\mu s$ for $\dot{\varepsilon} = 500\,s^{-1}$ [18]. Comparison of the theoretical data   received within the framework of MIR (fig.8, solid lines) according to formula (14)

$$\frac{1}{\bar{\rho}} = \frac{1}{\rho_0}\cdot[1+\dot{\varepsilon}(\xi)(t-\frac{\xi}{c_0})] \tag{14}$$

confirms the possibility for MIR to provide real estimates of the main characteristics of dynamic state of cavitation zone.

As mentioned, the cavitation fracture of free boundary liquid under pulse loading is a problem of unbounded inertial development of bubbly cluster and the inversion problem of two-phase state of cavitating liquids. The process of cavitation fracture of a liquid can be presented by the following main stages: 1. the formation and dynamics of cavitation clusters, 2. their transformation into a foam-like structure, 3. transition from a foam-like structure to a gas-droplet system. All these stages are characterized by relaxation effects, namely: the relaxation of tensile stresses (lots of microseconds), relaxation of an average density of cavitation zone to a bulk density (from tens microseconds to millisecond, depending on strain velocity) and relaxation process of transition from a foam-like structure to a droplet structure (milliseconds) .

An inversion of two-phase state of cavitating liquid as a physical mechanism of transition from a state of foam-type to a gas-droplet system is one of the principal problems in the investigation of the process of cavitation fracture. The experimental set-up for its studies includes an electromagnetic shock tube, a pulse light source, a system of delay and a high-speed camera. A drop of distilled water (about 1cm in size) was placed on a diaphragm of a shock tube (fig.9(1)). A shock wave (5-6 MPa amplitude and 3-4 μs duration) is generated in drop by pulse magnetic field which pushes the diaphragm. In a result of shock wave reflection from free surface of drop the pulse field of negative pressure arises and dense cavitation zone develops inside the drop (fig.9, (3)) over 50-60 μs .

Subsequently the cavitating drop is transformed into the structure of the liquid mesh type (fig.9b). The decomposition of the mesh cells by separate jets and then by drops can be considered as an inversion process of two-phase state of liquid.

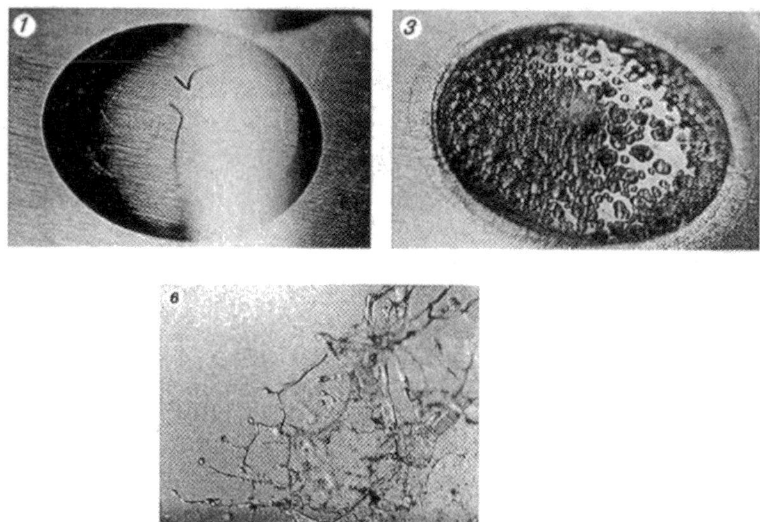

Figure 9. Dynamics of liquid drop structure at shoch wave loading: 1 – initial state, 3- state of "cavitating boiling", b-fragment of structure of liquid drop after its disintegration

Application of two-phase models to describe and to estimate the wave field parameters as well as the cavitation development makes it possible to understand the mechanism governing transformation process. The cavitation turned out to develop mainly inside a central region of the drop. This fact allows us to conclude that a drop is "exploded" from inside. In a result of this process the drop takes the form of a cupola on the stage of foam structure formation and then the form of a space mesh as a result of foam cell expansion.

Conclusion

A number of physical phenomena arising in a result of pulse loading of liquids as well as the mechanisms of their development can be explained with viewpoint of wave processes in cavitating media. The notion of real liquids containing microinhomogeneities as two-phase media has allowed to use corresponding mathematical model and to understand the main features of the process mentioned, to estimate their characteristic parameters and the timescales of relaxation effects as well as to simulate its different stages. With other hand very short times of tensile stress relaxation in cavitating media give the possibility to simplify mathematical simulation of second stage of cavitation development (MIR). The principle effect of rarefaction wave front steepness on limit maximum amplitudes of rarefaction waves in cavitating liquids was discovered.

The results of shock tube experiments, using X-ray technique, computer processing and mathematical simulations of different stages of inversion process of two-phase liquid state have allowed to understand main mechanisms governing by process of liquid fracture.

References

1.   Besov A.S. et al (1984) Investigation of initial stage of cavitation by diffraction optic method, ed. M.Pichal, *Proceedings of International Symposium "Optical Methods in Dynamics of Fluids and Solids"*, 129-135, Springer-Verlag, Berlin, Heidelberg .

2.   Kedrinskii V.K. (1975) Dynamics of cavitation zone at underwater explosion near free surface, *Journal of Applied Mechanics and Technical Physics*, **5**, 68-78.

3.   Kedrinskii V.K. (1976) Negative pressure profile in cavitation zone at underwater explosion near free surface, *Acta Astronautica* , July-August **3** (7-8) , 623-632.

4.   Kedrinskii V.K. (1989) On relaxation of tensile stresses in cavitating liquid, ed. P.Pravica, *Proceedings of 13th Int Congress on Acoustics*, Belgrade , Yugoslavia, SAVA CENTAR, Novi Beograd, v.1, 327-330.

5.   Kedrinskii V.K. (1978) Surface effects at underwater explosion (review), *Journal of Applied Mechanics and Technical Physics*, **4**, 66-78.

6.   Berngardt A.R. et al (1984) Optic and X-ray Investigation of Water Fracture in Rarefaction Wave at Later Stages, ed. M.Pichal, *Proceedings of International Symposium "Optical Methods in Dynamics of Fluids and Solids"*, 137-142, Springer-Verlag, Berlin, Heidelberg

7.   Chernobaev N.N. (1989) Modeling of Shock-Wave Loading of Liquid Volumes, eds. G.Meier, Ph.Thompson, *Proceedings of International Symposium "Adiabatic Waves in Liquid-Vapor Systems"*, 361-370, Springer-Verlag, Berlin, Heidelberg

8.   Berngardt A.R., Kedrinskii V.K., Pal'chikov E.I. (1995) Evolution of inner structure of liquid fracture zone at pulse loading, *Journal of Applied Mechanics and Technical Physics*, **2**, 99-105.

9.   Kedrinskii V.K., Besov A.S., Gutnik I.E. (1997) Inversion of two-phase state of liquid at pulse loading, *Doklady Russian Academy of Sciences*, **352**(4), 477-479.

10. Kedrinskii V.K. (1973) The experimental research and hydrodynamic model of "sultan", *Archives of Mechanics*, **26**(3),

11. Kedrinskii V.K. (1989) Hydrodynamics of Explosion: experiment and models, eds. G.Meier, Ph.Thompson, *Proceedings of International Symposium "Adiabatic Waves in Liquid-Vapor Systems"*, 395-405, Springer-Verlag, Berlin, Heidelberg

12. Gavrilov R.L. (1970) *Power Ultrasonic Fields*, Moscow, Nauka.

13. Hammitt F.G. et al (1974) Cavitation threshold and superheat in various fluids, *Proc. Conf. on Cavitation.*, Edinburg, 3--5 Sept., 1974. London, N.~Y.: Mech. Eng. Publ. Ltd.

14. Sirotyuk M.G. (1968) Experimental study of ultrasound cavitation, *Power ultrasonic fields*, Moscow, Nauka, (5), 167-220.

15. Strasberg M. (1956) Undissolved air cavities as cavitation nuclei, *Cavitation in Hydrodynamics*. London: National Phys. Lab.

16. Kedrinskii V.K. (1986) On multiplication mechanism of cavitation nuclei, *Proceedings, 12th Intern. Congress on Acoustics*. Toronto

17. Kedrinskii V.K., Berngardt A.R., Chernobaev N.N. (1994) Behaviour of a liquid at dynamic loading , eds. S.Morioka, L. Wijngaarden, *Proc. IUTAM Symposium on Waves in Liquid/Gas and Liquid/Vapor Two-phase Systems,* Kyoto, Japan, Kluwer Ac.Publ., Dordrecht, 429-438.

18. Kedrinskii V.K. (2000) *Hydrodynamics of Explosion: experiments and models*, Siberian Branch of the Russian Academy of Sciences Pbl., Novosibirsk

MEASUREMENTS OF THE CAVITATION THRESHOLD OF LIQUIDS UNDER DYNAMIC STRESSING BY PULSES OF TENSION.

P.R. WILLIAMS and R. L. WILLIAMS
Centre for Complex Fluids Processing,
School of Engineering,
University of Wales Swansea, Singleton Park, Swansea SA2 8PP (UK)

Abstract. This paper addresses discrepancies between the results of different measurements of the effective tensile strength (F_c) of liquids, in experiments in which a pulse of tension (or 'negative pressure') is created by the reflection of a pressure pulse at a suitable boundary. We show that a key feature of the pressure records previously reported in experiments such as the 'Bullet-Piston' (B-P) pulse-reflection apparatus [1] may have been misinterpreted. The first complete account of such pressure records is reported here. We also report a new method of estimating F_c in a modified B-P apparatus and the results obtained indicate that samples of degassed, deionised water can sustain tensions which are an order-of-magnitude greater than previously reported in B-P work. Results are also reported for work involving samples of Newtonian silicone oils, for which the dependence of F_c on shear viscosity, μ, found in this work confirms that of an earlier study although the absolute values of F_c are found to be considerably greater than previously reported.

1. Introduction

The maximum tension which a liquid can withstand (its effective tensile strength, F_c) has long been a matter of scientific interest. The pioneering work of Berthelot [2] involved subjecting liquids to a gradually increasing (quasi-static) tension but many important cavitation phenomena result from the dynamic stressing of liquids [3]. While many dynamic techniques for measuring F_c have been reported, several are associated with anomalously low results, particularly in the case of water [1], the values of F_c reported for this liquid falling in three main groups (see Table 1).

The first group is associated with 'quartz inclusion' work, such as that reported by Alvarenga *et al.* [22] who measured the speed of sound in water liquid-vapour inclusions within specially prepared specimens of quartz. These measurements suggest values of F_c close to the homogeneous nucleation limit of *ca.* 1300 bar [23] but involve long extrapolations into the negative pressure regime using equations of state which are determined in the range of positive pressures [4].

The values of F_c found in the second main group of results, which include those produced by both dynamic and quasi-static stressing techniques, fall in the range 100 bar to 300 bar. In a Berthelot tube, distilled water has been found to sustain a tension of 160 bar [11]. Briggs [9] reported a value of *ca.* 250 bar when a centrifugally induced tension was applied to a column of water. In later work involving Briggs' technique, Strube and Lauterborn [10] reported a value of 175 bar for degassed water. Greenspan and Tschiegg [12] and Galloway [13] have reported that suitably clean (i.e. nucleus free)

A.R. Imre et al. (eds.), Liquids Under Negative Pressure, 285–300.
© 2002 *Kluwer Academic Publishers. Printed in the Netherlands.*

TABLE 1. The tensile strength of water as determined in various experiments

1. 'Quartz-inclusion' work
Estimates of F_c up to the homogeneous nucleation limit: 800 - 1300 bar
(Zheng *et al* [4]; Green *et al* [5]; Roedder [6])

2. Static and Dynamic work

Degassed, distilled water: (pulse) Marston & Unger [7]	100 bar
Degassed, deionised water: (pulse) Williams *et al* [8]	300 bar
Degassed water: (centrifugal technique) Briggs [9]	250 bar
Degassed water: Strube & Lauterborn [10]	175 bar
Distilled water: Henderson & Speedy [11]	160 bar [B-T]
water ultrasound Greenspan & Tschiegg [12]	160-210 bar
Degassed water: 0.05% saturation ultrasound Galloway [13]	200 bar

3. Pulse reflection techniques

Degassed distilled water: Davies *et al* [14]	10 bar[B-P]
Degassed deionised water: Couzens & Trevena [15]	15 bar[B-P]
Deionised water: Trevena *et al* [16, 17]	10 bar[B-P]
Deionised water: Richards *et al* [18]	12 bar
Distilled water: Crum & Fowlkes [19]	12 bar
Untreated tap water: Couzens & Trevena [15]	8.5 bar[B-P]
Untreated tap water: Bull [20]	15-16 bar [B-P]
Untreated tap water: Wilson *et al* [21]	8 bar

[B-P denotes work involving the Bullet-Piston technique and its derivatives]
[B-T denotes work involving the Berthelot Tube technique and its derivatives]

water can sustain ultrasonically-induced tensions of *ca.* 200 bars.

The third main group of results, which contains anomalously low values of F_c , represents the outcome of dynamic stressing experiments in which a pulse of tension is generated within a liquid by the reflection of a compressional wave at a suitable boundary. In the 'shock tube' work of Richards *et al* [18], this reflection took place at a flexible membrane while in the 'Bullet-Piston' (B-P) technique reflection occurs at the free surface of the liquid. Early B-P work suggested that degassed, distilled water was incapable of sustaining a tension greater than 10 bar [14]. In subsequent B-P work, Couzens and Trevena [15] found a value of 8.5 bar for ordinary tap water and 15 bar for degassed, deionised water, while Couzens and Trevena [16] and Sedgewick and Trevena [17] reported values of 10 bar for deionised water. In similar work, Bull [20] found a value of 16 bar for tap water, with a second pulse reflection technique used by Bull [20] yielding an estimate of 15 bar. This latter technique was similar to that of Wilson *et al* [21] and involved the detonation of an explosive charge below the surface of a liquid. From measurements of the initial velocity of the spray dome formed above the original undisturbed free surface, F_c was estimated to be 8.0 bar for 'ordinary' water.

An unexpected feature of these low values of F_c is that they result from dynamic stressing. The rate of stressing a liquid experiences is an important factor determining F_c: usually, the higher the rate of stressing, the higher the value of F_c obtained [24]. In a Berthelot tube, tension usually develops over a period of several minutes, whereas in dynamic experiments the time of stress development is typically 10^{-6} s to 10^{-3} s.

Previously we explained how F_c might be underestimated in experiments involving tension pulses due to inadequacies in pressure transduction [8] and how, by overcoming these inadequacies, previous low estimates of F_c for water in dynamic stressing work could be reconciled with the higher values (50 bar to 60 bar) obtained using a steel Berthelot tube [25]. However, this explanation did not account for the much larger discrepancies in F_c discussed above and, to this extent, the question of the B-P technique's reliability has remained open. This is an important issue given that the B-P technique has been used to investigate several aspects of cavitation phenomena, including the dependence of F_c on shear viscosity [16]; and this issue is of no little practical concern given that the characteristics of the tension pulses involved in B-P work correspond to some forms of therapeutic ultrasound [26]. In the latter context low values of F_c are disquieting, as some biomedical applications of ultrasound can involve transient tensions *in vivo* of *ca.* 40 bar [7].

The explanation proposed by Temperley and Trevena [1] for the anomalously low values of F_c in B-P experiments invoked the concept of a metastable transition layer, between liquid and vapour. The transition layer makes the free surface an imperfect reflector, and accounts for the distortion of reflected pulses found in B-P work. This implies *inter alia* that the amplitude of such a reflected pulse is not a reliable indication of the tensile strength of the bulk liquid. Temperley and Trevena [1] considered that this explanation accounted for the anomalously low values of F_c but some points remain unresolved. Firstly, the ratio of (peak tension)/(peak pressure) reported in B-P work varies from 0.36 for deionised water, to 0.5 for boiled deionised water. Given that the peak amplitude of the incident pressure pulse could vary between 100 bar and 300 bar [16] it follows that tensions of between 40 bar and 100 bar would be expected to ensue. Furthermore, in order to account for the anomalously low values of F_c, Temperley and Trevena's theory involving the transition layer between liquid and vapour requires a very large difference between the speed of sound in liquid water, c, and that in water-vapour, c_v: this is necessary in order that the wavelength of the pulses undergoing reflection becomes commensurate with the thickness of the transition layer, which is of the order of only a few intermolecular distances [27]. However, the ratio c_v/c is around one-third [28], which is far too large to provide a satisfactory explanation. Moreover, the same, anomalously low value of F_c was found by Richards *et al* [18] in their pulse reflection experiments in which no free surface was involved.

It is clear from the foregoing that an additional, hitherto unsuspected mechanism plays a role in the recording of anomalously low tensions in B-P work. Such a mechanism is considered herein.

2. Experimental

It is appropriate to begin by describing the main features of previous B-P experiments which involve a vertical tube fitted with a piston at its lower end. A pressure pulse generated by a bullet striking the piston travels up a liquid column within the tube, to be reflected at the free surface as tension (due to the large difference in acoustic impedance between liquid and air). By varying the peak pressure, P, in the incident pulse, the

288

values of P and the peak tension, F, of the reflected pulse are recorded. The (F, P) curve shows a limiting value which provides an estimate of F_c [25].

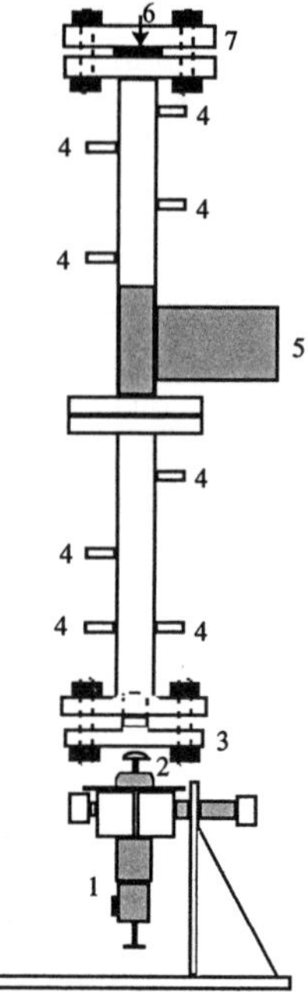

1   Magnum model 7000 cattle stun gun
2   Mushroom-headed cosh
3   Steel piston
4   Transducer ports
5   Support arm
6   Sealing gasket
7   Sealing flange

Figure 1. Schematic of the B-P apparatus

The present apparatus consists of a cylindrical, stainless steel tube (length 1.4 m and internal diameter 0.0243 m) closed at its lower end by a piston (see figure 1). The piston's lower surface is coupled to the bolt of a cartridge-driven stun-gun (Magnum model 7000, Shelby and Vokes, UK) which generates a pressure pulse in a column of liquid within the tube. The topmost steel flange permits connection of the tube to a high pressure steam line, a vacuum line (−1 bar) and a regulated pressure line connected to an oxygen-free nitrogen supply and a pressure gauge (PSITronix, USA). The lower flange houses the piston and provides a connection to a steam condensing system and a liquid drain point. Prior to experiments, superheated steam is blown through the tube, the walls of the open tube being subsequently heated by a flexible heating jacket wrapped around the tube's external surface (not shown in figure 1) and maintained at 100°C until dry. The tube is then allowed to cool prior to filling to the required depth with liquid, which is subsequently heated to the test temperature and degassed (the upper flange being connected to the vacuum line). The temperature of the liquid is measured using a thermocouple at the tube's inner surface. All experiments were conducted at a temperature of 20°C ($\pm$ 1°C).

Pressure changes within the liquid were monitored using three dynamic pressure transducers (Kistler type 603B, Kistler UK) mounted in mechanically isolated ports in the wall of the tube. The 603B, which has a short rise-time (1 μsec), a high natural frequency (> 450 kHz) and a 200 bar dynamic range, has been used in other cavitation studies [29]. The 603B outputs were sampled at rates between 1 MHz and 20 MHz by 12-bit analog to digital convertors, ADC's (Gage Inc., Can.). In addition, a KP-136 hydrophone (Ktech Corp., USA) was used. The output from the KP-136 was connected via BNC connectors and RF coaxial cables to a Gage 8012A, 12-bit ADC (Gage Inc., Can.) and converted to pressure in the following way.

Assuming one dimensional planar shock propagation normal to the transducer's active area, the pressure (in kbar) is related to the charge/active area, x (μC/cm^2), of the transducer as follows (after calibration by Sandia National Laboratories, USA):

$$\text{Pressure (kbar)} = a\text{x} + b\text{x}^2 + c\text{x}^3 + d\text{x}^4 + e\text{x}^5 + f\text{x}^6 \tag{1}$$

where a = 4.2695164; b = 12.413068; c = -13.229975; d = 8.7742958; e = -2.6080845; f = 0.27130771; and x =CV/A, where C is the total capacitance of the recording circuit (comprising transducer (41 pF) + lead (178 pF) + CS8012A input (25pF) to give a total of 0.000244 μF); V is the voltage output and A is the active area of the transducer (0.0905 cm^2).

3. Results

3.1 THE 'PRIMARY' AND 'SECONDARY' PRESSURE-TENSION CYCLES.

The main features of a typical pressure record obtained from a 603B transducer in an experiment on a sample of silicone oil (shear viscosity 1 cS) are shown in figure 2, in which the data are presented in terms of transducer output in unscaled ADC units (positive values correspond to positive pressure and vice versa). In this experiment the length, L , of the liquid column was 0.945 m. A pressure pulse (feature '1' in figure 2) is followed immediately by a tension pulse (feature '2'). We refer to cycle '1–2' as the 'primary' pressure-tension cycle. Thereafter, the record comprises 'secondary' pressure-tension cycles ('3–4', '5–6' etc.) which have been associated with cavitational activity [30].

In what follows we seek to provide a complete explanation of the pressure record and begin by establishing that cavitation may result from the development of the 'primary' pressure-tension cycle. That the response of a bubble nucleus to a single pressure-tension cycle, such as that generated in the present experiments, may result in several cycles of cavitational activity may be illustrated as follows. We begin by

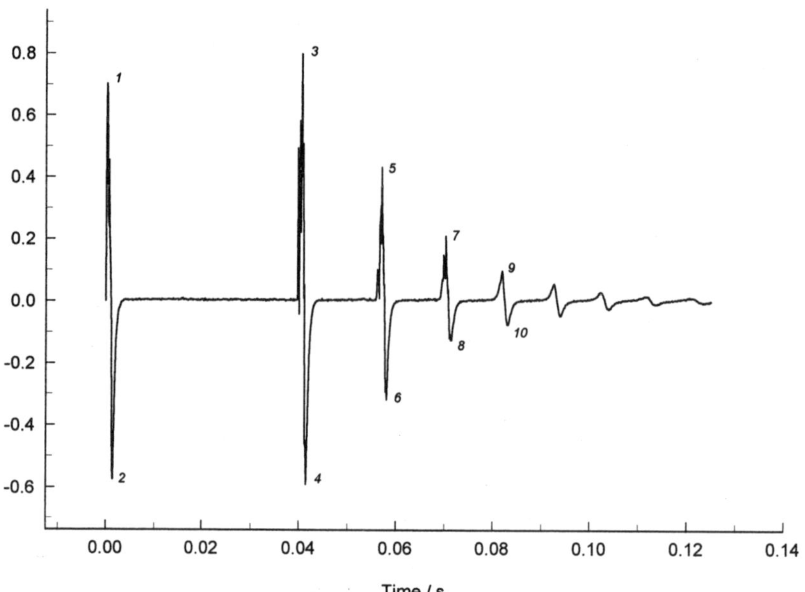

Figure 2. A typical pressure record obtained from a 603B transducer in an experiment on silicone oil. The pressure transducer output is in ADC units, positive values of which correspond to positive pressure and vice versa. The record contains the 'primary' ('1–2') and 'secondary' pressure-tension cycles ('3–4', '5–6' etc).

considering the collapse of a cavity in a large mass of incompressible liquid, in which case the cavity boundary $R(t)$ obeys the relation

$$R\ddot{R} + \frac{3}{2}(\dot{R})^2 = \frac{1}{\rho}\{p_i - p_\infty - \frac{2\sigma}{R} - \frac{4\mu}{R}\dot{R}\} \tag{2}$$

in which ρ is the liquid density and p_∞ is the pressure in the liquid at a large distance from the bubble. The generalized 'Rayleigh-Plessett' equation (equation (2)) includes the effects of surface tension, σ, and shear viscosity, μ, and the pressure in the gas at the bubble wall, p_i (and p_∞) may be a function of time [31].

Figure 3 shows the result of an integration of equation (2) (using a fourth-order adaptive Runge-Kutta method) in which $p_\infty(t)$ takes the form of a 'primary' pressure-tension cycle such as '1-2' in figure 2 and is represented by :

$$P(t) = \frac{d}{dt} \frac{a_0}{\left(1 + \frac{(t-a_1)^2}{a_2^2 \times a_3}\right)^{\left(\frac{a_3}{2} + \frac{1}{2}\right)}} \tag{3}$$

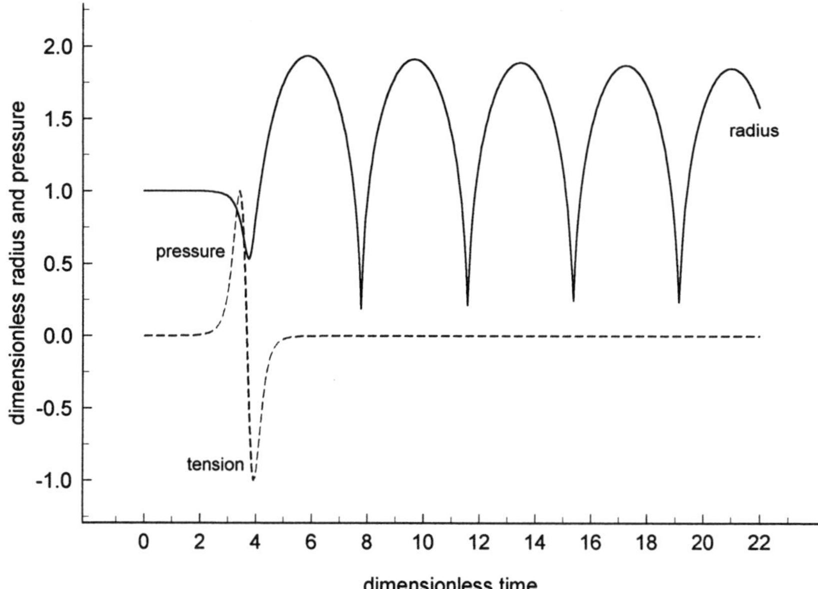

Figure 3. Results of a numerical integration of equation (2) illustrating the response of a cavity nucleus (initially at rest) to a pressure variation which represents the 'primary' pressure-tension cycle generated in the experiments. The resulting change in cavity radius involves several oscillatory cycles, of gradually diminishing amplitude and period.

in which a_0 is the amplitude, a_1 is the center, a_2 is the width, and a_3 is a shape parameter. In this example the representative values of σ, μ, ρ and p_i are taken to be those of water at 20 °C [28], the vapour pressure of the liquid at this temperature representing p_i and the initial radius of the cavity (at rest) being 10 μm. The resulting change in the cavity radius, $R(t)$ in response to the 'primary' pressure-tension cycle involves a series of oscillatory cycles of bubble growth followed by rapid collapse and rebound, as shown in figure 3 in which time is made dimensionless by the appropriate value of the 'Rayleigh' collapse time, τ [32] where:

$$\tau = 0.915 R_m \left(\frac{\rho}{p_0 - p_v} \right)^{\frac{1}{2}} \tag{4}$$

and the cavity radius is represented by R/R_m where R_m is the maximum radius of the cavity (note that the pressure p_0 is atmospheric pressure, the values of pressure in the 'primary' cycle being made dimensionless by the peak pressure, P, in the incident pulse).

In the corresponding experiments, the hydrodynamic pressure variation produced about the oscillating cavitation bubble(s) is recorded as the 'secondary' pressure-tension cycles shown in pressure records such as figure 2 [8, 33].

In previous B-P work, the 'primary' cycle '1-2' has been reported as a single pressure pulse which is immediately followed by a *single* pulse of tension [16] but in the following we show that *two* tension pulses are actually recorded within this cycle and explain their origin. This point has significance in estimates of the tension sustained by the liquid in B-P work (Section 3.2).

The 603B transducers were used to monitor the development of pressures in a column of silicone oil ($L = 0.945$ m). For the sake of clarity, only the records from the lowest and uppermost transducers are shown, their positions above the face of the piston being 33.6 cm and 77.8 cm, respectively (see figure 4). Following the firing of the stun-gun (at time $t = 0$) the pressure pulse, a', is recorded by the lower transducer, followed by pulse a'' which is recorded by the upper transducer. This confirms the assumption made by previous workers that the initial pulse (produced by the piston) travels upwards through the liquid. Pulses a' and a'' occur at 0.373 ms and 0.85 ms, respectively, in the record. The distance travelled is 44.2 cm, at a velocity of 926.6 ($\pm$9.9)ms^{-1}.

The next feature in the record is the tension pulse b'' which is first sensed by the upper transducer: this is followed by the corresponding tension pulse b' which is recorded by the lower transducer. We conclude that these pulses move downwards through the liquid column. Their velocity, calculated from the time delay (0.46 ms) between b'' and b' and the distance travelled (44.2 cm), is 961 ms^{-1}.

From the foregoing we predict the time of occurrence of b'' and b' in the pressure record if (as has been assumed in previous B-P work) they correspond to the reflection of a'' and a' from the free surface. In the case of b'' the total distance travelled is 33.4 cm and its predicted occurrence is at 0.354 ms following pulse a'' i.e. it should occur at 1.204 ms (0.354 ms + 0.85 ms). The actual occurrence of b'' is at 1.200 ms. We next calculate when b' should be found if it occurs as a result of the reflection of a' from the free surface. The total distance travelled is 121.8 cm and thus the predicted time of occurrence of b' in the pressure record is 1.291 ms after a' i.e. at 1.664 ms. The actual time of occurrence of b' is at 1.660 ms.

We next consider a feature which is absent from previous descriptions of the pressure records in B-P work, namely the tension pulses c' and c'' (which propagate upwards through the liquid column). The former occurs first in the pressure record, being sensed by the lower transducer and is followed by the corresponding pulse c'' recorded by the upper transducer. Their velocity, calculated from the time delay (0.49 ms) between c'' and c' and the distance travelled (44.2 cm), is 902 ms^{-1}.

Finally we calculate the times of occurrence of c' and c'' in the pressure record on the assumption that they represent the reflections of b' and b'', respectively, from the face of the piston. The time required for b' to travel the 33.6 cm to the face of the piston at 961 ms^{-1} is 0.35 ms; and the time required by c' to return, at 902 ms^{-1}, is 0.37 ms. On this basis, the predicted time of occurrence of c' in the pressure record is 0.72 ms after b', corresponding to a time of 2.38 ms. The actual time of occurrence of c' is 2.37 ms (within 0.5 % of the predicted position) confirming that it is the reflection (as tension) of the downward travelling tension pulse b'.

The time required for b'' to travel the 77.8 cm to the face of the piston at 961 ms^{-1} is 0.81 ms; and the time require by c'' to return at a velocity of 902 ms^{-1} is 0.86 ms. On this basis, the predicted time of occurrence of c'' in the pressure record is 1.67 ms after

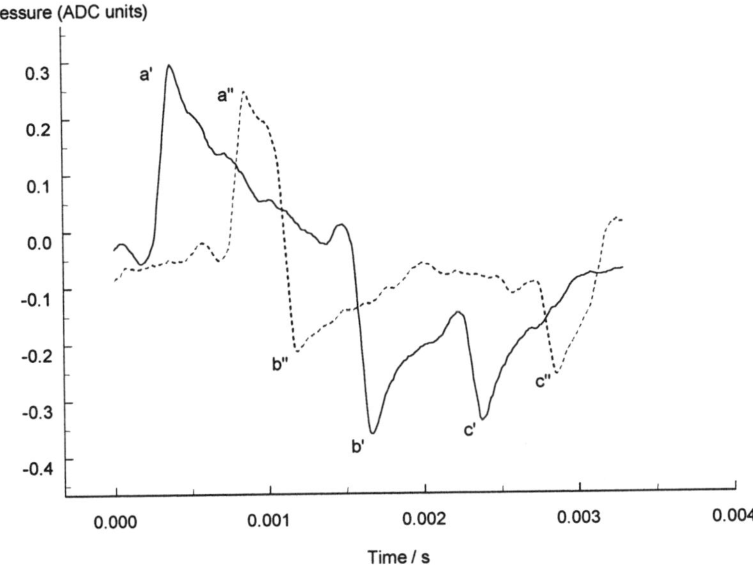

Figure 4. Details of a 'primary' pressure-tension cycle recorded in experiments on silicone oil. For the sake of clarity, only the records from the lowest (*prime symbols*) and the uppermost transducer (*double prime symbols*) are shown. The tension pulses *b'* and *b''* originate in the reflection (at the free surface) of the upward travelling pressure pulses *a'* and *a''*, respectively; and the upward travelling tension pulses *c'* and *c''* originate in the reflection of tension from the face of the piston.

b'' i.e. at 2.87 ms. The actual time of occurrence of *c''* is 2.86 ms, confirming that it is the reflection of the downward travelling pulse *b''*.

We summarise these findings as follows. The tension pulses *b'* and *b''* originate in the reflection (at the free surface) of the upward travelling pressure pulses *a'* and *a''*, respectively; and the upward travelling tension pulses *c'* and *c''* originate in the reflection of the downward travelling tension pulse from the face of the piston. This latter mechanism of tension pulse reflection has not been reported in previous B-P work.

A further series of experiments was conducted on samples of deionised water obtained from a two-stage, reverse-osmosis, ion-exchange purification system (Elga, UK) incorporating a carbon filter (for organics), a 'nuclear grade' deionisation stage and a final membrane filter (0.2 μm). The length of the liquid column was 0.95 m, the position of the free surface being 61.4 cm above the location of a KP-136 hydrophone. A typical record of dynamic pressures within the water is shown in figure 5 in terms of the KP-136's voltage output which was recorded at a sampling rate of 1 MHz (solid line). The corresponding record obtained at a simulated 10 kHz sampling rate is also shown, this sampling rate corresponding to that employed in previous B-P work. Note that the pulses F_1 and F_2 recorded at 10 kHz appear as a *single*, broad tension pulse which corresponds to descriptions of this feature in previous B-P work.

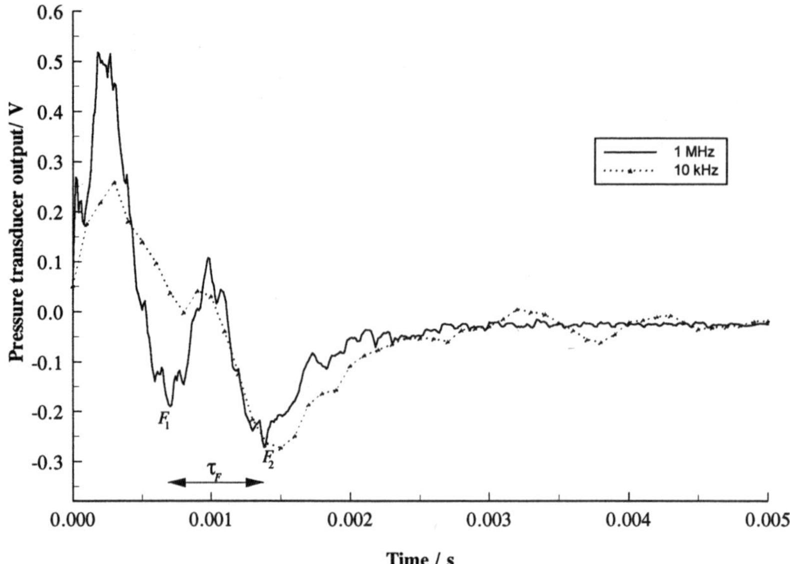

Figure 5. Results of experiments on degassed, deionised water. The pressure-tension cycles were recorded using the KP-136 at a sampling rate of 1 MHz (solid line). The pulses F_1 and F_2 recorded at 1 MHz appear as a single tension pulse at a (simulated) 10 kHz sampling rate.

As in the tests conducted on silicone oil, the upward-travelling pressure pulse is reflected at the free surface as a downward-travelling tension pulse (F_1). This is followed, after a time τ_F, by an upward-travelling tension pulse (F_2) due to reflection from the face of the piston. Using the relationship given in Section 2 for the calibration of the KP-136 hydrophone, the maximum dynamic tension sustained by degassed deionised water was found to be 90 bar ($\pm$ 5 bar).

3.2 CONSEQUENCES OF TENSION PULSE REFLECTION IN MEASUREMENTS OF F_C.

Pulse reflection has long been recognised as a factor in determining the results obtained in B-P experiments [15]. In previous work this has involved the following considerations. A pulse of duration τ and peak pressure P propagates upwards through the liquid at a velocity c. At the free surface this pulse is reflected as tension. The resultant pressure variation below the free surface is then determined by the superposition of the incident and reflected pulses whose overlap may be avoided at a depth D by satisfying the condition $2D/c > \tau$. In practice this entails positioning transducers at a sufficiently large distance from the free surface but this measure increases the likelihood of encountering overlap due to any reflection at the *base* of the tube.

To properly assess the influence of the latter mechanism on measurements of tension, a knowledge of the velocities of pulses such as F_1 and F_2 is required. Given that the amplitude F of the tension within the liquid near the base will increase due to pulse reflection, a situation may arise where the magnitude of tension pulse F_1 is less than F_c and, accordingly it travels at a velocity c appropriate to liquid water (some 1435 ms^{-1}). However, if as a result of reflection the tension increases sufficiently, such that F_c is exceeded, then cavitation ensues and the velocity of the upward travelling pulse may be closer to that (494 ms^{-1}) appropriate to the propagation of sound in water-*vapour*, c_v [8]. It follows from basic considerations of acoustic theory that such a difference in velocity between pulses such as F_1 and F_2 will influence estimates of F_c, resulting in an underestimate of tension, the underestimate being commensurate with the discrepancy which exists between previous (anomalously low) B-P results and expected values of F_c (ie. those within the second group of results discussed in Section 1).

3.3 EXPERIMENTS INVOLVING THE APPLICATION OF A REGULATED STATIC PRESSURE.

The B-P technique used in previous work has the disadvantage that the means by which P is varied (using different combinations of piston mass, piston length and bullet momentum) necessitates frequent dismantling of the apparatus and removal of the liquid. This was overcome in the present work by a new method which allows F_c to be estimated with the same sample *in situ* throughout.

This method involves regulating a static pressure, P_s , in the space above the liquid column, P_s being increased gradually in a series of dynamic stressing experiments. From the dynamic pressure records obtained in these experiments a record is made of the time delay, τ_i , between the peak incident pressure (corresponding to '1' in figure 2) and the first pressure pulse arising from cavitation bubble collapse (corresponding to '3' in figure 2). It is useful to recall here that, under tension, cavitation bubbles grow from pre-existing nuclei within the liquid and eventually collapse and rebound, emitting a pressure wave into the liquid as they do so. Hence the interval τ_i , which encompasses the attainment of maximum cavity radius and its subsequent decrease to a minimum value, is reduced by increasing P_s (τ_i therefore provides a convenient measure of cavitational activity).

From the work described in Section 3.1, it is clear that the B-P experiment involves the transmission of a pulse of tension by the liquid to the face of the piston; and that cavitation may result from the reflection of this pulse. It follows that in the case of experiments in which cavitational activity is detected, the magnitude of the tension transmitted by the liquid is sufficient to result in the development of a transient net negative pressure in the presence of a background static pressure (P_s). Thus an estimate of the magnitude of tension capable of being transmitted by the liquid can be obtained from a knowledge of P_s.

The results of such experiments on degassed, deionised water are shown in figure 6, in which τ_i is plotted as a function of P_s (absolute, in p.s.i.) and from which F_c is estimated in the following way. The time delay, τ_o , between pulses corresponding to '1' and '2' in figure 2 represents the time required for the upward travelling pressure wave to return, as tension, to the lower transducer's location: it also represents the smallest time interval for which a cavity growth-collapse cycle could occur given that a bubble would have to grow and collapse infinitely quickly in order that $\tau_i = \tau_o$. Thus F_c is estimated by extrapolation of the data in figure 6 to that pressure P_s at which $\tau_i = \tau_o$, this condition representing the complete suppression of cavitation. This procedure yields a value of F_c of 96 bar ($\pm$ 5 bar), in good agreement with the maximum value of 90 bar ($\pm$ 5 bar) recorded by the KP-136 hydrophone.

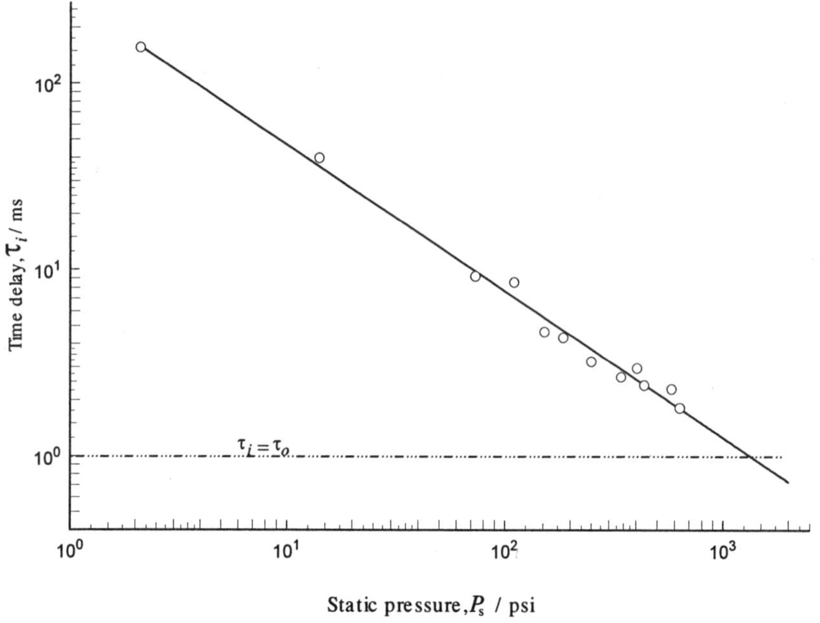

Figure 6. The time interval τ_i as a function of applied static pressure, P_s (in p.s.i.) for degassed, deionised water. Also shown is the value of the time interval τ_o used to estimate F_c .

This procedure is distinct from that used to estimate F_c by Zheng *et al* [4], which was based on long extrapolations into the negative pressure regime using equations of state determined in the range of positive pressures. The present method involves estimates of dynamic tensions which are based on measurements of static positive pressure and extrapolation in the range of positive pressures.

It is appropriate to compare our results with values obtained by other dynamic stressing techniques which do not rely upon the direct measurement of negative pressures using pressure transducers. Such a technique has been reported by Marston and Unger [7] and involved a compressive pulse of duration 1.7 μs which was reflected at a flexible membrane to create a pulse of tension in the liquid. Their values of F_c (100-110 bar) for degassed, distilled water are in good agreement with the results of the present work which yield a value of 96 bar ($\pm$ 5%). Our slightly lower value may be attributed to the longer time of stress development involved in the present work (50 μs - 100 μs).

3.4 THE DEPENDENCE OF F_C ON SHEAR VISCOSITY.

Few studies of the dependence of F_c on shear viscosity μ have been conducted but among the work reported, pulse reflection methods have figured prominently, the B-P experiment being the most widely employed. Bull [20] and Couzens and Trevena [16] reported that their results could be represented by an expression of the form $F_c \propto \mu^n$, the value of n being 0.2 in the former study whereas the latter workers reported a value of 0.1. In work involving a pulse reflection technique, Carlson and Levine [34] reported a power-law relationship for samples of glycerol whose viscosity was varied by changing the test temperature, the value of n found in their work being 0.3.

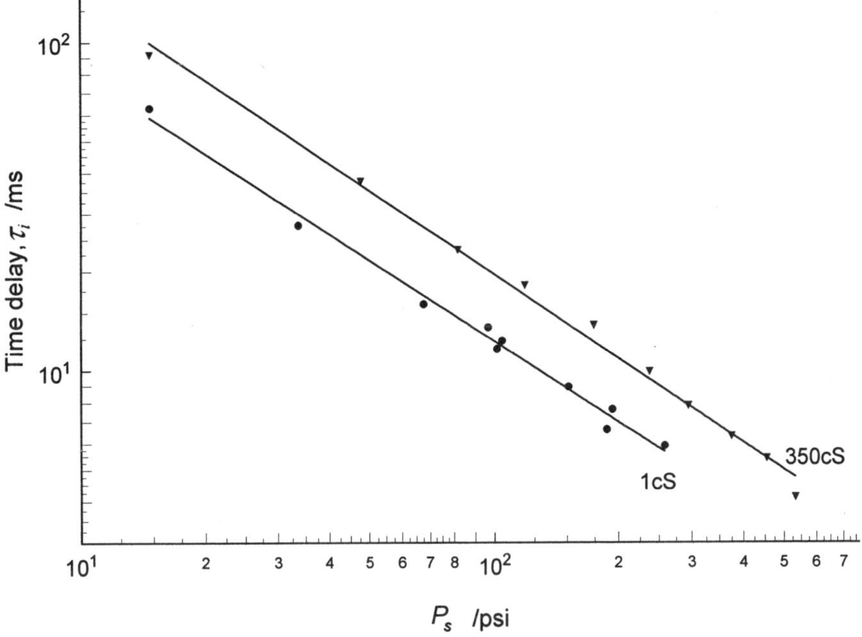

Figure 7. The time interval τ_i as a function of applied static pressure, P_s (in p.s.i.) for two silicone oils of different shear viscosities (1 cS and 350 cS).

In the light of the different results found in the aforementioned studies, experiments were conducted using the new B-P method on a range of Newtonian silicone oils (Dow Corning 200 series, BDH Chemicals, UK) of different shear viscosities (1, 50, 100 and 350 cS). For the sake of clarity the results obtained in experiments on two silicone oils (viscosity 1 cS and 350 cS, respectively, measured using a Rheometrics Scientific (USA) ARES rheometer fitted with cone-and-plate measuring geometry) are shown in figure 7, the values of F_c obtained for the oils being shown in Table 2. The results may be represented by an expression of the form $F_c \propto \mu^n$, where $n = 0.09$ ($\pm$ 0.005). This finding corresponds closely to that ($n = 0.1$) reported by Couzens and Trevena [16] for silicone oils, in a similar viscosity range (although the absolute values of F_c found in the present work are significantly higher).

TABLE 2. Tensile strength F_c for Dow Corning® 200 silicone oils at 20°C

Shear viscosity of oil / cS	F_c / bar
1	69
50	72
100	81
350	125

The relation $F_c \propto \mu^{0.2}$ obtained by Bull [20] in pulse reflection work on samples of water, olive oil, glycerol and syrup has been attributed as due more to a coincidental distribution of nuclei in these different liquids than to other, more basic features of cavitation [35]. In view of the agreement found between the results of the present study and those of Couzens and Trevena [16], it is noteworthy that both involved a series of silicone oils with a range of viscosities, rather than a range of different fluids, with different viscosities, as in the work of Bull [20].

4. Discussion

It is pertinent to consider whether the mechanism of tension pulse reflection reported herein arises in other cavitation experiments. In the work reported by Fujikawa and Akamatsu [36], a rarefaction wave impinged on the free surface of a column of water. The rarefaction transmitted into the water was reflected (also as a rarefaction) at the closed, lower end of the water column, thereby increasing the magnitude of the tension. In the work reported by Richards et al [18], deionised water was held in the lower part of a vertical shock-tube. Ignition of an oxyacetylene mixture in the upper part of the tube caused a compression wave to propagate in the liquid. This wave was reflected upwards (as tension) from a diaphragm supporting the liquid. In order to increase the magnitude of the reflected tension, a converging-diverging nozzle was placed in the test section. When no cavitation occurred, the velocity of both the downward-moving compression wave and the upward-moving tension pulse corresponded to the velocity of sound in water. When cavitation was observed, an acoustic field radiated by the collapse of bubbles, was found to be associated with *two* velocity components whose values were 1550 ms^{-1} and 550 ms^{-1} respectively. The former corresponds closely to the expected value for liquid water, c, whereas the latter is commensurate with that expected in water-vapour, c_v.

From the foregoing we conclude that tension pulse reflection is a feature of several different cavitation experiments, and is encountered both in water 'shock-tube' and B-P work. Wave interference may accompany this reflection and, under a level of tension sufficient to cause cavitation, a velocity of the associated tension wave appropriate to that in water-vapour may result.

5. Concluding remarks

The results of the present study establish that samples of degassed, deionised water can sustain a transient tension which is an order-of-magnitude greater than previously reported in pulse reflection experiments in the B-P apparatus or in a water 'shock tube' and reconcile the outcome of B-P experiments with those reported in the second group of results discussed in Section 1 of this paper. Moreover, the results of a new method of estimating F_c based on B-P experiments which involve the regulation of a static pressure are shown to accord with the results of direct measurements of dynamic tension made at adequately high sampling rates; and with the results obtained by Marston and Unger [7] in pulse reflection work. Previous B-P work has only involved consideration of the overlap of pressure and tension pulses following reflection of the former at the free surface of the liquid. The explanation advanced herein for the anomalously low values of F_c recorded in previous work B-P involves the overlap of tension pulses at the base of a column of water. When cavitation of the liquid accompanies this pulse overlap, the velocities of the pulses may differ significantly, with the result that the tension recorded within the liquid appears too low.

The work reported here suggests that previous explanations for such low values of F_c are incomplete and offers an explanation for the discrepancy between reported and expected values of F_c in the B-P apparatus, and in similar experiments. Further experiments are now planned in which the new method of estimating F_c in the new B-P apparatus reported here will be employed to study the cavitation properties of liquids such as motor oils over a range of temperatures.

Acknowledgements

This work was carried out under EPSRC Grant GR/L62160. The authors are grateful to Dr. P.M. Williams for technical assistance during the construction of the apparatus.

References

1. Temperley, H.N.V., Trevena, D.H. (1987) Why is the tensile strength of water measured dynamically less than that measured staticall, *J. Phys. D: Appl. Phys.*, **20**, 1080-1081.
2. Berthelot M. (1850) Sur quelquesphenomenes de dilation force de liquids, *Ann. Chim. Phys.*, **30**, 232-237.
3. Joseph, D.D. (1998) Cavitation and the state of stress in a flowing liquid, *J. Fluid Mech.*, **366**, 367-378.
4. Zheng, Q., Durben, D. J., Wolf, G. H. and Angell, C. A. (1991) Liquids at large negative pressure: water at the homogeneous nucleation limit, *Science*, **254**, 829-832.
5. Green, J. L., Durben, D. J., Wolf, G. H. & Angell, C. A. (1990) Water and solutions at negative pressure: Raman spectroscopic s tudy to – 80 Megapascals, *Science*, **249** , 649-652.
6. Roedder, E. (1967) Metastable superheated ice in liquid water inclusions under high negative pressures. *Science* 155, 1413-1417.
7. Marston, P.L. and Unger, B.T. (1986) Rapid cavitation induced by the reflection of shock waves, in *Shock Waves in Condensed Matter* Plenum, New York 401.
8. Williams, P.R., Williams, P.M. and Brown, S.W.J. (1997) Pressure waves arising from the oscillation of cavitation bubbles under dynamic stressing, *J. Phys. D: Appl. Phys.*, **30**, 1197-1206.
9. Briggs, L.J. (1950) Limiting negative pressure of water, *J. Appl. Phys.*, **21**, 721-722.
10. Strube, H.W. and Lauterborn, W. (1970) Investigation by the centrifuge method of cavitation nuclei at the interface between glass and water, *Z.Angew. Phys.*, **29**, 349-357.
11. Henderson, S.J., Speedy. R.J. (1980) A Berthelot-Bourdon tube method for studying water under tension, *J. Phys. E: Sci. Instrum.*, **13**, 778-782.

12. Greenspan, M., Tschiegg, C.E. (1967) Radiation-induced acoustic cavitation; apparatus and some results, *J. Res. Natl. Bur. Stand.* Sect. C **71**, 299-312.

13. Galloway, W.J. (1954) An experimental study of acoustically induced cavitation in liquids, *J. Acoust. Soc. Am.*, **26**, 849-857.

14. Davies, R.M., Trevena, D.H., Rees, N.J.M. and Lewis, G.M. (1956) The tensile strength of liquids under dynamic stressing. *Proceedings of the National Physical Laboratory Symposium on Cavitation in Hydrodynamics.* **5**, 1-20.

15. Couzens, D.C.F. and Trevena, D.H. (1969) Critical tension in a liquid under dynamic conditions of stressing, *Nature*, **222**, 473-474.

16. Couzens, D.C.F. and Trevena, D.H. (1974) Tensile failure of liquids under dynamic stressing, *J. Phys. D: Appl. Phys.*, **7**, 2277-2287.

17. Sedgewick, S.A. and Trevena, D.H. (1976) Limiting negative pressure of water under dynamic stressing, *J. Phys. D: Appl. Phys.*, **9**, 1983-1990.

18. Richards, B.E., Trevena, D.H. and Edwards, D.H. (1980) Cavitation experiments using a water shock tube, *J. Phys. D: Appl. Phys.*, **13**, 1315-1323.

19. Crum, L.A., Fowlkes, J.B. (1986) Acoustic cavitation generated by microsecond pulses of ultrasound, *Nature*, **319**, 52-54.

20. Bull, T.H. (1956) The tensile strengths of liquids under dynamic loading, *Phil. Mag.*, **8**, 153-165.

21. Wilson, D. A., Hoyt, J. W. and McKune, J. W. (1975) Measurement of tensile strength of liquids by an explosion technique, *Nature*, **253**, 723-725.

22. Alvarenga, A.D., Grimsditch, M. and Bodnar, R.J. (1993) Elastic properties of water under negative pressures. *J. Chem. Phys.*, **98**, 8392-8396.

23. Fisher, J. C. (1948) The fracture of liquids, *J. Appl. Phys.*, **19**, 1062-1067.

24. Trevena, D.H. (1982) Time effects in cavitation experiments, *J. Phys. D: Appl. Phys.*, **15**, L111-114.

25. Trevena, D.H. (1987) *Cavitation and Tension in Liquids*, Adam Hilger, Bristol.

26. Rosenschein, U. and Rassin, T. (1998) Ultrasound Thrombolysis, *Science and Medicine*, **5**, 36-43.

27. Als-Nielsen, J. (1985) The Liquid Vapour Interface, *Z. Phys. B*, **61**, 411-414.

28. Bolz, R and Tuve, G. (Eds.) (1973) *CRC Handbook of Tables for Applied Engineering Science*, 2nd Edn, CRC Press.

29. Tomita, Y. and Shima, A. (1986) Mechanisms of impulsive pressure generation and damage pit formation by bubble collapse, *J. Fluid Mech.*, **169**, 535-564.

30. Overton, G. D. N. & Trevena, D. H. (1981) Cavitation phenomena and the occurrence of pressure-tension cycles under dynamic stressing, *J.Phys.D: Appl.Phys.*, **14**, 241-250.

31. Plessett, M.S. and Prosperetti, A. (1977) Bubble Dynamics and Cavitation, *Ann. Rev. Fluid Mech.*, **9**, 145-185.

32. Rayleigh, Lord. (1917) On the pressure developed in a liquid during the collapse of a spherical cavity, *Phil. Mag.*, **34**, 94-98.

33. Williams, P.R. and Williams, P.M. (1996) Pressure-tension cycles induced by dynamic stressing and cavitation in liquids, *J.Phys.D: Appl.Phys.*, **29**, 1904-1909.

34. Carlson, G.A. and Levine, H.S. (1975) Dynamic tensile strength of glycerol, *J.Appl.Phys.*, **46**, 1594-1601.

35. Hsieh, D.Y. (1970) Bubble growth in a viscous liquid due to a transient pulse, *J. Basic Eng.*, **92**, 815-818.

36. Fujikawa, S. and Akamatsu, T. (1980) Effects of the non-equilibrium condensation of vapour on the pressure wave produced by the collapse of a bubble in a liquid, *J. Fluid Mech.*, **97**, 481-512.

NEGATIVE PRESSURE TAIL OF A REFLECTED PRESSURE PULSE: A LATTICE BOLTZMANN STUDY

GÁBOR HÁZI

KFKI Atomic Energy Research Institute, Simulator Development Department, H-1525 Budapest, Hungary
E-mail: gah@sunserv.kfki.hu

ATTILA R. IMRE

KFKI Atomic Energy Research Institute, Materials Department, H-1525 Budapest, Hungary
E-mail: and imre@sunserv.kfki.hu

Abstract. In this paper, a numerical pressure wave reflection experiment in a two-dimensional liquid is presented. The liquid is simulated by the pseudo-potential extension of the lattice-Boltzmann method. In our experiment a pressure pulse is produced by a point source and the resulting pressure wave is reflected back by a wettable rigid wall. Negative pressure tail can be observed at the vicinity of the wall/liquid interface.

1. Introduction

There are several situations where solid-liquid interfaces are subjects of strong positive pressure pulse coming from the liquid. Behind the positive peak, a rarefaction wave can be find with pressure lower than the surrounding hydrostatic pressure. Reflecting back from a free surface or from a solid wall, the reflected pressure pulse can produce negative pressure in the liquid [1-6].

In this paper we would like to present a numerical simulation, where – after a reflection from a rigid wall - the pressure of the rarefaction wave will be lowered below its initial value. This pressure can become easily even negative, i.e. the liquid can be stretched, causing mechanical and thermodynamical metastability [2,6].

This phenomenon is very important: most underwater structures (like oil-lines, submarines, etc.) are designed to withstand big outer pressure, including pressure waves. These structures are cylindrical; a cylindrical object can withstand relatively large positive pressure differences (p_{out}-p_{in}, when the p_{out}>p_{in}; p_{out} and p_{in} is the outer and inner pressure, respectively) but they can open up experiencing large negative pressure differences (p_{out}<<p_{in}). Being p_{in} is constant, a cylindrical structure can survive a large positive pressure wave (like an underwater detonation or pressure wave generated by an earthquake) but it can open up when p_{out} turns to be negative. Therefore it would be crucial to avoid – or at least to reduce – the reflected negative pressure waves. In our simulation, the phenomenon is modeled by the so-called lattice-Boltzmann method. A short overview about this method can be seen in the next chapter.

A.R. Imre et al. (eds.), Liquids Under Negative Pressure, 301–306.
© 2002 *Kluwer Academic Publishers. Printed in the Netherlands.*

302

2. The lattice-Boltzmann method for non-ideal gases and liquids

As it is well known, in the microscopic level discrete atoms or molecules build up any fluid, but in the macroscopic level it shows continuous behavior and partial differential equations (PDEs) can describe its dynamics. The form of these PDEs does not depend on the microscopic details. Indeed, if the interactions between molecules satisfy some conservation principles then these interactions can affect only on transport coefficients (e.g. viscosity) of macroscopic equations, but not their form. Consequently, we can produce realistic results in macroscopic level without the complete knowledge of molecular interactions.

The lattice-Boltzmann method (LBM) exploits this observation. It simulates the behavior of particle populations by a simple stream-and-collide procedure forcing the movement of molecules to a lattice. At each time step, populations propagate to a neighboring lattice link where a local collision takes place redistributing the populations (for more details see e.g. our recent review [7]).

Using a lattice with b number of links and discretizing in the time domain, the evolution of distribution functions of a lattice site can be written as:

$$f_i(\mathbf{r}+\mathbf{c}_i, t+1) - f_i(\mathbf{r}, t) = \Omega(f_i),\qquad(1)$$

where $\mathbf{c}(\mathbf{r},t)$ is the particle velocity and $f(\mathbf{r},\mathbf{c},t)$ is defined in such a way that $f(\mathbf{r},\mathbf{c},t)d\mathbf{r}d\mathbf{c}$ is the number of molecules at time t positioned between $\mathbf{r}$ and $\mathbf{r}+d\mathbf{r}$ which have velocities in the range $\mathbf{c}$ and $\mathbf{c}+d\mathbf{c}$ and the index i=1..b is used to identify the lattice links.

Equation (1) is the so-called lattice-Boltzmann equation, which name is due to the collision operator Ω, since it is selected by following Boltzmann's assumptions.

In the simplest model, we can assume that the distribution function relaxes to its equilibrium by a constant rate which gives the so-called *BGK* (Bhatnagar, Gross and Krook) [8] or *linear collision operator* [9]:

$$\Omega_i = -\frac{1}{\tau}(f_i - f_{i,eq}).\qquad(2)$$

The relaxation parameter $\omega=1/\tau$ determines the kinematics viscosity.

Since it is known from kinetic theory that using Boltzmann's equation the single-particle distribution function reduces to the Maxwell-Boltzmann distribution at equilibrium, as a discrete approximation one can take the second-order expansion in the local velocity of the Maxwell-Boltzmann distribution function assuming that $|u|\ll 1$.

To recover the correct fluid equations, the following form of equilibrium distribution can be adopted:

$$f_{i,eq} = \rho\left[\frac{1-d_0}{b} + \frac{D}{c^2 b}(\mathbf{c}_i \cdot \mathbf{u}) + \frac{D(D+2)}{2c^4 b}(\mathbf{c}_i \cdot \mathbf{u})^2 - \frac{D}{2bc^2}u^2\right],\qquad(3)$$

where d_0 is a constant and determines the compressibility of the fluid, $c=\Delta x/\Delta t$ is the reference speed and D is the dimension number (=2 in our case). The coefficients of Eq.

(3) need to be found subject to isotropy, conservation principles and Galileian invariance in order to obtain the required equilibrium distribution (to accuracy $O(u^2)$).

The macroscopic quantities can be obtained after summing the distribution functions at lattice sites:

$$\rho = \sum_i f_i \qquad \rho u_\alpha = \sum_i f_i (\mathbf{c}_i)_\alpha \,, \qquad (4)$$

where ρ is the density, u is the local hydrodynamic velocity and the Greek subscript α denotes the space directions in Cartesian coordinate.

Using this model, the Navier-Stokes equations - with the equation of state of an ideal gas - in the limit of low Mach number can be recovered via Chapman-Enskog expansion.

LBM has several advances over traditional computational fluid dynamics methods, e.g. complex solid boundaries can be introduced into the model easily by implementing the so-called bounce-back boundary condition.

LBM was extended to model two-phase flow in several ways. The most popular two-phase flow model was derived originally by Shan and Chen [11] and later extended by many others. To model two-phase flow we have to achieve that the equation of state to be non-ideal. In the model of Shan and Chen it is achieved by incorporating nearest-neighbor interactions into the model by defining interaction potential at each site:

$$V(\mathbf{x}, \mathbf{x}') = G(\mathbf{x}, \mathbf{x}')\psi(\mathbf{x})\psi(\mathbf{x}') \,, \qquad (5)$$

where G is a Green function and ψ defines the form of the equation of state resulting. The neighboring fluid particles exchange momentum through a short-range force:

$$F(\mathbf{r}) = -\psi(\mathbf{r}) \sum_i G_i \psi(\mathbf{r} + \mathbf{c}_i) \mathbf{c}_i \,. \qquad (6)$$

Taking into account only local interparticle interactions between neighbors, for a two-dimensional eight-link with one resting population (D2Q9) model the Green function can be written as [11]:

$$G_i = \begin{cases} 4\Gamma & |\mathbf{c}_i| = 1 \\ \Gamma & |\mathbf{c}_i| = \sqrt{2} \\ 0 & \text{otherwise} \end{cases} \,. \qquad (7)$$

The absolute value of the Green function Γ determines the strength of the interaction and its sign selects between attractive and repulsive forces. The parameter Γ appears directly in the equation of state and plays the role of „temperature".

This model can be used to model multi-phase flows with phase transition by the suitable choice of the ψ function. A suitable function assures that the pressure not be a monotone function of density in the equation of state. Shan and Chen proposed the following function (others can be found in [12]):

304

$$\psi = \rho_0 \left[1 - e^{\rho/\rho_0} \right],\tag{8}$$

where ρ_0 is a constant.

The critical value of Γ, where phase transition takes place can be determined from the equation of state ($\partial p/\partial\rho = 0$ and $\partial^2 p/\partial\rho^2 = 0$):

$$p = \frac{c^2}{D}\left[(1-d_0)\rho + \frac{b}{2}\Gamma\psi^2 \right].\tag{9}$$

This method was later extended to three dimensions and local interaction between solid and liquid was introduced. Using the solid-liquid model proposed by Martys and Chen it was demonstrated that we can control the surface wetting characteristic of the solid [13,14]. The model was analyzed carefully by Shan, Martys and their co-workers [15-17] and applied for different problems [11,14,18].

3. Numerical shock wave experiment in liquid

In our numerical experiment we used the D2Q9 LBM model with the above-introduced Shan-Chen two-phase extension. The simulation was carried out on a 512x512 lattice and a wettable solid was placed at the middle of the geometrical domain. Periodical boundary conditions were applied at each side of the domain. To model solid-liquid interaction we used the method proposed by Martys and Chen [13], which uses Eq. (6) where now the parameter Γ_w controls the wettability of the solid. The connection between Γ_w and experimentally determinable quantities (like surface tension, contact angle) can be seen in [14,18].

In the equilibrium distribution function Eq. (3) the compressibility parameter d_0 was set to 0.5. For this model the critical value of the temperature like parameter is $\Gamma \cong -0.111$ [11]. To avoid phase transition we selected $\Gamma = -0.180$. The shape and the magnitude of the pressure profile strongly depends on the value of the parameter, i.e. on the wettability of the wall. The surface wetting parameter (Γ_w) of the wall model was chosen for from -0.05695, -0.0570 and -0.05705. Initially the density field was homogenous $<\rho> = 2.9$ with a small (1%) random perturbation and the velocity components were set to zero in the overall system. At the beginning of the simulation a pressure shock leaves the wall due to the lack of interaction potential balance at the solid. After 350 steps this pressure shock propagated far enough from the wall and we started our experiment. We introduced a pressure pulse into the system on the horizontal symmetry axis near to the wall and we followed its effect at the wall sites. In Fig. 1/a,b the minimum pressure of the overall system and the pressure at the wall surface are shown.

In Fig. 1/a the first, sharp minimum can be observed at the neighborhood of the pulse source as a direct effect of the source and the second negative pressure peak develops at the wall surface as it can be seen in Fig. 1/b. In Fig. 1/b the positive peak

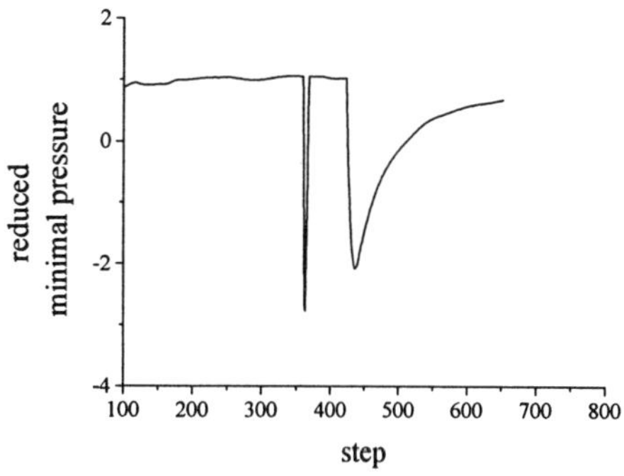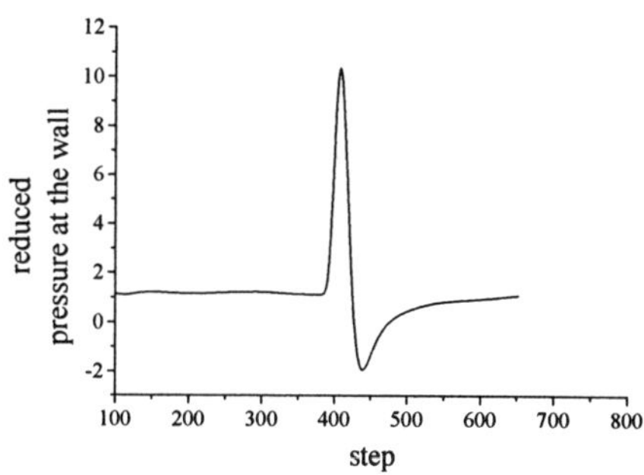

Figure. 1. a (left): Reduced minimal pressure of the overall system. The first negative peak can be observed at the position of pressure pulse and the second can be detected at the wall. b (right): Reduced pressure at the solid surface. Value for the wettability parameter parameter was $\Gamma_w=0\text{-}0.05695$. In both case the reduction factor was the overall pressure after 300 step.

corresponds to the reflected wave and a second negative peak follows it. Comparing to our simulations to experimental results [20], we can see good qualitative agreement.

Liquids under negative pressure are metastable for liquid/vapour phase transition (see e.g. [2,6]) i.e. the stretched liquid tends to equilibrate by producing bubbles. Being the LBM a little bit inaccurate with phase transition phenomena [7], cavitation (i.e. liquid-vapour phase transition) was neglected in our simulation. Cavitation can be avoided in real liquids when the negative pressure is small or its duration is very short [2,21], therefore our model - in its present form - can be applied only for very short and/or very small pressure waves.

4. Conclusions

In this paper we demonstrated that the development of negative pressure can be observed easily in lattice-Boltzmann models. The numerical method used for this simulation can be implemented easily and can help us to design new, real experiments. Both the optimal arrangements of reflectors and their materials (rigidity, surface treatment) can be found in a straightforward manner by such numerical simulations. These results might help us to design underwater structures, which can resist to the negative rarefaction waves.

Acknowledgement

This work was partially supported by the Hungarian Research Foundation (OTKA) under contract number F034333. One of the authors (A.R.I.) was supported also by the Bolyai Research Fellowship.

References

[1] Kedrinskii, V.K. (1976) Negative pressure profile in cavitation zone at underwater explosion near free surface, *Acta Astronautica* **3**, 623-632
[2] Trevena, D.H. (1987) *Cavitation and Tension in Liquids*, Adam Hilger, Bristol
[3] Vinogradov V.E. and Pavlov, P.A. (2000) The Bounday of Limiting Superheats of n-Heptane, ethanol, benzene and Toluene in the Region of Negative Pressures, *High Temperature* **38**, 379-383
[4] Eisenmenger, W., Köhler, M., Pecha, R. and Wurster, C. (1997) Negative pressure amplitudes in water measured with the fiber optic hydrophone, *Prog. Nat. Sci.* **7**, 499-501
[5] Carnell, M.T., Gentry, T.P. and Emmony, D.C. (1998) The generation of negative pressure waves for cavitation studies, *Ulrasonics* **36**, 689-693
[6] Imre, A., Martinás, K., and Rebelo, L.P.N. (1998) Thermodynamics of Negative Pressures in Liquids, *J. Non-Equilib. Thermodyn.* **23**, 351-375
[7] Házi G., Imre R. A., Mayer G. and Farkas I. (2002) Lattice Boltzmann nethods for two-phase flow modeling, *Ann. Nucl. Energy*, **29**, 1421-1453
[8] Bhatnagar P. L., Gross E. P., Krook M. (1954) A model for collision processes in gases. I. small amplitude processes in charged and neutral one-component systems, *Phys. Rev.*, **94**, 511-525
[9] Qian Y.H., d'Humiéres, Lallemand P., (1992) Lattice BGK for Navier-Stokes equation, *Europhys. Letters*, **17**, 479-484
[10] Shan, X., Chen, H. (1993) Lattice Boltzmann model for simulating flows with multiple phases and components, *Phys. Rev. E*, **47**, 1815-1819
[11] Sehgal B. R., Nourgaliev R. R., Dinh T. N. (1999) Numerical simulation of droplet deformation and break-up by lattice-Boltzmann method, *Prog. Nucl. Energy*, **34**, 471-488
[12] Qian Y. H. and Chen S. (1997) Finite size effect in lattice-BGK models, *Int. J. Mod. Phys. C* **8**, 763-771
[13] Martys N. S., Chen, H. (1996) Simulation of multicomponent fluids in complex three-dimensional geometries by the lattice Boltzmann method, *Phys. Rev. E*, **53**, 743–750
[14] Yang Z. L., Dinh T. N., Nourgaliev R. R. and Sehgal B. R. (2001) Numerical Investigation of bubble growth and detachment by the lattice-Boltzmann method, *Int. J. Heat and Mass Transfer* **44**, 195-206
[15] Shan, X. and Chen , H. (1994) Simulation of nonideal gases and liquid-gas phase transitions by the lattice Boltzmann equation, *Phys. Rev. E* **49**, 2941-2948
[16] Shan X., Doolen G. (1996) Diffusion in a multicomponent lattice Boltzmann equation model, *Phys. Rev. E*, **54**, 3614-3620
[17] Martys N. S. and Douglas J. F. (2001) Critical properties and phase separation in lattice Boltzmann fluid mixtures, *Phys. Rev. E* **63**, 1205-1218
[18] Langaas K. and Grubert D. (1999) Lattice Boltzmann simulations of wetting and its application to disproportionate permeability reducing gels, *J. Petr. Sci. Eng.* **24**, 199-211
[19] Hazlett, R.D. and Vaidya, R.N. (2002) Lattice-Boltzmann simulations and contact angle hysteresis in convergent-divergent media, *J. Petrol. Sci. Eng.*, **20**, 167-175
[20] Kedrinskii, V.K. (2002) Relaxation effects and disintegration problems of cavitating liquids at pulse loading, this book
[21] Šponer, J. (1990) The Dependence of Cavitation Threshold on Ultrasonic Frequency, *Czech. J. Phys. B* **40**, 1123-1132

ULTRASONIC CAVITATION IN FREON AT ROOM TEMPERATURE

FRÉDÉRIC CAUPIN AND VINCENT FOURMOND
Laboratoire de Physique Statistique
de l'Ecole Normale Supérieure
associé aux Universités Paris 6 et Paris 7 et au CNRS
24 rue Lhomond 75231 Paris Cedex 05, France

Abstract. We report preliminary results on ultrasonic cavitation in freon (1,1,2-trichloro 1,2,2-trifluoro ethane). We use a high intensity 1 MHz acoustic wave produced by a hemispherical transducer to quench a small volume of liquid in the negative pressure region during a short time, far from any wall. For a sufficiently large pressure oscillation, we observe the nucleation of bubbles. We describe the three different methods we use to detect cavitation: diffusion of light, optical imaging and acoustic detection by the emitting transducer itself. We present our first results on the statistics of cavitation. We finally address the questions of calibrating the negative pressure reached at the focus and of the nature of cavitation in our experiment (homogeneous vs. heterogeneous).

1. Introduction

Any liquid may be kept for some time at negative pressure in a metastable state. However, there is an absolute limit of metastability, called the spinodal pressure, at which the liquid becomes macroscopically unstable against long wavelength fluctuations. In most liquids, the spinodal pressure increases monotonically with temperature. In the case of water, this usual behaviour is predicted by several theories, but another theory expects the spinodal pressure to reach a minimum around 40°C [1]. This second scenario is related to the existence of a line of density maxima in water, that would intersect the spinodal line at its minimum.

Experimentally, the liquid-gas spinodal cannot be reached, because vapour bubbles nucleate at less negative pressures. In the absence of nucleation seeds like dissolved gases or solid surfaces, this pressure is called the homogeneous cavitation pressure P_h. It is related to the spinodal pressure. Therefore, measuring the temperature dependence of the homogeneous cavitation pressure in water would be of great interest, because experimental data are lacking in the range from 0 to 40°C, and they would allow to decide between the two scenarios mentioned above.

As a preliminary study for this purpose, we have investigated cavitation in a freon: 1,1,2-trichloro 1,2,2-trifluoro ethane. This substance does not exhibit any line of density maxima, but allows us to check the efficiency at room temperature of a method we have previously used to study cavitation in liquid helium [2]. This

307

A.R. Imre et al. (eds.), Liquids Under Negative Pressure, 307–313.

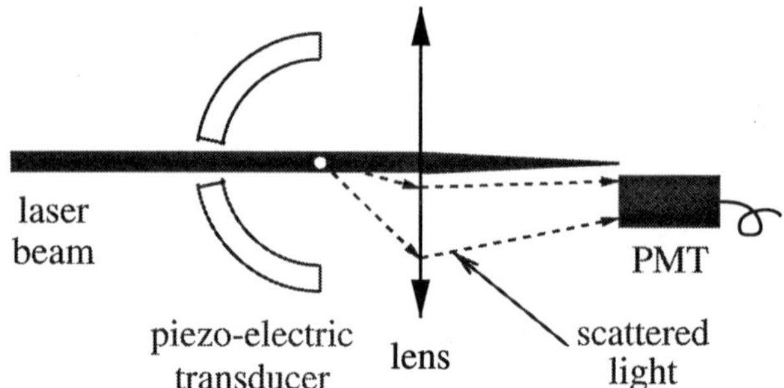

Figure 1. Schematic drawing of the experimental setup.

method allowed us to obtain some experimental evidence for the existence of a minimum in the spinodal line of liquid helium 3, for which we have also given theoretical arguments [3, 4]. We have chosen this freon because it is weaker than water. According to Zel'dovich [5] or Fisher's theory [6], we have:

$$P_{\mathrm{h}} = -\sqrt{\frac{16\pi}{3k_{\mathrm{B}}T}\frac{\sigma^3}{\ln\left(\mathcal{N}_{\mathrm{A}}k_{\mathrm{B}}T/h\right)}} \qquad (1)$$

where σ is the liquid-gas interfacial tension; $\mathcal{N}_{\mathrm{A}}$ is Avogadro number, here taken as the number of nucleation sites, and the attempt frequency of nucleation has been estimated by a thermal frequency $k_{\mathrm{B}}T/h$. Eq. 1 gives $P_{\mathrm{h}} = -1320$ bar for water and -179 bar for freon at 20°C.

2. Experimental methods

In our experiments, we quench the liquid in the negative pressure region with a high amplitude acoustic wave. This wave is focused by a hemispherical piezo-electric transducer, driven with short bursts of a 1 MHz sine wave (typically 1 to 30 cycles; all the figures shown in this paper were obtained for 8 cycles). These electric bursts are produced by a homemade amplifier that reaches up to 1600 W of instantaneous power (200 V on 25 Ω).

In this room-temperature experiment, the transducer is immersed in freon inside a sealed stainless steel cell. A laser beam is shined on the acoustic focus, through the two windows of the cell and through a hole drilled in the transducer (see Fig. 1). The scattered light is collected with a lens (50 mm focal length) in a photomultiplier tube. For a low driving voltage of the transducer, we observe a modulation due to the acoustic wave (grey curve on Fig. 2). At sufficiently high voltage, a peaked structure appears (black curve on Fig. 2) around 10 μs after the largest oscillation of the driving voltage (at time $\tau_{\mathrm{b}} = 8.5\,\mu$s); this corresponds to a bubble. Because of the finite quality factor of the transducer, its oscillation builds up over several cycles, and the acoustic wave near its surface is largest at τ_{b}; the 10 μs delay comes from the flight time of the sound wave (across 8 mm

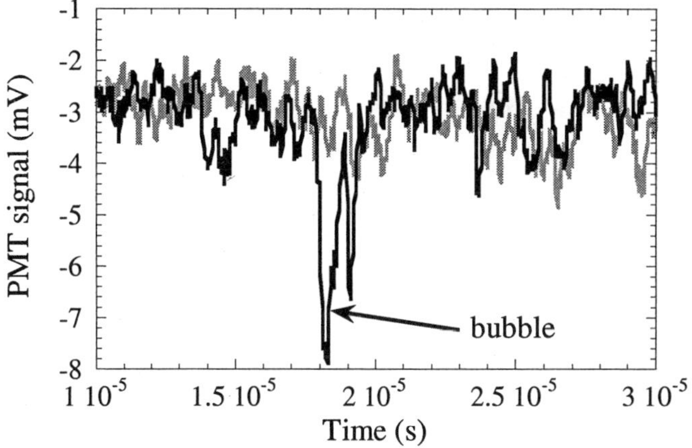

Figure 2. Two signals corresponding to successive bursts produced under the same experimental conditions. Only the lower curve exhibits a cavitation event.

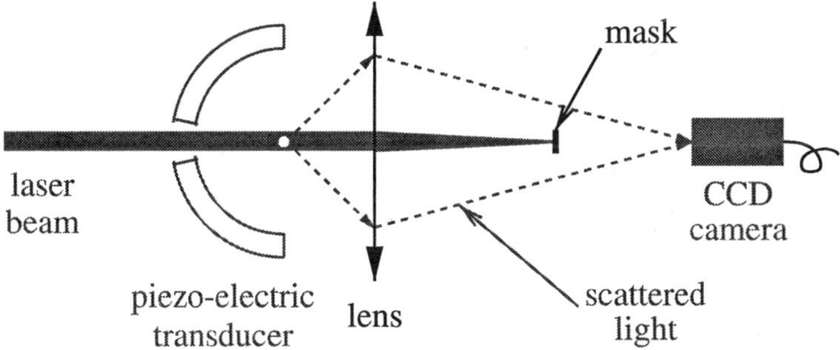

Figure 3. Setup for imaging of the acoustic focal area.

at $783.7\,\mathrm{m\,s^{-1}}$). The two signals shown on Fig. 2 were actually obtained with the same driving voltage. Only some of the bursts triggered cavitation, which allowed us to define and measure a cavitation probability (see Sec. 3).

We now describe two other methods of detection that were not used in our experiments on liquid helium.

Using the setup shown on Fig. 3, we can make images of the acoustic focal area on a CCD camera. The mask is a small disk made of black tape placed in the focal plane of the lens; its purpose is to remove the direct beam and thus suppress the white background on the image. This standard optical filtering technique allows us to image the sources of light scattering with an enhanced contrast.

A typical image sequence showing cavitation is displayed on Fig. 4. For the sake of clarity, the contrast was inverted and the static background due to dusts and defects in the optics was removed. The magnification was not yet calibrated: we estimate the actual width of the picture to be of the order of $1\,\mathrm{mm}$. The first image gives the time origin; it corresponds to the arrival of the sound wave at the acoustic focus. As explained above, this produces some light scattering, which can be seen here as a diffuse grey zone; many tiny bubbles might also be present here.

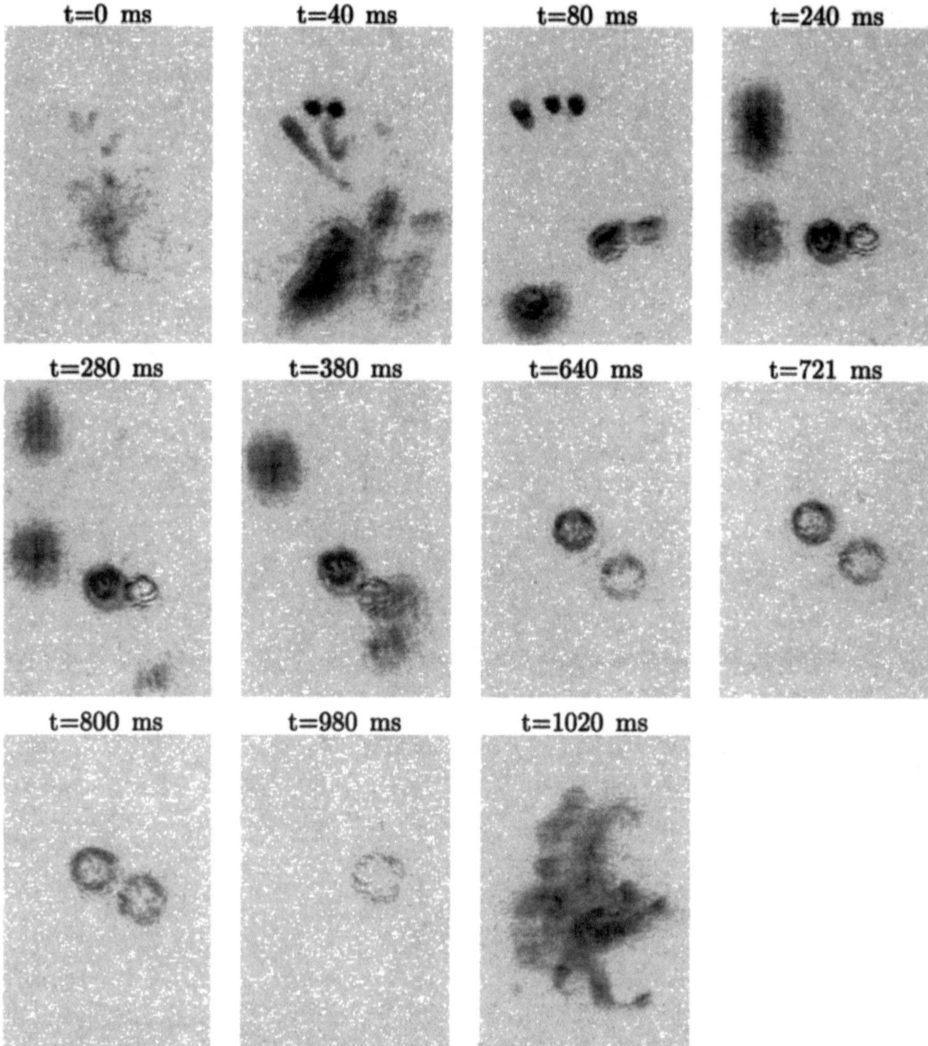

Figure 4. Sequence of images obtained with the setup of Fig. 3. Each picture is labeled above with the time; the time origin corresponds to an acoustic impact at the focus.

The eight following images show the evolution of several large bubbles: first they diverge from the acoustic focus; then they move upward because of the buoyancy forces; finally, they collapse in the liquid at atmospheric pressure. The last two images are around time 1 s: the first shows that the disappearance of the last bubble, and the second one shows the halo and bubbles created by the next burst.

The third method we used is based on the echo phenomenon. When the acoustic wave travels back to the transducer surface, the corresponding displacement is converted into voltage; this signal is shown on Fig. 5. In the absence of bubble, one observes the dashed grey curve: there is a small 1 MHz signal, which we

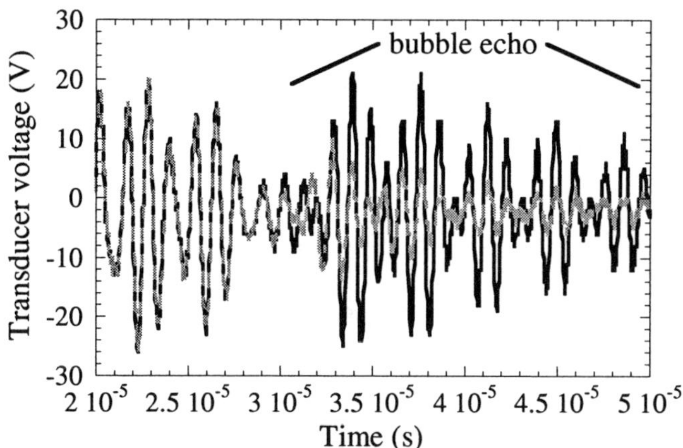

Figure 5. Observation of the echo on the transducer for two successive bursts produced under the same experimental conditions.

attribute to some ringing of the transducer or transmission of the wave from one side to the other near its equator. When a bubble is present at the focus, one observes the solid black curve: it is undistinguishable from the dashed curve until a time $\tau_{\mathrm{echo}} = 30\,\mu$s. We have $\tau_{\mathrm{echo}} \simeq \tau_{\mathrm{b}} + 2\tau_{\mathrm{flight}}$: the wave reaching the bubble at the end of the burst is reflected by it and goes back to the transducer where it is detected.

To conclude this description of the experimental details, it is important to note that the three methods we used have always detected consistently the onset or the absence of cavitation.

3. Results and discussion

All three methods described in Sec. 2 lead to the same conclusion that cavitation is a stochastic phenomenon: for a certain voltage range, when we repeat the experiment under the same conditions, we obtain cavitation only in some of the bursts. This was also noticed in liquid helium [2]; following the same analysis, we define the cavitation probability as the fraction of bursts showing a nucleation event. We have performed a first measurement of the cavitation probability as a function of the driving voltage (see Fig. 6). Each probability was computed over one hundred bursts. One can see that the probability raises smoothly from 0 to 1, over a width which is larger than the experimental noise on the voltage; this means that the phenomenon is intrinsically random, and not an all-or-none process smoothed by noise.

Cavitation is a thermally activated process: there is a pressure-dependent energy barrier $E_{\mathrm{b}}(P)$ to overcome to nucleate the vapour phase. The cavitation rate Γ follows an Arrhenius law:

$$\Gamma = \Gamma_0 \exp\left(-\frac{E_{\mathrm{b}}(P)}{k_{\mathrm{B}}T}\right) \tag{2}$$

312

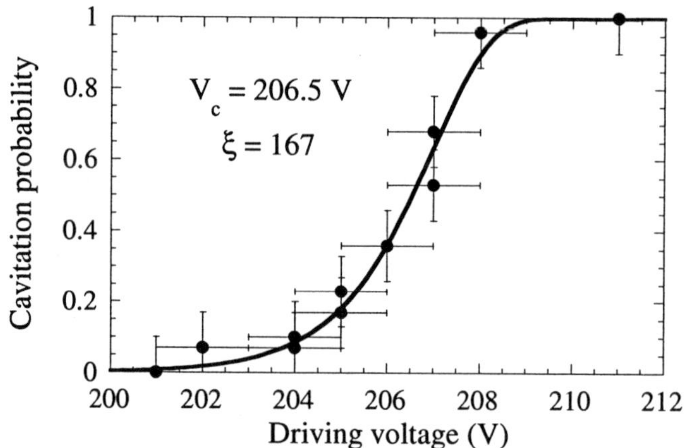

Figure 6. Cavitation probability as a function of driving voltage. The horizontal and vertical error bars respectively represent the noise on the voltage amplitude and the statistical uncertainty. The solid curve is a fit with Eq. 4.

where Γ_0 is a prefactor. The cavitation probability Σ in an experimental volume V_{exp} and during an experimental time τ_{exp} can then be written:

$$\Sigma = 1 - \exp\left(-\Gamma V_{\mathrm{exp}} \tau_{\mathrm{exp}}\right) \tag{3}$$

The largest negative pressure P reached in the wave is a function of the driving voltage V, and by using Eqs. 2 and 3 and expanding around the cavitation voltage V_{cav} where $\Sigma = 1/2$, we obtain:

$$\Sigma = 1 - \exp\left\{-\ln 2 \exp\left[\xi\left(\frac{V}{V_{\mathrm{cav}}} - 1\right)\right]\right\} \tag{4}$$

with $\xi = \Gamma_0 V_{\mathrm{exp}} \tau_{\mathrm{exp}} (\partial \ln P / \partial \ln V)_{V_{\mathrm{cav}}}$. Fig. 6 shows that the measured probability is satisfactorily fitted by this formula.

The main difficulty in our experiment is to estimate the negative pressure P_{cav} corresponding to V_{cav}. By comparing P_{cav} with $P_{\mathrm{h}} = -179$ bar (see Sec. 1), we will be able to decide whether we observed homogeneous or heterogeneous cavitation. We can estimate the relation between P and V in the case where the focusing of the wave is linear; the corresponding analysis is described elsewhere [2]. Among other parameters, we need to know the value of the quality factor Q of the transducer; it is determined by measuring the cavitation voltage as a function of the number of cycles in the burst [2]: we find Q to be around 20. Because the signal delivered by our homemade generator has not a square envelope, there are some difficulties in the analysis. A very rough estimate gives $P_{\mathrm{cav}} > -370$ bar. This is a lower bound because of the non-linearities; they are due to the large decrease in the sound velocity at negative pressure, and their effect is to reduce the efficiency of the focusing. Since it is more negative than P_{h}, but also of comparable amplitude to it, we consider our result as a preliminary indication of homogeneous nucleation. This regime is likely to be reached in our experiment because of the small experimental volume (tens of microns in diameter) and time (tens of nanoseconds) involved, which make the probability to find an impurity very low.

Of course, this point requires further investigation. We plan to improve the accuracy in the estimate of P_{cav} first by improving the quality of the amplifier to obtain a more reliable value of the lower bound; then we can also obtain an upper bound by measuring the variation of V_{cav} with the static pressure, as was done for liquid helium [2]. At room temperature, another solution is available, which would consist in measuring directly the pressure at the focus by inserting an hydrophone with an active area smaller than the sound wavelength.

4. Conclusion

We have reported preliminary measurements on cavitation in freon at room temperature. We have detailed three different methods used to detect cavitation, and presented our first measurements of the cavitation probability. We have discussed the homogeneous character of cavitation. By using a more powerful amplifier, we should be able to produce and study cavitation in water at large negative pressure.

Acknowledgements

We would like to thank Sébastien Balibar for helpful discussions, and Christophe Herrmann who built our amplifier.

References

1. Debenedetti, P.G. (1996) *Metastable liquids*, Princeton University Press, Princeton, and references therein.
2. Caupin, F. and Balibar, S. (2001) Cavitation pressure in liquid helium, *Phys. Rev. B* **64**, 064507 (1–10).
3. Caupin, F., Balibar, S., and Maris, H.J. (2001) Anomaly in the stability limit of liquid helium 3, *Phys. Rev. Lett.* **87**, 145302 (1–4).
4. Caupin, F. and Balibar, S. (2002) Quantum statistics of metastable liquid helium, *this conference*.
5. Zel'dovich, Ya. B. (1942) On the theory of the formation of a new phase: cavitation, *Zh. Eksp. Teor. Fiz.* **12**, 525-538.
6. Fisher, J.C. (1948) The fracture of liquids, *J. App. Phys.* **19**, 1062–1067.

DEPENDENCE ON KINDS OF IMPURITY GASES IN METALS OF NEGATIVE PRESSURES IN WATER/METAL BERTHELOT TUBE SYSTEMS

YOSHIHITO OHDE and YASUTOSHI TANZAWA

Department of Systems Engineering, Nagoya Institute of Technology, Gokiso, Shouwa-ku, Nagoya 466-8555, Japan

Abstract.   With an all-stainless-steel Berthelot tube having a cavitation-free range up to ca. -10 MPa, trends in negative pressure with repeated cavitation in water were investigated, I) when pre-degassed single-crystalline copper pieces were additionally exposed to either hydrogen gas or nitrogen gas and then co-sealed together with de-aerated water in the tube, respectively, and II) when its pre-degassed stainless steel plugs were additionally exposed to super-critical water just before their sealing uses. These additional pre-treatments for pre-degassed metals deteriorated trends in negative pressure in the tube. All the trends were found qualitatively consistent when we admit the earlier trends were dominated by hydrogen concentration in metals and later trends by the other kinds of impurity gas concentrations in the metals.

1. Introduction

For facilitating experimental studies of liquids under static negative pressure we have elaborated on Berthelot method using metal tubes since 1987. Our motivation came from failure of polymers known as 'Environmental Stress Cracking' that occurred when polymer test pieces were weakly loaded while immersed in their non-solvents. Increase in their mutual solubility under dilative stress (equivalent to negative pressure) was concluded to be the cause [1]. For checking the conclusion we searched for relevant data in the literature in vain. There had been only a few measurements of thermodynamic properties of water under negative pressure. Henderson and Speedy [2] measured temperature of maximum density of water to -22 MPa in the Berthelot method using glass capillaries of ca.10^{-3} cm^3 in volume. Their success came from the two principles for minimizing the cavitation probability, 1) use of small-sized sample, and 2) preparation of 'clean' liquid/container system [3]. In spite of their success, we selected metal tubes because of the mechanical sealing, since our targets were flammable organics including polymer solutions.

When a water/stainless steel tube/copper plug (Berthelot) system was subjected to a run of temperature cycles, negative pressure, though scattered wide, increased gradually with the number of temperature cycles. This cavitation history effect was a result of exhaustion of gaseous nuclei of cavitation on the tube wall [4]. Furthermore, it was consistently interpreted only when the gas-trapping crevice model [5-7] was supplemented with a gas-supply assumption [8]. The original models all assumed that gases (air) trapped within tiny 'crevices' on the container wall and in dust particles worked as cavitation nuclei. Our model additionally assumed that those surface crevices should be replenished with impurity gas from the metal bulk.

315

The extended model guided us to find that: 1) degassing pre-treatments for metals were essential for achieving negative pressures above −10 MPa, though they accompanied a retardation effect; that is, rises in negative pressure for earlier cycles were badly hindered. The retardation effect was attributed to annealing of metals owing to the high temperatures for their efficient degassing, and was partly removed by additional room temperature underwater pressurization at a few hundreds MPa for 100 min [9]. Our negative pressures were raised to −17 MPa for water at 59 ^{0}C [9] and to around −20 MPa for some organics at 15 ^{0}C [10] in ca. 1 cm^3 of a stainless steel tube sealed with pre-degassed Ni plugs, respectively.

Unfortunately, each maximum was achieved only after several thousands of temperature cycles, and negative pressures scattered wide. The physics involved in the two drawbacks, the time consumption and the wide scatter, has been grasped only recently. Trends in negative pressure in single-crystalline and poly-crystalline molybdenum (s-Mo and p-Mo) tubes sealed with single-crystalline copper (Cu) plugs, respectively, were studied 1) when all the metals were pre-degassed and 2) after both tubes alone were exposed to nitrogen gas temporarily [11]. The single-crystalline Mo tube yielded a narrow scatter of ca. 1MPa wide while the poly-crystalline Mo tube did scatter even wider than 10 MPa. The s-Mo tube's capability deteriorated by the gas exposure was recovered only partially after a total of ca. 5000 cycles while the p-Mo tube's capability partly even after 12000 cycles. The time consumption and the scatter are natural results of both slow gas transports in metals and high cavitation probability of the super-expanded liquid.

Figure 1 shows a combined trend of two trends investigated in the successive runs before and after its nitrogen gas exposure (Figs.3 & 5 in [11]) for the same s-Mo tube. The deterioration of the tube's capability is seen in two points, 1) the low negative pressures from −3 MPa for the onset a few cycles, and 2) the later steady falling trend. Nevertheless, the intermediate negative pressures were around −16 MPa higher than the maximum of −15 MPa before the gas-exposure. Careful readers may note the frequency of cycles yielding low cavitation negative pressures is much less after the gas-exposure than that before the gas-exposure. They are a puzzle to be solved, and are the direct motivation of the present work.

We investigate trends in negative pressure for water in an all-stainless-steel Berthelot tube, I) when pieces of s-Cu that were a) pre-degassed and additionally exposed to either b) hydrogen gas or c) nitrogen gas and respectively co-sealed together with de-aerated water in the tube, and II) after its stainless steel plugs were a) pre-degassed in ultra-high-vacuum range and b) additionally exposed to super-critical water in a metal autoclave. We shall point out the puzzle has a natural answer if we take into accounts thermodynamics and kinetics of impurity gases in the liquid/metal container (Berthelot) system, and of impurity gas transports in metals used.

2. Experiments

The apparatus and the experimental procedure employed here are similar to those reported before [11]. So here we describe points specific to this work in detail and the others briefly.

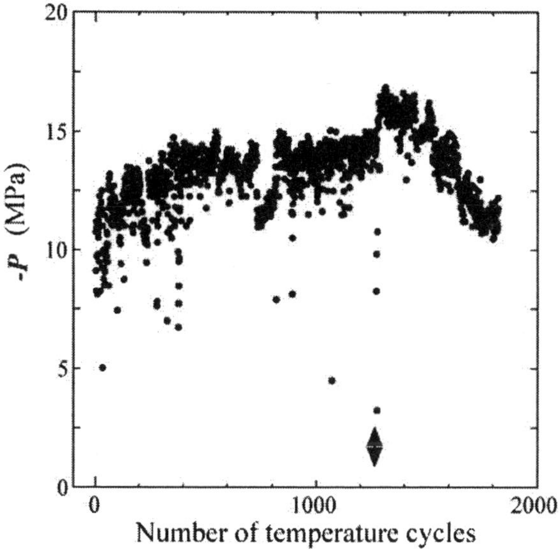

Figure 1. Combined trend in negative pressure for water in an all-single-crystalline Mo/Cu tube before and after nitrogen gas exposure of the s-Mo tube alone. The arrow indicates the onset cycle just after the gas exposure. Each point represents negative pressure at which cavitation occurred each cycle. © 2001 IOP Publishing Ltd. J. Phys. D: Appl. Phys. 34, 1711-1726 [11]

2.1. TUBE, PLUG, AND AUTOCLAVE

A tube was made of a precipitation-hardened grade of stainless steel, SUS630. It consisted of three parts, 1) top cap for sealing 2) specimen chamber and plug, and 3) bottom cap for waterproof of sensors. They were assembled with 8 pairs of bolts and nuts angled every $\pi/4$ radian at the outskirts of their flanges. The design principle was the same as reported before [11]. Its specimen chamber was a cylindrical hole of 3 mm diameter and 22mm in depth. Outside of its bottom, metal-film strain gauges were cemented to form a built-in pressure transducer of Wheatstone bridge type. The transducer was calibrated by filling the specimen chamber with nitrogen gas of known positive pressures up to +18 MPa at several temperatures between 20 and 80 ^{0}C. We therefore measured 'negative pressure' by extrapolating the linear relation between the elastic strain of the bottom diaphragm and positive pressure into the negative range. Near to the transducer a Chromel-Alumel thermocouple was fixed to monitor the liquid temperature.

Plugs were made of stainless steel, SUS316L. As-machined pieces of 5×5×3 mm in size were pre-degassed at 270 ^{0}C under vacuum of 5×10^{-4} Pa for 2 months at least. Some steel plugs were specifically pre-degassed at 1250 ^{0}C for 64 h under vacuum of 2×10^{-7} Pa, and additionally subjected to super-critical water at +180 MPa, 530 ^{0}C for 12 h in a metal autoclave. Unless otherwise noted, the former plugs were used in the present work.

The autoclave was made of heat-resistant Ni-rich steels supplied by Daido Steel Co. Its specimen chamber was a cylindrical hole of 60 mm in diameter and 60 mm in depth. It was filled with de-aerated water in which a pre-degassed steel plug was immersed. It

318

was sealed with a thick cap assisted by a Bridgman type gasket of graphite-asbestos composite. The chamber was connected at the center of the cap with a SUS304 thick pipe of which the upper end was a pressure manipulating part including a Bourdon pressure gage, a leak valve, and a rupture disk made of copper sustaining up to the safety limit, +200 MPa. The pressure manipulating part stood in air with the connecting pipe through a hole of heat-resistant ceramics boards above the autoclave in a furnace.

When the water/autoclave system was heated in the furnace with a heater of 1.2 kW, excess positive pressure in water was generated in the same thermodynamic principle as the Berthelot method, and some drops of water were leaked out through the leak valve to keep the pressure below the safety limit. The critical point of water, +22.1 MPa at 374 ^{0}C (647K) was easily exceeded below 600 ^{0}C, the temperature safety limit of the autoclave.

Just before its use, any plug was heated under boiling water and during the heating ultra-sonic waves were irradiated intermittently while immersed in the boiling hot water. It took ca. 30 min or a little longer time to exhaust almost all the continual boiling sites on the plug.

2.2 RUN OF TEMPERATURE CYCLES

Water, once distilled, was heated for 30 min in a Pyrex beaker to boiling, and irradiated intermittently by ultrasonic waves. The hot water was poured directly into the specimen chamber. The chamber was sealed with the prepared plug under compression by rotating the screw rod through the top cap. Then the tube was immersed in a thermostat bath, and by elevating its temperature step-wise we measured a pressure versus temperature relation by which we determined the characteristic temperature, T_0, of the current Berthelot system, where T_0 is the temperature at which pressure equals to zero.

Then we operated an automatic temperature cycle repeater (ATCR) for carrying out a run of temperature cycles for the Berthelot system. About the assembly of the ATCR and its operation, please consult our report elsewhere [11]. When the water/metal tube Berthelot system was first heated in its bath, excess positive pressure was generated in the sample liquid in the tube. A personal computer of the ATCR emitted a control pulse to a stepping motor for lifting up the tube in air for cooling. Through a 12 bit A/D converter the PC read voltage signal from the tube after being amplified by ca. 2000 times with a DC amp, and converted it to current pressure. The PC was so programmed as to monitor pressure signal every ca. 10msec. It compared the current pressure with that at the previous sampling time. And when a pressure jump-up was first greater than a threshold value (+0.5 MPa, a noise margin), the PC regarded the event as a result of cavitation, and emitted a control pulse to the stepping motor to immerse the tube back into the heating bath for the next cycle.

It allowed us with minimum attention to continue a run of temperature cycles day and night even over a month at cycle rates of around 60 to 140 cycle per day.

2.3. PIECES OF SINGLE-CRYSTALLINE COPPER

Small pieces of 2×2×1 mm in size were cut with a micro-cutter using a circular blade of solidified diamond powder out of a rod of single-crystalline copper purchased from

Goodfellow Co. They were pre-degassed at 270 ^{0}C for a month under vacuum of 5×10^{-4} Pa, and were kept at room temperature in the same vacuum chamber. Just before the co-sealing test, each piece, with or without its exposure to either hydrogen gas or nitrogen gas at $+10$ MPa at 75 ^{0}C for 15 h, was heated in boiling water for 30 min and intermittently irradiated by ultra-sonic waves while immersed in the hot water. We set T_0's ca. 75 ^{0}C for the runs to investigate effects of co-sealing of s-Cu pieces together with water in the tube and the maximum temperature of the temperature cycles ca. 90 ^{0}C throughout the present work.

3. Results

3.1. CO-SEALING OF GAS-EXPOSED SINGLE-CRYSTALLINE COPPER

Figure 2 shows a trend in negative pressure when a pre-degassed s-Cu piece was co-sealed together with de-aerated water in the all-stainless-steel tube. Negative pressure increases steeply to around -10 MPa in the initial 50 cycles and very gradually to around -11 MPa to the last cycle, number 630. The frequency of cycles yielding negative pressure below -10 MPa becomes less and less steadily. The trend shows a typical cavitation history effect.

It is good when we recall the second run trend for the pre-degassed poly-crystalline Mo tube (Figure 4 in [11]); the lower envelope remained to be as low as -4 MPa even after a total of 4000 cycles while the higher envelope rose to -14 MPa and negative pressures scattered uniformly in the width. Thus, the all-stainless-steel Berthelot tube has been so well conditioned that it have a negative pressure window opened up to ca. -10 MPa.

Since a liquid under negative pressure is in a super-expanded metastable state, a cavitation event is triggered by a single nucleus, by then the weakest one. Thus, the good trend in Fig. 2 is also a manifestation of the effectiveness for exhausting nuclei by the boiling pre-treatment (heating in boiling water and intermittent irradiations of ultrasonic waves) for the co-sealed s-Cu piece. We use this trend as a reference with which trends for gas-exposed s-Cu pieces are to be compared.

Figure 3 shows a trend for a hydrogen gas-exposed piece. Negative pressures scatter much wider towards lower values. They distribute almost uniformly in the wide range, but if an average negative pressure for a group of successive, say, 5 or 10 cycles were plotted against the number of temperature cycles, it would decrease steadily in the initial 300 cycles and level off for subsequent cycles. The steep initial increase may be due to the boiling pre-treatment after the gas-exposure of which effectiveness is pointed out just above. The intermediate steady falling in the average negative pressure was caused by the gas-exposure. Then what does it mean the final leveling in the average trend? Let see the trend shown in Fig. 4 obtained in the next run for de-aerated water alone in the same tube sealed with the same plug. Negative pressures exhibit a wider scatter owing to the steady increase of the higher envelope as a result of the cavitation history effect.

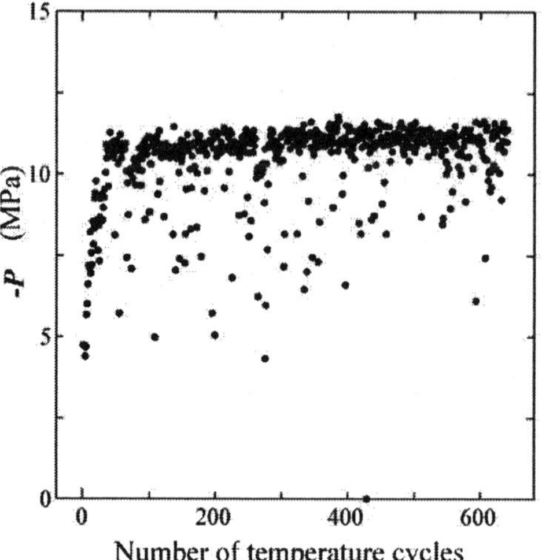

Figure 2 Trend in negative pressure for pre-degassed s-Cu

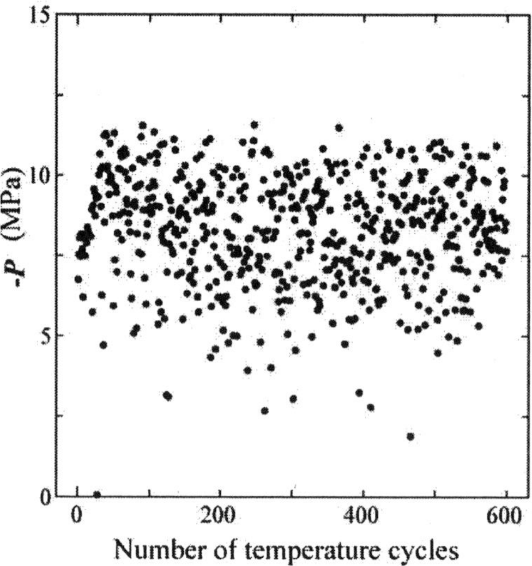

Figure 3. Trend in negative pressure for hydrogen-exposed s-Cu

Since the steel plug had also been subjected to the boiling treatment just before the run, the wide scatter was primarily caused by weak gaseous nuclei on the tube wall. The tube's capability had been badly deteriorated during the last run.

The last leveling at the low average negative pressure in Fig. 3 is not a result of diminishing of out-gassing from the hydrogen-gas exposed s-Cu piece but rather a result of kinetic balance of impurity gas in the liquid/container system. We assured this by carrying out successive two similar runs for another hydrogen-exposed piece, though not shown here.

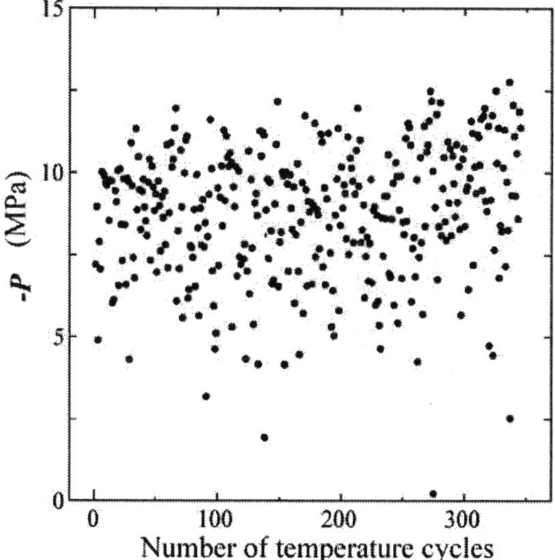

Figure 4. Trend in negative pressure for water in the same tube just after the hydrogen-exposed s-Cu run in Figure. 3.

Before achieving recovery to the original capability like shown in Figure 2, we similarly carried out the run for another nitrogen-gas exposed piece (see Figure 5). Negative pressure for the nitrogen-gas exposed piece scatters wide, too. But one can see a narrow band along the higher envelope of the scatter. Most of cycles yield negative pressures falling in the band. Though not shown here, we carried out a water-alone run for checking the tube's capability just after this run. We again saw a faint band along the higher envelope increasing gradually with the number of temperature cycles.

Again, the narrow band in the Fig. 5 increases steadily up to the cycle ca. 200 and then decreases steadily. It took 3.5 days to carry out those 250 cycles. When we then changed the cycle rate from 72 to 96 cycles per day, negative pressure temporarily recovered a similar increasing trend. However, it soon began to decrease steadily at the faster cycle rate. Such dependence of trend in negative pressure on the cycle rates is also seen as repeated staggering of the later steady falling trend in Fig. 1.

Careful readers may note the lower envelope of the scatter decreases steadily for later cycles, too. It is impossible to completely differentiate which kind of gas, hydrogen or nitrogen in s-Cu, was responsible for a specific cavitation event.

3.2. EXPOSURE TO SUPERCRITICAL WATER

Figure 6 shows a trend for a SUS316L plug specifically pre-degassed at 1250 ^{0}C under vacuum of 2×10^{-7} Pa in the range of ultra-high vacuum (UHV). Figure 7 shows a trend for another so-pre-degassed steel plug after its additional exposure to super-critical water at +180 MPa and 530 ^{0}C generated in the metal autoclave. Before the trials three successive water alone runs of a total of 4500 cycles were done to recover the tube's capability of having a cavitation-free range up to –10 MPa.

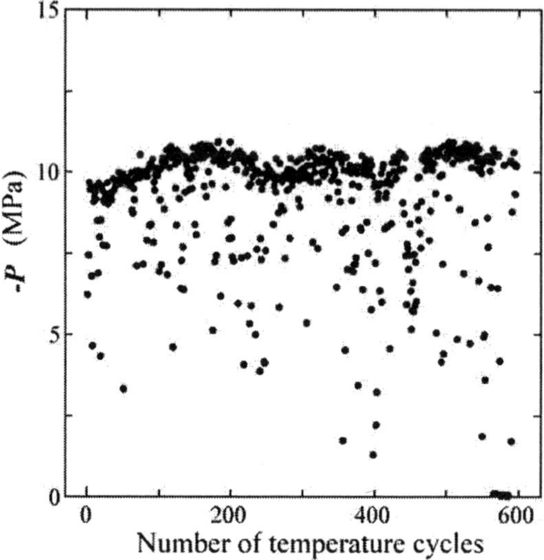

Figure 5. Trend in negative pressure for nitrogen-exposed s-Cu

Negative pressure for the UHV pre-degassed steel plug exhibits a typical two-stage increase, initial steep stage and subsequent gradual stage. It is not so good a trend as that shown in Fig.2 in spite of the UHV de-gassing. This is partly due to what we called the retardation effect of degassing pre-treatment [9]. Anyway, we can use the trend in Fig.6 as a reference with which the trend in Fig. 7 for another so pre-degassed steel plug that had been additionally exposed to super-critical condition of water in the autoclave just before the run.

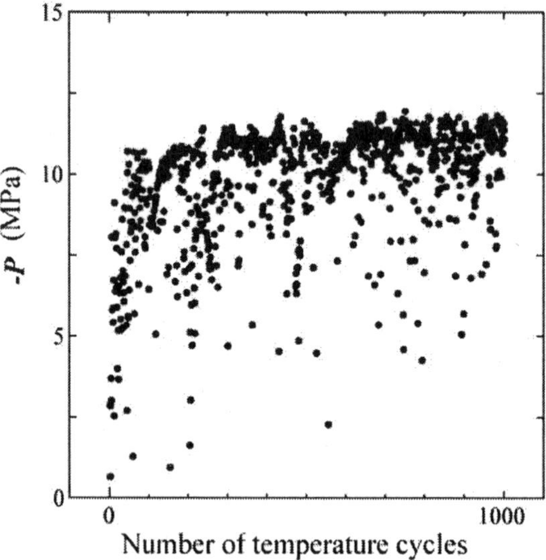

Figure 6. Trend in negative pressure for a pre-degassed steel plug

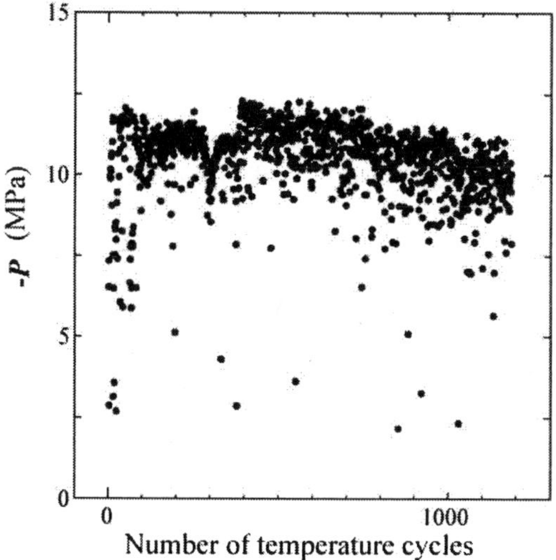

Figure 7. Trend in negative pressure for another pre-degassed steel plug after its additional exposure to super-critical water

The trend in Fig.7 is peculiar with respect to the two points. First, negative pressures levels around -12 MPa in initial ca. 650 cycles, and decreases steadily for later cycles.

Though not shown here, the plug's second run after renewing the water yielded similar but a little poorer trend, leveling around -11 MPa for earlier cycles, and subsequent steady and gradual falling to a final leveling of -9 MPa to the total of 2500 cycles.

Careful readers may note the scatter width becomes wider in the later steady falling trend. Thus, the later decreasing trend suggests concentrations of some kinds of impurity gas atoms had been increased in the steel plug during the autoclaving run. Second, nevertheless, the frequency of intermediate cycles yielding negative pressure below, say, -9 MPa is much less in Fig. 7 than the frequency of those intermediate cycles in the reference trend in Fig. 6 for the merely pre-degassed plug.

Thermodynamically the autoclave run is nothing other than a temperature cycle for a metal Berthelot tube. There should be many gas sources in the autoclave system. Those gases, mainly of nitrogen and oxygen, would have once dissolved into the water, and might have furthermore diffused into the steel plug in the autoclave of which temperature was kept high at 530 ^{0}C for 12h. We encounter another puzzle in the sense that earlier trend in negative pressure is better than the reference trend in spite of the increased impurity gas concentrations in the metal caused by its additional treatment, exposure either to a kind of gas or to super-critical water.

4. Discussion

We here point out that puzzles can have a simple answer, if we admit the two assumptions, first, thermodynamics of impurity gas in liquid/container system, and second, kinetics of impurity gas in metals. I) Impurity gases in metals re-distribute themselves at own rates from metals to water and vice versa in any thermo-mechanical processes for any closed liquid/container system to achieve thermodynamic equilibrium of each kind of gas among metals and the water. II) Hydrogen atoms diffuse much faster in metals at any temperature than the others including nitrogen atoms, and the difference in the rate will be greater in metals with higher melting points in the temperature range below 100 ^{0}C where our runs of temperature cycles were done.

At first we give an answer to the puzzle in Fig. 1. The hydrogen atoms concentrations both in the s-Mo tube and in the s-Cu plug in their pre-degassed states (the earlier part in Fig. 1) would be reduced after the run since the concentration of hydrogen molecules in the pre-boiled water should be very low. Thus, at the onset of the run just after the gas exposure to the s-Mo tube alone, the hydrogen atom concentration in the s-Cu plug was at a lower level than that at the onset of the run in the pre-degassed state. This may hold for the hydrogen atom concentration in the nitrogen gas exposed s-Mo tube, too, at least in its near-surface region.

It is reasonable to assume that initial negative pressures were dominated by hydrogen atom concentration in the metal with the lower melting point (Cu), and the later steady falling in negative pressure were dominated by slow replenishing of surface crevices on the Mo tube wall with nitrogen from the bulk. The puzzle is turned into a demonstrative example of dependence on kinds of impurity gases in the single-crystalline Mo and Cu of trends in negative pressure owing to the difference in transport rates of hydrogen and nitrogen atoms in the metals below 100 ^{0}C .

Similar answer to the peculiar trend for the additional exposure of the pre-degassed steel plug is possible. The hydrogen atom concentration of the steel plug having degassed under the ultra-high vacuum might be reduced through the dissolution into the de-aerated water during the autoclave run, while the concentrations of nitrogen and oxygen would be increased after the autoclaving run. The peculiar trend is not a puzzle but an example showing an overlapped effect of hydrogen and the other impurity gas atoms in the steel.

In the case of poly-crystalline metal Berthelot tubes the crevices were pores buried in as-machined layers [4] and long-thin pore-like defects that were occasionally incorporated in the poly-crystalline textures [12]. They were replenished with impurity gases through efficient gas transport along grain boundaries [11]. On the surfaces of single-crystalline metals such pore-like defects would be much less in amount. When those 'crevices' were in contact with de-aerated water, a kind of underwater pressurization effect would force the gases even then being trapped within crevices to dissolve into the water [9]. Though the pressure was rather low of ca. +30 MPa each cycle in the all-single-crystalline Mo/Cu tube in the last work, and of ca. +15 MPa for the co-sealed Cu in the present work, the positive pressure worked on all the crevices. Without replenishing of those crevices with impurity gases from in the metal bulk, they could not work long as cavitation nuclei.

Main out-gassing species from stainless steel after being baked (heated under vacuum) was hydrogen, and its mechanically stimulated increase in the out-gassing rate was directly measured [13]. As speculated before [11], mechanical sealing would cause its after-effect; temporarily enhanced out-gassing from the s-Cu plug might hinder the rise in negative pressure in the all-single-crystalline Mo/Cu tube even in its pre-degassed state. Super-expanded meta-stable liquid has a stronger de-gassing power to contacting metal than vacuum environment.

In conclusion, the experimental difficulty in achieving high negative pressure in metal Berthelot tubes lies in both impurity gas transports in metals and high cavitation probability of super-expanded meta-stable liquid state.

Pioneering applications of the Berthelot method to measure properties of liquids including polymer solubility at negative pressures have already been reported by using glass tubes [14,15]. We hope metal Berthelot tubes may contribute such studies deep in the negative pressure range, in particular about pressure dependence of thermodynamic properties of soft materials like polymers including bio-ones.

Acknowledgements

We thank Mr. Yoshikazu Muto for his commitment to the s-Cu co-sealing runs. We also thank Mr. Kaname Motosita in NIT for his advice about computer software.

References

[1] Okamoto H and Ohde Y (1986) Environmental Stress Failure: An Irreversible Thermodynamic Approach. in W. Brostow and R D Corneliussen (eds.), *Failure of Plastics,* Hanser Publishers, New York, pp. 330-344.

[2] Henderson, S. J. and Speedy, R. J. (1987) Temperature of maximum density of water at negative pressure, *J. Phys. Chem.* **91**, 3062-3068.

[3] Henderson, S. J. and Speedy, R. J. (1980) A Berthelot-Bourdon tube method for studying water under tension, *J. Phys. E.* **13**, 778-782.

[4] Ohde Y., Ikemizu M., Okamoto H., Hosokawa. W., and Ando T. (1988) The two-stage increase in negative pressure with repeated cavitation for water in a metal Berthelot tube, *J. Phys. D: Appl. Phys.* **21**, 1540-1542.

[5] Harvey E.N., Whiteley A.H., McLroy W.D., Pease D.C. and Cooper K.W. (1944) Bubble formation in animals I, Physical factors. *J. Cellular Comp. Physiol.* **24**,1-22.

[6] Apfel R.E. (1970) The role of impurities in cavitation-threshold determination, *J. Acoust. Soc. Am.,* **48**, 1179-1186.

[7] Winterton R.H.S. (1977) Nucleation of boiling and cavitation, *J. Phys. D: Appl. Phys.* **10**, 2041-2056.

[8] Ohde Y., Ikemizu M., Okamoto H., Yokoyama T. and Shibata S. (1989) Cavitation history effect of a water-metal Berthelot tube system interpreted by an elaborated gas-trapping crevice model, *J. Phys. D: Appl. Phys.* **22**, 1721-1727.

[9] Ohde Y., Hiro K., Ouchi T., Takeshita T. and Hayashi N. (1991) Pre- and post-treatment conditioning approach for a metal Berthelot tube system, *J. Phys. D: Appl. Phys.* **24**, 859-864.

[10] Ohde Y., Watanebe H., Hiro K., Motoshita K. and Tanzawa Y. (1993) Raising of negative pressure to around −200 bar for some organic liquids in a metal Berthelot tube, *J. Phys. D: Appl. Phys.* **26**, 1188-1191.

[11] Ohde Y.,Komori K. Nakamura T.,Tanzawa Y., Nishino Y. and Hiro K. J. (2001) Effects of gas transports in metals on negative pressures in water in Mo/Cu Berthelot tubes, *J. Phys. D: Appl. Phys.* **34**, 1717-1726.

[12] Ohde Y., Hiro K., Ono M., Isono H and Watanbe H. (1992) Effects on trends in negative pressure of surface pre-treatments for the sealing plugs of a water-metal Berhtelot tube system, *J. Phys. D: Appl. Phys.* **25**, 1098-1101.

[13] Ishikawa Y. and Yoshimura T. (1991) Mechanically stimulated outgassing from stainless steel surface. *J. Vac. Sci. Technol.* **A 9**, 2021-2024

[14] Imre A. and Alexander Van Hook W. (1997) Continuity of solvent quality in polymer solutions. Poor-solvent to Θ-solvent continuity in some polystyrene solutions. *J. Polym. Sci. B.* **35**, 1251-1259

[15] Rebelo L.P.N. and Szylowski J. (2002) Measuring the properties of liquids and liquids mixture at absolute negative pressures. (in this volume)

HETEROGENEOUS NUCLEATION, INFLUENCE OF HETEROPHASE FLUCTUATIONS ON THE CAVITATION STRENGTH, AND THE SUPERCOOLING OF LIQUIDS WITH PHASE INCLUSIONS

V.A. BULANOV

Institute of Marine Technology Problems,
Far Eastern Branch Russian Academy of Sciences,
5a, Sukhanov St, Vladivostok, 690600, RUSSIA

1. Introduction

Disagreement between the experimental cavitation strength of liquids and theoretical values calculated for a pure liquid [1] is explained by the presence of different inclusions [1,2] and is referred to as heterogeneous nucleation. Let us consider the basics of heterogeneous nucleation in a liquid in the presence of a sound field. Usually the cavitation strength of a liquid is calculated on the assumption of a loss of mechanical stability between the balance of internal and external pressure and this method does not take into account thermodynamic aspect of a problem at all. In the present work, it is shown that additionally heterophase fluctuations should be taken into account. This reduces the obtained cavitation strength. Taking into account this mechanism is very important for homogeneous nucleation, and turns out to be essential for heterogeneous nucleation.

The following interesting question is discussed in this work: through what states do the critical parameters of a nucleus of a new phase pass during the growth of the nucleus of a new phase (bubbles or crystallization centers) under the influence of an external field, for example, under the influence of external acoustic fields? In the paper, it is shown that under the influence of an acoustic field the cavitation nuclei cavitation grow, passing always through two critical sizes. One of them corresponds to the traditional unstable position of equilibrium, which is connected to a top of a potential barrier for minimal work, separating the steady and the metastable phase. The other size corresponds to a steady equilibrium position, which is formed under the influence of an external rapidly oscillating acoustic field, resulting in the appearance of additional effective potential forming this new position of equilibrium. From the physics viewpoint, the appearance of the just mentioned steady position equilibrium corresponds to the appearance of new balanced slowly-varying parameters, which occur under the influence of a rapidly varying external field [3,4]. As a consequence of the influence of the rapidly varying field, there is the appearance of nuclei of a new phase inside the metastable liquid with a steady-state narrow size distribution.

A.R. Imre et al. (eds.), Liquids Under Negative Pressure, 327–340.
© 2002 *Kluwer Academic Publishers. Printed in the Netherlands.*

2. Heterogeneous Nucleation Without Taking Into Account of Heterophase Fluctuations

2.1. FEATURES OF RUPTURE OF A LIQUID IN THE PRESENCE OF SOLID INCLUSIONS OR VAPOR BUBBLES

Let us consider rupture of a liquid in the presence of solid particles. It is possible to write down the condition of equilibrium of a bubble with radius R_k :

$$P_k' = P_k + \frac{2\sigma}{R_k} \tag{1}$$

where σ is the surface tension coefficient, the primes hereinafter refer to the bubble, and P is the pressure. Cavitation or rupture of a liquid is the formation of bubbles with radius R larger than the critical R_k : $R > R_k$. Assuming that in the presence of solid cavitation nuclei, the rupture of the liquid occurs on these nuclei, i.e., believing that $R_k \cong R_0$ where R_0 is the radius of the solid particle, it is possible to arrive at the following approximate formula [2, 5]:

$$\Delta P_k \approx \frac{2\sigma}{R_0}\left(1 - \frac{\rho'}{\rho}\right)^{-1}, \tag{2}$$

where $\Delta P_k \approx P_\sigma - P_k$ is the cavitation strength, $P_\sigma \equiv P_\sigma(T)$ is the equilibrium vapor pressure above the flat surface at temperature T .

2.2. NUCLEATION IN LIQUIDS CONTAINING GAS BUBBLES

Cavitation strength of liquids containing inclusions, such as gas bubbles, is usually calculated as follows. The pressure inside the cavitation bubble is the sum of the pressure of gas P_g and pressure of vapor P_σ , i.e., $P' = P_\sigma - P_0$. The pressure of the gas in the bubble depends on its size. In the usual case, it is possible to write down P_g in a bubble of radius R as:

$$P_g = P_g^{(0)}\left(\frac{R^{(0)}}{R}\right)^{3\gamma} = \left(\frac{2\sigma}{R_0} - \Delta P_0\right)\left(\frac{R^{(0)}}{R}\right)^{3\gamma} \tag{3}$$

where $P_g^{(0)}$ is the pressure of gas in the initial bubble of radius $R^{(0)} \cong R_0$, γ is the adiabatic exponent of the gas, $\Delta P_0 = P_\sigma - P_0$, and P_0 is the hydrostatic pressure. Substituting the formula (3) in (1), and taking into account of (2), it is possible to obtain the condition of equilibrium of the critical gas bubble in the liquid as:

$$\Delta P_k = \left(\frac{2\sigma}{R_k} - \frac{a}{R_k^n}\right)\left(1 - \frac{\rho'}{\rho}\right)^{-1} \tag{4}$$

where $a = P_g^{(0)} R^{(0)n} = R_0^n (1 - \Delta P_0 R_0 / 2\sigma)$, $n = 3\gamma$. Further, it is usually assumed that the rupture occurs only when the size of the negative pressure on the absolute scale exceeds the maximal value ΔP_k, found from Eq. (4) [5, 6]. For this purpose, let us differentiate (4) and find a point $R_* = R_0 n^{1/(n-1)} (1 - \Delta P_0 R_0 / 2\sigma)^{1/(n-1)}$ or $R_* = (na / 2\sigma)^{1/(n-1)}$ giving a maximum of the function ΔP_k. The following expression for the cavitation strength of a liquid with a gas bubble can be obtained:

$$\Delta P_k = \alpha_0 \frac{2\sigma}{R_0} \left(1 - \frac{\rho'}{\rho}\right)^{-1}, \qquad \alpha_0 = \frac{n-1}{n} n^{-1/(n-1)}, \tag{5}$$

where the coefficient α_0, which is taking into account the influence of the "internal pressure" in the gas bubble is equal to $[(3\gamma - 1)/3\gamma](3\gamma)^{-1/(3\gamma-1)} < 1$. The last inequality testifies that the rupture of liquids in the presence of gas bubbles is easier in comparison with solid inclusions and occurs at smaller values ΔP_k.

As is obvious from Eqs. (3), (4), in the present work we suppose that the cavitation occurs on a nucleus in equilibrium with the liquid environment, i.e., we believe that R_0 and the hydrostatic pressure P_0 are connected by the relation

$$\Delta P_0 = \left(\frac{2\sigma}{R_0} - \frac{a}{R_0^n}\right)\left(1 - \frac{\rho'}{\rho}\right)^{-1}. \tag{6}$$

On the phase equilibrium curve, we have $\Delta P_0 = P_\sigma - P_0 = 0$, whence $R_0 = (a/2\sigma)^{n-1}$.

2.3. NUCLEATION IN LIQUID HELIUM, CONTAINING ELECTRONIC BUBBLES

By a similar method, the cavitation strength of liquid helium containing electron bubbles [6, 7] was calculated. In this case, the role of the partial pressure of the gas was taken by the pressure P_e caused by the presence of an electron in a cavity [7]. The value P_e equals [7]

$$P_e = \frac{a_e}{R_{0e}^5}, \qquad a_e = \frac{\pi \hbar^2}{4m_e}, \qquad R_{0e} = \left(\frac{a_e}{2\sigma}\right)^4, \tag{7}$$

where $\hbar$ is Planck's constant, and m_e is the mass of the electron. The cavitation strength was calculated in the same way as for a liquid with a gas bubble also using the formula (5), but taking into account of $n = 5$, and the value $a = a_e$. Then the value of α_0 will be $4 \cdot 5^{-5/4}$ giving the following value for ΔP_k

$$\Delta P_k = \alpha_{0e} \frac{2\sigma}{R_{0e}}, \qquad \alpha_{e0} = \frac{4}{5^{5/4}}. \tag{8}$$

3. Influence of Heterophase Fluctuations on Heterogeneous Nucleation of Liquids Containing Inclusions with "Internal Pressure"

3.1. MECHANISM OF HETEROPHASE FLUCTUATIONS IN HETEROGENEOUS NUCLEATION

As was remarked above, Eq. (5) used for finding a maximum of the function $\Delta P_k(R_k)$ does not take into account the effect of heterophase fluctuations. To take into account this effect, we shall write an expression for the minimal work of formation of cavitation bubble on nuclei in the form of gas or electronic bubbles:

$$W = -\frac{4}{3}\pi(R^3 - R_0^3)\Delta P\left(1 - \frac{\rho'}{\rho}\right) + 4\pi\sigma\left(R^2 - R_0^2\right) - \frac{4\pi a}{3-n}(R^{3-n} - R_0^{3-n}). \quad (9)$$

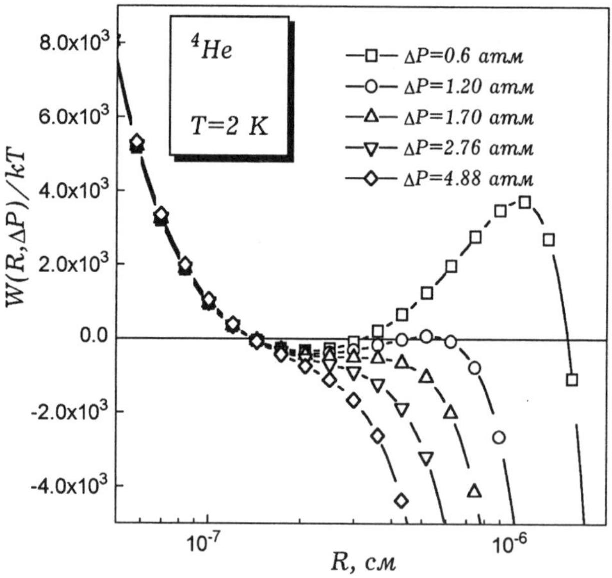

Fig. 1. Dependence of $W(R)$ for liquid helium containing electronic bubbles.

In Fig. 1, the dependence of $W(R)$ is presented for liquid helium containing electron bubbles. In the case of the curve $\Delta P_k = 1.7$ atm., the critical radius R_k corresponds simultaneously to the conditions: $dW/dR = 0$ and $d^2W/dR^2 = 0$. By direct check it is possible to be convinced, that from the condition $dW/dR = 0$ it is possible to arrive at the Eq. (4), and from the condition $d^2W/dR^2 = 0$ Eqs. (5) and (8). Thus, this case corresponds to formation of nuclei of new phases without activation. It is also possible to have activated nucleus formation. The curve $\Delta P_k = 1.20$ atm. exhibits a potential barrier which can be overcome by thermal or quantum heterophase fluctuations [9]. In this case, the quantity of nuclei of a new phase capable of growing to macroscopic size (so-called supercritical nuclei with the sizes $R > R_k$) per unit of time per unit of liquid volume, and which is named as a flow of nuclei J through top of potential barrier formed by the minimal work W, is given by the following expression:

$$J = C\exp(-W_k / kT),\tag{10}$$

where $W_k(\Delta P_k)$ is the minimum work of formation of a critical nucleus of radius R_k, and the preexponential factor C is a weak function of ΔP_k and depends on the kinetics of the transition through the potential barrier [6,9,10]. We shall designate through y the ratio R_0 / R_k, then from Eq. (5) it is possible to write down:

$$\Delta P_k = \frac{2\sigma}{R_0} y(1 - Dy^{n-1}), \qquad D = 1 - \frac{\Delta P_0 R_0}{2\sigma}.\tag{11}$$

Using the formulas (9), (11) we can obtain the following equation to define y:

$$1 - \left(3 + b - \frac{6D}{3-n}\right)y^2 + 2y^3 - D\left(\frac{2n}{3-n}y^{n-1} - 2y^{n+2}\right) = 0,\tag{12}$$

where $b = 3kT\ln(C/J)/4\pi\sigma R_0^2$. From the formula (12), it is evident that the equation that defines y is nonlinear and its analysis is possible only after further simplification with reference to a concrete model of heterogeneous nucleation.

3.2. CAVITATION STRENGTH OF LIQUID HELIUM WITH ELECTRON BUBBLE

In many cases of practical importance, it is possible to make the assumption $y \ll 1$, which corresponds to the condition that the initial nucleus R_0 is small in comparison with the size of a critical nucleus R_k, i.e., the condition $R_0 << R_k$. Equation (12) then is easily solved and for $n \neq 3$ it is possible to write down its solution in closed form as

$$y = \left(3 + b - \frac{6}{3-n}\right)^{-1/2}.\tag{13}$$

For liquid helium containing an electron bubble, it is possible to calculate the value of b which is included in this result for y using the expression for C from refs. [6, 9]. It is found that $\ln(C/J) \cong 70$ and for electron bubbles the value $b \approx 0.4$ at $T = 2$ K, the temperature at which the experiments of ref. [8] were carried out. Substituting Eq. (13) into Eq. (12), we find the final expression for cavitation strength of liquid helium with electron bubble:

$$\Delta P_k = \alpha_{ef}\frac{2\sigma}{R_0}, \qquad \alpha_{ef} = (6+b)^{-1/2}\left(1 - \frac{\rho'}{\rho}\right)^{-1}.\tag{14}$$

The factors α_{ef} and α_{e0} take into account the decrease of the cavitation strength of liquid helium with electron bubbles due to the electronic "internal pressure" [7, 10] in comparison with the capillary pressure $2\sigma/R_0$. Their values are $\alpha_{ef} = 0.39$ and $\alpha_{e0} = 0.53$. Thus, it is shown that the taking into account of heterophase fluctuations results in a decrease of cavitation strength of $\alpha_{fe}/\alpha_{e0} \cong 4(6+b)^{1/2}/5^{5/4} \approx 0.74$.

3.3. CAVITATION STRENGTH OF LIQUIDS WITH GAS BUBBLES

For derivation of an expression for the cavitation strength of a liquid with a gas bubble, let us note that as a general rule, gas bubbles nuclei have large enough size and so for them the inequality $b \ll 1$ is carried out. Then in the case of adiabatic gas bubbles, when $n \neq 3$, the value y following from (13), is approximately equal to $[3+2D/(\gamma-1)]^{-1/2}$. In the case of isothermal gas bubbles, for which $n = 3$, when solving Eq. (13), we should first make the limiting transition $n \to 3$ and put $b = 1$, then obtaining $y \approx 0.386$. From here it is possible to define cavitation strength of a liquid with gas bubbles at small hydrostatic pressure, which in most cases can be calculated under the following formula:

$$\Delta P_k = \frac{2\sigma}{R_0}\left(3+\frac{2D}{\gamma-1}\right)^{-1/2} = \alpha_f \frac{2\sigma}{R_0}, \qquad \alpha_f \approx \left(3+\frac{2}{\gamma-1}\right)^{-1/2} . \qquad (15)$$

For a liquid with isothermal bubbles, we obtain $\Delta P_k = 0.386(2\sigma/R_0)$.

It is of interest to compare results for the cavitation strength of a liquid with gas bubbles when taking into account (by the factor α_f), and without taking into account of heterophase fluctuations (by the factor α_0). Then depending on the value of the adiabatic exponent γ we have at $\gamma = 7/5, 5/3, 1, 1.3, 1.5$ the following values of $\alpha_f/\alpha_0 = 0.70, 0.74, 0.85, 0.62, 0.64$, respectively. Thus, it is shown, that the downturn of cavitation strength of a liquid with gas bubbles caused by heterophase fluctuations, reaches on the average 30-40 %, i.e., it appears significant.

3.4. SUPERCOOLING OF LIQUIDS CONTAINING CHARGED PARTICLES

In liquid helium, alongside with electron bubbles, others specific nuclei can also exist – charged crystallization centers, formed in a liquid near to charged extraneous particles as a result of electrical energy of a polarizing origin [11]. Their characteristic size is usually 8-10 Å and is thus more than 2 times less than electron bubbles. Nevertheless the presence of such nuclei results in the appearance of heterogeneous nucleation, which can lead to the crystallization of liquid helium. It is necessary to note, that such charged nuclei are observed also in other low temperature liquids.

We shall consider supercooling liquids in view of heterogeneous nucleation, caused by presence of the positively charged particles forming solid nuclei in a liquid. This problem in many respects is similar to a problem about cavitation strength of helium with electron bubble. The value of supercooling can be determined under the formula

$$\Delta T_k = T_\sigma - T_k = \left(\frac{dT}{dP}\right)_\sigma \Delta P_k \tag{16}$$

where the value ΔP_k is connected to the radius of a critical nucleus $\Delta P_k = \Delta P_k(R_k)$, and in general is defined by Eq. (4), and the derivative $(dT/dP)_\sigma$ is taken along the curve of liquid–solid phase equilibrium, $(dT/dP)_\sigma = T(\rho'-\rho)/L\rho'$, L is the latent heat. Let us consider some physical values describing a particular crystallization nucleus. Because of the additional polarization in the presence of an electrical field E of the charged particle, the free energy of a nucleus $\Delta F = [(\varepsilon-1)/8\pi]\int E^2\,dV$ is changed. This results in a change of the minimal work W for critical nucleus formation. It is possible to write down the expression for the work W of formation of a nucleus of radius R in the metastable liquid in the presence of a solid nucleus of radius R_{ch} as

$$W = -\frac{4}{3}\pi(R^3 - R_{ch}^3)\frac{\rho' L}{T}\Delta T + 4\pi\sigma(R^2 - R_{ch}^3) - \frac{4\pi\,a_{ch}}{3-n}\left(\frac{1}{R} - \frac{1}{R_{ch}}\right), \tag{17}$$

where $a_{ch} = (Ze)^2(\varepsilon'-e)/8\pi$, the equilibrium radius R_{ch} of a nucleus is equal $R_{ch} = a_{ch}/R_{ch}^4$, and the "internal pressure" is defined by the formula $P_{ch}(R) = a_{ch}/R_{ch}^4$.

Considering the task of formation of a new phase without taking into account of heterophase fluctuations on the basis of an unactivated transition through a potential barrier similar to Eq. (5), we shall obtain the following expression for supercooling $(\Delta T_k)_0$:

$$(\Delta T_k)_0 = \left(\frac{dT}{dP}\right)_\sigma \frac{3}{4^{4/3}}\frac{2\sigma}{R_{ch}}\frac{\rho}{\rho'-\rho}. \tag{18}$$

The taking into account of heterophase fluctuations results in the following value of supercooling in the presence of charged particles:

$$(\Delta T_k)_f = \left(\frac{dT}{dP}\right)_\sigma \frac{2\sigma}{R_{ch}}\frac{\rho}{\rho'-\rho}\frac{1}{\sqrt{9+b}}. \tag{19}$$

The calculations under the Eqs. (18), (19) give the following value of the relation

$$(\Delta T_k)_f /(\Delta T_k)_0 \approx 0.675, \tag{20}$$

which shows that taking into account of heterophase fluctuations has a large effect on the calculation of supercooling of liquids in the presence of solid nuclei formed by positive particles in liquids.

3.5. FEATURES OF HETEROGENEOUS NUCLEATIONS IN AN ACOUSTIC FIELD

The above mentioned results concern heterogeneous nucleation in a field of static tensile strain or slow supercooling. Alongside this important direction is the study of the nucleation process when the liquid is exposed to a rapidly oscillating acoustic field. For definiteness, we shall consider the case of cavitation. Under the influence of an acoustic field, the growth of cavitation nuclei occurs in two regimes. The first regime is characterized by the fast rupture of the liquid during the half cycle when the pressure is negative. This regime corresponds to, in essence, the situation already considered above, and is referred to as nonstationary or transient cavitation. The second regime, on the contrary, is characterized by a rather small growth rate of the nuclei, and growth taking place during many cycles of an acoustic field owing to the effect of rectified gas diffusion or rectified heat transfer [6,12]. Let us consider the second regime, i.e., stationary cavitation. In contrast to what happens in the first regime, the influence of the acoustic field results in the appearance of additional critical sizes of nuclei, which essentially change the general picture of growth of nuclei in the metastable liquid. In essence, an additional parameter controlling growth of nuclei in the liquid appears. We have already mentioned above, the critical size of nuclei corresponding to the well known unstable position of equilibrium, at the top of the potential barrier for the minimum work separating the steady and metastable phase. The other size is completely new. It does not have an analog in the traditional scheme of heterogeneous nucleation, and corresponds to a <u>steady</u> position of equilibrium of nucleus in metastable liquid.

The origin of the just mentioned steady position of equilibrium corresponds to the appearance of new equilibrium slowly varying parameters under the influence of rapidly varying external fields. As is known [3,4], the nonlinearity of such fields results in a new effective potential which essentially changes the potential previously existing without an external field that determines the usual position of equilibrium in system.

For study of the specified mechanism, we shall consider briefly the process of rectified heat transfer for the growth of a vapor nucleus [6, 12] or crystallization nucleus [12] in metastable, i.e., superheated or supercooled liquid. The simplest result is obtained for the dynamics of a vapor nucleus. Let us consider the Rayleigh equation for a bubble in an incompressible liquid:

$$\rho\left(R\ddot{R}+\frac{3}{2}\dot{R}^2\right)=\left[P_\sigma-P_\infty-\left(\frac{2\sigma}{R}-P_i\right)\right]=\Delta P-\left[\frac{2\sigma}{P}-P_i\right], \qquad (21)$$

where $\dot{R}=dR/dt$, $\ddot{R}=d^2R/dt^2$. The pressure P_i inside of the bubble consists of two components – the equilibrium vapor pressure P_σ and the additional pressure $\overline{P}'$ caused by the rectified action of the sound owing to the additional rectified change of temperature, and also owing to the additional rectified changes of vapor pressure which are connected with the inertial effects of the liquid adjacent to the bubble and the nonequilibrium of the phase transformation dynamics of the vapor nucleus in the liquid. Thus it is possible to write that $P_i(R)=P_\sigma+\overline{P}'$. The pressure $\overline{P}'$ depends quadratically on the amplitude of the acoustic field P_m and depends on the frequency ω, i.e.,

$\overline{P'} \approx B(R,\omega)P_m^2$. As a result, the growth of a nucleus is observed with the average speed given by the formula [6, 12]

$$\dot{\overline{R}} = \frac{G}{R}\left[\Delta P - \frac{2\sigma}{R} + B(R,\omega)P_m^2\right] = \frac{A(R,\omega)}{R}\left[P_m^2 - P_{mk}(R,\omega)^2\right], \qquad (22)$$

where G is some function, dependent on the thermodynamic and kinetic characteristics of the liquid, $A(R,\omega) = GB(R,\omega)$, and at the same time the value P_{mk} represents the cavitation strength of a liquid with nuclei, which it is possible to define as

$$P_{mk}(R,\omega) = \left[\left(\frac{2\sigma}{R} - \Delta P\right)/B(R,\omega)\right]^{1/2} . \qquad (23)$$

From Eq. (22) it is seen that at when the amplitude of the acoustic field P_m exceeds a critical value P_{mk} the growth of nuclei is observed.

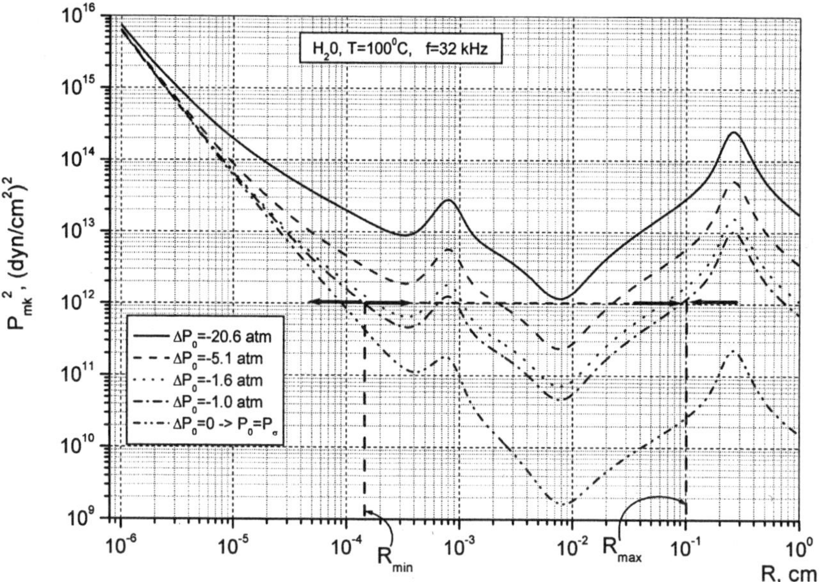

Fig. 2. Dependence of the threshold pressure of acoustic cavitation on the radius of bubbles.

In Fig. 2 the dependence of $P_{mk}^2(R)$ at frequency 32 kHz is shown for the case of vapor-gas bubbles in water at $T = 80°C$. Here, the different curves correspond to various external hydrostatic pressures $\Delta P_0 = P_\sigma - P_0$. From Fig. 2, it is seen that rupture of the liquid in the sound field can occur not only at downturn of pressure P_0 below the value P_s, but also at $\Delta P_0 < 0$, i.e., when the liquid is exposed to external pressure. The arrows in Fig. 2 specify positions of the critical sizes $R_{\min}$ and $R_{\max}$, formed from crossing of curve $P_{mk}(R)$ and constant line P_m, to the appropriate amplitude of an acoustic field. It is seen, that the appearance of two critical sizes for vapor bubbles is a completely new phenomenon in the mechanism of heterogeneous nucleation and it is connected exclusively with action of an acoustic field. Under the action of a sound field, a rectified heat transfer to the inside of the bubble occurs and the bubble grows steadily during many periods of the field. At the same time, some distribution of bubbles with size between the limits from $R_{\min}$ up to $R_{\max}$ is formed. The point $R_{\min}$ corresponds to the usual unstable critical radius, known from the theory of nucleation. However, the appearance of the steady radius $R_{\max}$ is a new effect, connected to the action of the sound.

The picture considered above is observed also when the static stretching ΔP exceeds some critical value dependent on the amplitude of the acoustic field:

$$\Delta P_k = \frac{2\sigma}{R} - B(R,\omega)P_m^2 . \tag{24}$$

From the formula (24), it is evident that the value $B(R,\omega)P_m^2$ can be considered as additional "internal" pressure, which influences heterogeneous nucleation in an essential way. From (28) it is seen, that at large enough ΔP_k each bubble grows up to macroscopic size and at the same time there will be no restriction to bubble growth in the form of the appearance of $R_{\max}$. To find the condition of unactivated nucleation, we shall make use of the recipe considered in 2.2. We find a condition of an extremum from the equation

$$\left[\frac{2\sigma}{R^2} + B'(R,\omega)P_m^2\right]_{R_*} = 0, \quad B'(R,\omega) \equiv \frac{\partial B(R,\omega)}{\partial R} . \tag{25}$$

Substituting Eq. (25) in Eq. (24) we can obtain

$$\left(\Delta P_k\right)_{R_*} = \frac{\partial B(R,\omega)}{\partial R}\bigg|_{R_*} P_m^2 . \tag{26}$$

From the condition (26) it is seen that the value $\left(\Delta P_k\right)_{R_*}$ is directly connected with the amplitude of the acoustic field. At the same time, the condition (26), as is seen from the above mentioned analysis for gas and electronic bubbles, gives an overestimate since it does not take into account the mechanism of heterophase fluctuations and corresponds to unactivated nucleation. For taking into account thermally-activated nucleation, it is necessary to write an expression for the minimum work with the account of the rectified action of an acoustic field.

We turn again to the equation (21). Let us consider the average nucleus parameters during their dynamics in an acoustic field, so that Eq. (21) does not depend on the time in explicit form. It allows us to transform the left part of Eq. (21) into the form

$$\rho\left(R\ddot{R} + \frac{3}{2}\dot{R}^2\right) = \frac{\rho}{2R^2}\frac{d}{dR}(R^3\dot{R}^2). \tag{27}$$

Let us notice that the kinetic energy $E = m_b\dot{R}^2/2$ of radially-symmetric movement of a nucleus in an incompressible liquid is equal to $E = \rho \cdot 3V_b\dot{R}^2/2 = 2\pi\rho R^3\dot{R}^2$. Hence, $m_b = 4\pi R^3\rho$. Then from (21) and (27), we obtain

$$E(R) - E(R_0) = 4\pi\int_{R_0}^{R} R^2\left[B(R,\omega)P_m^2 - \left(\frac{2\sigma}{R} - \Delta P\right)\right]dR. \tag{28}$$

Further, taking into account that for a critical nucleus at $R = R_k$ we have $E(R_k) = 0$, from (28) we can determine the kinetic energy $E(R_0)$ as

$$E(R_0) = -4\pi\int_{R_0}^{R_k} R^2\left[B(R,\omega)P_m^2 - \left(\frac{2\sigma}{R} - \Delta P\right)\right]dR. \tag{29}$$

Taking into account that the minimal work $W(R)$ is equal to the loss in kinetic energy $E(R_0)$, we find that $W(R) = -E(R_0)$. Thus, it is possible to conclude that the kinetic energy of radial growth of the nucleus and the minimal work are connected by the equation of energy conservation of the kind

$$E(R) + W(R) = W_k, \quad W(R) = -4\pi\int_{R_0}^{R} R^2\left[B(R,\omega)P_m^2 - \left(\frac{2\sigma}{R} - \Delta P\right)\right]dR. \tag{30}$$

Now taking into account Eq. (10), it is possible to write down the following equation for heterogeneous nucleation in the acoustic field:

$$1 + \frac{B(R_0/y)R_0}{\sigma y}P_m^2 = \left[3 + b - \frac{B(R_0/y)R_0}{\sigma}P_m^2 - \frac{3P_m^2}{\sigma R_0^2}\int_{R_0}^{R_0/y}\xi^2 B(\xi)d\xi\right]y^2. \tag{31}$$

The important circumstance complicating the further analysis of Eq. (31) is the complexity of the explicit form of the function $B(\xi)$. Let us consider some limiting cases which are amenable to simple analysis. The simplest case is when $B(R) = B_0 = \text{constant}$. It is easy to see that from Eq. (31) under the condition of $y \ll 1$, we obtain the expression $y \approx (B_0 R_0 P_m^2/\sigma)/[1 - B_0 R_0 P_m^2/\sigma] \approx B_0 R_0 P_m^2/\sigma$. Substituting this into Eq. (24), written down as $\Delta P_k \approx (2\sigma y/R_0)[1 - B_0 R_0 P_m^2/2\sigma y]$, we have $\Delta P_k \approx B_0 P_m^2$.

Let us consider more complex cases. In Fig. 3 and Fig. 4 the dependences of the function $B(R,\omega)$ and the modulus of compressibility $|K(R,\omega)|$ are shown. For a monochromatic wave the compressibility $|K(R,\omega)|$ is defined by

$$\delta R/R = -(1/3)K(R,\omega)P_m\exp(-i\omega t),$$

338

where $\delta R / R$ is the relative pulsation of the radius of the bubble, and P_m is the amplitude of the sound. The results in Fig. 4 take account of the phase transformation.

From this we can see that for radius smaller than the radius of a thermal maximum R_T [6, 12], the function $B(R) \approx B_1 R$. In an interval of radius $R_Y << R << R_\omega$, where $R_\omega \approx (3 / \rho \, \mathrm{Re}(K))^{1/2} / \omega$ is the resonant radius of bubble for the given frequency [6, 12], the function $B(R, \omega)$ has a maximum. And then at $R >> R_\omega$, there is a decrease of the function $B(R)$ under the law $B(R) \approx B_2 R^{-2}$.

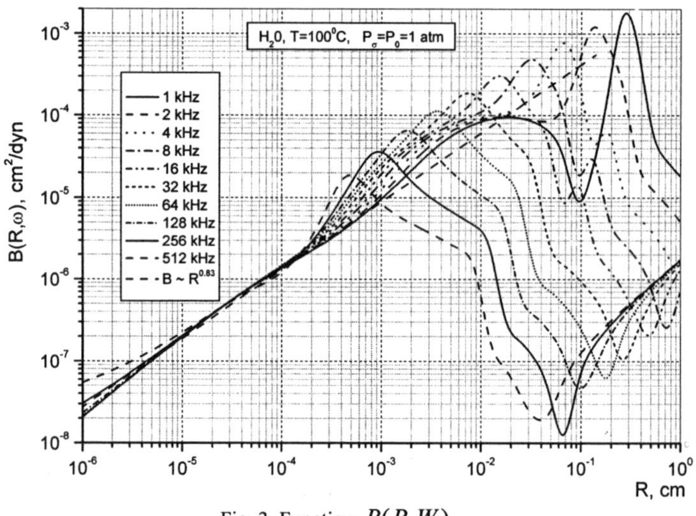

Fig. 3. Function $B(R, W)$.

Fig. 4. The function $| K(R, W) |$.

Let us consider the case of small bubbles $R < R_T$, thus $B(R) \approx B_1 R$. Then from (31), we obtain $y \approx (5B_1 R_0^2 / \sigma)^{1/2} P_m$, whence we conclude $\Delta P_k \approx -2\sigma / 5R_0$. The obtained result shows that cavitation in the presence of a field arises in a liquid which not only is subject to a stretching, but, on the contrary, is compressed by external hydrostatic pressure, which in our case is equal to $2\sigma / 5R_0$. The specified case is shown in Fig. 2.

Let us consider now the case $R > R_0$, when $B(R) \approx B_2 R^{-2}$. In this case, from Eq. (31) the following the equation can be obtained

$$\left[3 + b + \frac{3B_2 P_m^2}{\sigma R_0} \right] y^2 - \frac{4B_2 P_m^2}{\sigma R_0} y - 1 = 0 , \tag{32}$$

the positive root of which is

$$y = \frac{2B_2 P_m^2}{\sigma R_0} \left[3 + b + \frac{3B_2 P_m^2}{\sigma R_0} \right]^{-1} \left\{ 1 + \left[1 + \frac{3}{4} \left(1 + \frac{B_2 P_m^2}{\sigma R_0} \right) \right] \left(\frac{\sigma R_0}{B_2 P_m^2} \right)^2 \right\}^{1/2} . \tag{33}$$

There are two limiting cases corresponding to sound with small and large amplitude $y \approx 1/\sqrt{3}$ at $P_m^2 << \sigma R_0 / B_2$ and $y \approx 4/3$ at $P_m^2 >> \sigma R_0 / B_2$. In the first case we can obtain

$$\Delta P_k \approx 2\sigma / \sqrt{3} \, R_0 , \quad P_m^2 << \sigma R_0 / B_2 . \tag{34}$$

The second limiting case actually does not satisfy the condition $y < 1$ and so should not be considered.

4. Conclusion

The basic principles of heterogeneous nucleation in a liquid are considered in the presence of a sound field. The results of investigations of cavitation strength of liquids are affected by the presence of inclusions such as gas and vapor bubbles, and electron bubbles in liquid helium. The "internal pressure" due to gas pressure within gas bubbles results in a decrease of the cavitation strength of the liquid up to factor $\alpha_0 = 0.4$ to 0.5 in comparison with the capillary pressure. A similar result is obtained for the cavitation strength of liquid helium with electronic bubbles: $\alpha_{e0} = 0.53$. Moreover, it is also shown, that additional decrease of cavitation strength of a liquid with gas or electron bubbles caused by heterophase fluctuations reaches on the average 30-40 %, i.e., it appears to be significant. For the case of freezing of liquids, it is shown that the presence of positively charged particles in liquid helium forming solid nuclei reduces the supercooling threshold and additional reducing due to heterophase fluctuations amounts to a reduction of 43 %.

It is shown, that heterogeneous nucleation in an acoustic field in the presence of vapor nuclei has features that are inherent in nucleation at the presence of gas bubbles. The "internal" pressure in this case plays the role of the average rectified action of a sound field. The presence of a field results in the occurrence of asymptotic steady radius R_{max} and forming of bubbles distribution on the sizes in area from R_{min} up to R_{max} as a result.

5. Acknowledgments

The author is grateful to Victor A. Akulichev for numerous useful discussions of results.

The work was maintained by the Russian Foundation for Basic Researches, project 00-02-16913.

6. References

1. Zeldovich, Ya. B. (1942) On the theory of new phase formation. Cavitation, *Zh. Eksp. Teor. Phys.* **12**, 525-538. (in Russian)
2. Frenkel, Ya. I. (1975) *Kinetic theory of liquids*, Nauka, Leningrad. (in Russian)
3. Kapitza, P. L. (1951) Pendulum with vibrating pendant, *Usp. Phys. Nauk* **44,** 7-20 (in Russian).
4. Landau, L. D., Lifshitz, E. M. (1973) *Mechanics*, Nauka, Moscow (in Russian).
5. Knapp, R.T., Daily, J.W., Hammitt, F.G. (1970) *Cavitation*, McGraw-Hill Book Company, N.-Y.
6. Akulichev, V.A. (1978) *Cavitation in cryogenic and boiling liquids*, Nauka, Moscow (in Russian).
7. Akulichev, V. A. Boguslavskii, Y. Y. (1972) The cavitation strength of liquid helium caused by electronic bubbles, *Sov. Phys. JETP* **35**, 1012.
8. Marston, P.L. (1976) Tensile strength and visible ultrasonic cavitation of superfluid 4He, *J. Low Temp. Phys.* **25**, 383-407.
9. Akulichev, V.A., Bulanov, V.A. (1975) On the tensile strength of quantum liquids, *Sov.Phys. Acoustics*, **20**, 502-505.
10. Akulichev, V.A., Bulanov, V.A. (1996) About the influence of heterophase fluctuations on the cavitation strength and nonlinear parameter of liquids, in M.D. Ageev (ed.), *Marine technology*, Dalnauka, Vladivostok, pp. 148-168 (in Russian).
11. Atkins, K.R. (1959) Ions in Liquid Helium, *Phys. Rev.* **116**, 1339-1343
12. Akulichev, V.A., Alekseev, V.N., Bulanov, V.A. (1986) *Periodic phase transformations in liquids*, Nauka, Moscow (in Russian).